KNOWLEDGE, COMMUNICATION AND CREATIVITY

KNOWLEDGE, COMMUNICATION AND CREATIVITY

Editor

Carran King

Socialy
PRESS
AN IMPRINT OF SCITUS ACADEMICS

Knowledge, Communication and Creativity
Edited by **Carran King**

This Book is Printed on Acid-Free Paper. ♾

ISBN: 978-1-68117-844-8
Library of Congress Control Number: 2017938876

Socialy Press
(An Imprint of Scitus Academics),
616, Corporate Way, Suite 2, 4766,
Valley Cottage, NY 10989

www.scitusacademics.com

Notice

Printed in United States of America.

Preface

Computer technology is becoming more efficient, productive, and cheaper. Advances in technology are producing more powerful computing devices to create a dynamic virtual network that allows people all over the world to communicate and share information with each other. Information and communication technologies (ICTs) are electronic tools used to convey, manipulate and store information. The exponential growth of Internet access and ICTs greatly influenced social, political, and economic processes worldwide. However, the implementation of ICTs is occurring in a context where the cultural and institutional barriers are not well addressed. Knowledge, communication, and creativity are passions of contemporary modern societies. Conversely, little attention has been paid to the sociological relationships among these elements, let alone their impacts as processes driving social change. With the growth of the Internet, a wide range of ICTs have transformed social relationships, education, and the dissemination of information. It is argued that online relationships can have properties of intimacy, richness, and liberation that rival or exceed offline relationships, as online relationships tend to be based more on mutual interest rather than physical proximity. Despite this interest in technology, the attention that the field of social work has given to ICTs in research, education, and practice does not match the efforts of other national and international organizations that view ICTs as critical to improving the lives of disadvantaged and disenfranchised persons, and necessary for all forms of civil engagement.

Knowledge, Communication and Creativity provides dominant explorations into the creative developments that are connected to knowledge production, impact of a general increase in knowledge on individuals, lifestyles, institutions and technologies It also highlights how new communication and information technologies are transforming social relationships, communities and the international public sphere. This book will be of valuable for all those interested in understanding the Knowledge Society and how its new technologies of communication are transforming contemporary society.

Table of Contents

1

CHAPTER

The Methodology of a Multi-Model Project Examining How Facebook Infrastructures Social Relations

Beverley Skeggs and Simon Yuill

Department of Sociology, Goldsmiths, University of London, London, UK

ABSTRACT

It is the purpose of this paper to make explicit the methodology (the theory of the methods) by which we conducted research for an Economic and Social Research Council-funded research project on the relationship of values to value. Specifically, we wanted to study the imperative of Facebook to monetize social relationships, what happens when one of our significant forms of communication is driven by the search for profit, by the logic of capital. We therefore wanted to 'get inside' and understand what capital's new lines of flight, informationally driven models of economic expansion, do to social relations. Taking up the challenge to develop methods appropriate to the challenges of 'big data', we applied four different methods to investigate the interface that is Facebook: we designed custom software tools, generated an online survey, developed data visualizations, and conducted interviews with participants to discuss their understandings of our analysis. We used Lefebvre's [(2004). *Rhythmnanalysis: Space, time and everyday life.* London: Continuum] rhythmanalysis and Kember and Zylinska's [(2012). *Life after new media: Mediation as a vital process.* Cambridge, MA: MIT Press] ideas about 'lifeness' to inform our methodology. This paper reports on a research process that was not entirely straightforward. We were thwarted in a variety of ways, especially by challenge to use software to study software and had to develop our project in unanticipated directions, but we also found much more than we initially imagined possible. As so few academic researchers are able to study Facebook through its own tools (as Tufekci [(2014). Big questions for social media big data: Representativeness, validity and other methodological pitfalls. In ICWSM '14: Proceedings of the 8th International AAAI Conference on Weblogs and Social Media, pp. 505–514.). Big questions for social media big data: Representativeness, validity and other methodological pitfalls. In ICWSM '14: Proceedings of the 8th International AAAI Conference on Weblogs and Social Media (pp. 505–514)] notes how, unsurprisingly, at the 2013 ICWSM only about 5% of papers were about Facebook and nearly all of these were co-authored with Facebook data scientists), we hope that our methodology is useful for other researchers seeking to develop less conventional research on Facebook.

Keywords: Political economy, facebook, rythmanalysis, big data, interactivity, research methodology

INTRODUCTION

The Facebook project is part of a more general investigation of the relationship between values and value (see https://values.doc.gold.ac.uk/), funded by the Economic and Social Research Council, UK. We were concerned to investigate if anything at all was beyond the logic and analytic of capital (Skeggs, 2014a, 2014b). We were therefore interested in the places where the imposition of capital's logic, such as friendship (and in a parallel project, faith), could possibly be resisted. In particular, we were concerned to see what happens when affective social relationships, replete with values, are monetized.

Following Andrejevic (2013) and Morozov (2014), we believe that the foremost asset for monetization today is our personal data. But if all 'data is credit data' as suggested by Douglas Merrill of ZestFinance, does this mean that all life beats to the rhythms of monetization? And, if Facebook monetizes by developing an infrastructure to sell audiences to advertisers through massive online data aggregation, how do we live the relations of that connectivity?

Facebook was chosen because it is *the* 'relational' site for social networking in comparison to the more political and opinion orientation of Twitter or sites for self-promotion. Facebook is still remarkably successful, globally expanding daily. The first quarterly earning of Facebook in 2014 were recorded as $2.5 billion in revenue, $10 billion annually, of which 85% of Facebook revenue comes from advertising. Eight per cent of all global advertising now takes place on Facebook. Facebook has 1.28 billion total monthly users, 802 million daily users, and 609 million daily mobile users. As Van Dijck (2012) notes, Facebook *brokers* sociality.

Our original research questions about the significance of technological infrastructuring of relations were framed with a static model of the human and the technological, which assumes that we know in advance what the social, the personal and the technical actually are. Kember and Zylinska (2012), however, note: 'It is not simply the case that "we" – that is, autonomously existing humans – live in a complex technological environment, and it makes no more sense to talk of us using it, than it does of it using us.' (p. 13). We thus moved to a more entangled understanding, but how do we know what Facebook does in the world if we cannot untangle its mechanisms? So, our research is an attempt to use multiple models and methods to address the porosity of boundaries, to untangle and investigate what Facebook does across different domains of capital, labour, code, communication and social relations.

From our survey of existing research, we felt that many studies took for granted, and often reinforced, Facebook's own claims as to the platform's significance, both as a primarily 'social' entity, as opposed to a primarily technological or mediatized one, and to the significance of 'networks' to represent and engage in such sociality. Google as a search engine foregrounds its relation to the technology it operates on, whilst earlier platforms such as MySpace presented themselves as DIY media tools. From its very beginning, Facebook had openly modelled itself on ideas from Social Network Analysis.[1] Light and Cassidy (2014) outline how the majority of research on social network studies (SNS)'s focus on various conceptions of network, including networked individualism (Wellman & Gulia, 1999), networked sociality (Wittel, 2001), networked collectivism (Baym, 2007), networked publics (boyd, 2008; Ito, 2007), the networked imaginary (Lievrouw & Nguyen, 2007) and networked literacies[2] (Lievrouw, 2012). Much of the existing software tools used in studies of Facebook, such as Netvizz, accept Facebook's self-description as a 'social network' at face value and map out the data in a form that represents a network, most commonly adopting the now familiar, node and line diagram. These are static diagrams that present a network frozen in time. Munster (2013) argues, however, that such visualizations create a form of anaesthesia, numbing us to the very particularities and specificities of how such structures might come into being and evolve and proposes that:

Instead of seeing networks everywhere, perhaps we should look, think, and sense more thoroughly the patchiness of the network field. We need to immerse ourselves in the particularities of network forces and the ways in which these give rise to the form and deformation of conjunctions – the closures and openings of relations to one another. It is at this level of imperceptible flux – of things informing and reforming relationally – that we discover the real experience of networks. (p. 3)

We argue that we cannot assume what such socio-technical structures do in advance, e.g. do they even operate as a network for instance? For as Miller (2008) notes, socio-technical arrangements are indeterminate, complex, heterogeneous and mutable. SNS's are not simply phatic objects, or communications that have purely social (networking) intents. They are more likely to be what Langlois and Elmer (2013) characterize as thick digital objects, a combination of phatic, relational, networked and media objects that operate as an interface between different materialities (Galloway, 2012).

Network studies also tend to treat Facebook as a self-contained entity set apart from the rest of the Internet. Through the development of technologies such as the Like buttons placed on web pages and the Open Graph API, Facebook has, however, steadily expanded its presence well beyond the boundaries of its own domain, tracking users as they browse. What a given user receives on their Facebook account page, in the newsfeed or surrounding adverts, may be as much, or more, driven by activities outside of Facebook as within. Our own software and forms of visualization were designed to map these complex aspects of relational reforming and entanglement both across the Web and between Facebook and users' daily lives.

Rather than assuming that we knew what Facebook does, we wanted to get inside of Facebook to examine, as far as we could, its algorithmic behaviour and how that relates to the particular forms of sociality they construct and instrumentalize. Understanding Facebook algorithms, however, relies upon using tools and services that Facebook itself makes available, those which often reinforce the very assumptions and paradigms we were seeking to question.

Given the debates around how sociology can respond to both the opportunities and threats of research in the age of social networks and Big Data (Burrows & Savage, 2014; Savage & Burrows, 2007), we believe that it is vital that researchers continue to develop new methods and tools of analysis and not simply rely on off-the-shelf systems that may restrict the scope of investigation and how we frame the questions we put to it. We therefore developed our own tools expanding our existing methodologies to incorporate the new forms of knowledge such tools might enable. This process, however, is a complex one and so, in this paper, we wish to set out a frank exposition of the problems we encountered and how these were addressed so that our experience may help inform and encourage other researchers to continue developing new approaches to the field.

We sought to confront prevailing modes of analysis based on networks and sociality with an alternative conception of lifeness and rhythm. Following Kember and Zylinska (2012), we understand 'lifeness' as ongoing encounters that move in and out of synchronization and resonance with all aspects of our lives: rhythms of life rather than networks, rhythms which may come into conflict with other rhythmic structures imposed upon or encountered, such as work, parenting, relationships, care, or different paces of information exchange arising in different media (Weltevrede, Helmond, & Gerlitz, 2014). We used software to develop a 'life method' (Kember & Zylinska, 2012) rather than just a 'live method' (Back & Puwar, 2013) to investigate the dynamics, the moments and rhythms of engagement and entanglement with the Facebook platform. Did the platform produce the user's response, as proposed by Marres and Weltevrede (2013), determining what we even understand as time, producing distinct temporalities not external from but specific to devices (Weltevrede et al., 2014)? Or was engagement a form of intra-action (Barad, 2007) organized with other social, ecological and biological relationships such as energy, affect, attention and work rhythms? How does the time of lived experience, captured and communicated on a platform such as Facebook, relate to the temporal ordering and sequencing of data that forms the basis of every algorithm? Mackenzie (2006) describes 'algorithmic time' as the coming together of the step-by-step clock process through which code operates and the 'accretion of past biological knowledge production' (Mackenzie, 2006, p. 51) that led to the creation of that code and the data upon which it operates. If our own actions are part of the algorhythms (Miyazaki, 2012) of Facebook's newsfeed, are we merely enslaved within its beat or, if we also can determine aspects of algorithmic time through our own actions, is something more complex going on?

Hence, we used Lefebvre's (2004) rhythmanalysis as a methodology that links theory to our methods: 'Rhythm appears as regulated time, governed by rational laws, but in contact with what is least rational in human being: the lived, the carnal, the body' (p. 9). Rhythmanalysis emphasizes overdetermination – focusing on the many different features that co-join to determine use, drawing attention to the conjectural, and focusing on moments and elements that appear instinctual, impulsive, irresolvable and incommensurable, alongside those that we identify as habitual and repetitive. Rhythmanalysis allows us to interrogate the various moments when energy materializes and is monetized by Facebook. Our software enabled us to visualize this process, drawing out moments of engagement alongside rhythms of use, elaborating both the quality and quantity of use, with many interaction rhythms taking place simultaneously (see visualizations later) through the entanglement of lines that have different routes. Murphie (2007) and DeLyser and Sui (2013) have advocated using rhythmanalysis for understanding digital media, but only one project so far (Hochman & Manovich, 2013) has employed the technique in visualization.

Our software was thus designed to speak to these issues. We developed code that could capture 'internet time' (Karpf, 2012), entanglements, practices, attention, rhythms, and importantly, also the *contradictions* of temporal connectivity for, as Lefebvre (2004) emphasizes, not all rhythms merge seamlessly into one singular duration, but rather they may conflict with and interrupt one another and slip out of synchronicity.

Initial challenges: software development

To achieve our methodological aims, we combined data from two different sources. Firstly, that available through the official Facebook API[3] and, secondly, data relating to Facebook activity but not available in the API such as advertising shown to users on Facebook and the presence of Facebook tracking users as they browsed elsewhere. This required building two data collecting tools: one as a Facebook App for the API data and one as a browser plugin, which we named Admonitor, for the other sources. As these both produced a continuously updating stream of data, a custom website was built to collate material during the data collection phase. This site also provided 'live' visualizations of the data analysis enabling researchers and participants to review material as it was collected; they could follow their own rhythms.[4] During the course of the data collection period, a number of new visualization tools were added in response to the activity we were able to monitor.[5]

A number of challenges were encountered in the technical realization of the project, which created delays in the schedule and caused aspects of the original aims to be rethought. Whilst some of these were due to common problems facing the development of custom software systems, such as scaling up a system from prototype to production use, others reflected the inherent politics of conducting online research in the contemporary academic environment and engaging with powerful platforms shaped by business interests such as Facebook. As Mair, Greiffenhagen, and Sharrock (2015) point out, it is common practice to omit any discussion of difficulties encountered in the development of technical components in a research project; yet, these 'worldly troubles' and the ways in which a given team responds to them can often be significant factors in shaping how a given piece of research puts social relations on display, or even whether a given phenomenon can be studied.

Many platforms, such as Facebook and Twitter, provide access to their data through an API that is often available for free through registration. Whilst this does provide an entry point to information sources that would be beyond the means of a university project to generate itself, and has been commended by many researchers, it also has to be recognized that every API is a carefully controlled gateway which, as Bucher (2013) notes in relation to Twitter, has its own politics expressed through 'protocols that structure and exercise control over the specific social situations on which they are brought to bear'. An API will only ever give partial access to data that is carefully controlled by its providers to enable sufficient openness to encourage innovative spin-offs by outside developers, which can be harvested for future internal product lines, and, at the same time, ensure sufficient closure around that data so as to protect the provider's competitive interests (Bucher, 2013). As Bucher argues, control over external data access is also exercised through the inherent instability of the API. APIs are subject not only to internal processes of structural change, shutting off some resources whilst opening others and sometimes rendering outside services obsolete overnight, but also to contextual alterations in ever-changing licence agreements and developer documentation. Facebook is well known, notorious even, amongst both users and external developers for changing features of its services without prior notice and for the patchiness of its developer documentation. Whilst our project was relatively brief (one year in total) and thus short in its implementation period, we nevertheless encountered problems as a result of these factors forcing quite a substantial rewrite of the code at a number of points during the project. What was originally planned as a three-month development time for the software quickly expanded into six months and the project had to be extended to 18 months in total.

Due to our combination of data from the Facebook API and scraping, we also found ourselves in a grey area in relation to Facebook's terms of service, as these forbid the use of gathering data through alternative sources from the API itself. Due to this, we were unable to release our Facebook App publicly and had to rely on participants signing up as Facebook test developers in order to install it.[6] The process of registering people as test developers proved to be difficult and unreliable, a situation not helped by the lack of documentation on the process in Facebook's own developer site. This severely restricted the number of participants we were able to recruit. In our own experience, we would therefore say that the use of such propriety APIs has to be treated with caution by researchers. They are not always as reliable as might be assumed and the restrictions on their use placed by the platform owners can have significant impacts on how a research project is able to operate. This also had an impact on the kind of sampling strategy we were able to implement. We had initially hoped for a broad sweep of participants combining both an opportunistic response through online recruitment on Facebook and other online platforms along with purposively targeted groups such as teenagers and individuals who used Facebook as part of a commercial operation such as marketing. In order to access as wide as field as possible we asked people to forward requests to as many people as possible. This meant that our sample's social location, even geographical location was initially unknown, and as we found out from the survey below was widely spread (New Zealand, Canada, Australia, Sweden, Brazil and USA). In the end, however, the hurdles posed by Facebook's test developer registration process forced us to drastically reduce the number of participants and focus on those we could support in manually singing up and guiding via Skype and phone through the process, relying on those who had the patience to stay with us.

Using Facebook messenger presented particular problems as we were alerted to the fact that many of our requests for participation and return offers to participate did not get through. We still have no idea why many enthusiastic and even tenacious would-be participants were lost on the way. We finally captured data from 33 testers. Even though getting these data was an exacting process, we actually ended up with much more than we had imagined and were able to obtain more detailed analysis from this small group than we had originally expected.

Whilst scraping is frequently considered to be more prone to the instabilities of online media, in our experience, the changes encountered in scraping were less of an issue, involving only minor rewrites of some code.[7] We were fortunate that these occurred during the testing phase, as had they happened mid-way through data collection, they could have had greater impact. Due to the timescale of the project, however, we were limited to only developing a scraper for the Firefox desktop browser as both the mobile version of Firefox and the other widely used browsers (Internet Explorer, Chrome and Safari) would each require a bespoke version of the tool.

Another restriction was that devices such as iPhones and iPads would not let us through, as Apple do not approve the use of Firefox on their mobile systems. This meant that a large proportion of willing participants could not participate as they only accessed Facebook via their mobile devices or did not use Firefox. Our sample of research participants were therefore also determined by the devices and software they use.

The Admonitor browser plugin proved to be more successful than anticipated, however, and harvested advertising data from ambient background sites outside of Facebook to a degree of detail which we did not initially think possible, providing useful information on the specific kinds of data being exchanged by tracking and advertising companies during a live tracking process. It was through the plugin that we became aware of the scale of the tracking conducted by Facebook beyond a participant's Facebook page (see later).

METHODOLOGICAL ISSUES: THE SURVEY

More conventional studies of use through behavioural quantification for example, '75% of people do ….'. are just one element of a very complex picture. Yet, most research, even Facebook's own, is often based on quantitative rather than qualitative behavioural analysis. Rains and Brunner (2014) found, in their content analysis of SNS research published in six interdisciplinary journals between 1997 and 2013, that 62% of a total of 327 SNS studies used traditional survey research and that whilst many applied multiple methods of analysis, only 6% used some form of qualitative research.

Although aware of all the issues from previous media research of asking people about 'use', where participants notoriously underreport, and provide the responses they think the researchers want, we wanted to untangle forms of engagement, types of connectivity and intra-face-action. Did the Facebook platform make people do things? What do people think they are doing with Facebook? Did Facebook incite an 'anticipatory logic', as proposed by Kitchin and Dodge (2011), one that structures time and action through algorithms, protocols and prescriptions that sequence the possibilities for action and modulate experience, organizing the temporality of action and hence any form of use? Kitchin and Dodge describe platforms such as Facebook as 'transducted space', one which transforms that which passes through it, where code and software organize the spatiality and temporality of everyday life and one is produced through the other.[8]

We thus designed questions for our research respondents around how, when and what they did with Facebook: usage, relationship to Facebook, imagined audience, performance of self, type of attention and privacy. We addressed these particular issues because our previous research on audience responses to TV in mainstream media alerted us to the significance of affects and performance, especially the anxiety people feel when anticipating judgement of one's behaviour in public (Skeggs & Wood, 2012). We were therefore also interested in genre production, in how a specific digital form could operate as a highly charged affective medium that encourages people to divulge in a variety of ways, from the traditional biographical format to the immediate response mode. We see Facebook offering a communicative structure for generating a performance of a coherent persona (Skeggs, 2004, 2009) used to incite users to reveal information about themselves which can then be algorithmically disaggregated for the purposes of monetization.[9] Facebook is not just a 'data processor' as described by Peters (2013), part of a 'logistical media' whose content is not representational but rather organizational, instead, we maintain, it works with traditional forms of narrative and discourse to produce a particular genre of self-revelation, whilst simultaneously doing all the data processing that allocates data into saleable segments. Facebook as a platform with *both* genre and technical specifications offers possibilities for shaping our environment not just through code but also through the curating of discourse (Gillespie, 2010).

Therefore, in a similar way to that of Rogers (2013), we recognize Facebook as an epistemological platform, one that captures, processes, analyses, ranks, recommends formats and aggregates data before we even engage with it. We therefore see platforms as performative. They have what Hands (2013) calls 'platformativity'. Yet, most studies have only speculated about what this platformativity brings into effect, what it enables people, technology, organizations and capital to do or not do. Hence, rather than just focusing on the capacity of the platform itself (Galloway, 2012), we want to understand the types of communicative activity and personhood that Facebook mobilizes.

Conscious of ethics from the start (we had previously worked on legality and privacy, and as a longitudinal ethnographer on 'sensitive' topics), we asked people to sign an online consent form (see https://values.doc.gold.ac.uk/survey/). This created some confusion with the sentence 'I understand that confidentiality will be breached if I reveal matters that I later request to be withdrawn.' However, doing Facebook research immanently through the platform meant that all participants' queries could be answered directly and a debate ensued about the ethics of digital research and our phrasing of survey questions on our Facebook timeline. This dialogue generated a much more immediate sense of interaction than any surveys we had previously conducted (although still not in the zone of the intense ethnographic encounter). Surprisingly, the survey elicited some exceedingly long and detailed personal responses. Paradoxically, this suggested to us that people felt safe to discuss some of their performance anxieties digitally.

Methodologically, it would be very difficult for social media researchers to accurately identify participants via social categories due to the design of the medium. It would depend on honest declaration of detailed information (such as education, housing, occupation, etc.) and self-identification. As SNS's have promoted and supported dissimulation in the past, many users have become skilled in subterfuge (Turkle, 2011).[10] We think classifications and categories need studying in their making (gender, race, class and sexuality are after all performative). Some of the issues associated with classification became apparent through responses to the survey that revealed anxiety about revelation and judgement.

We did ask respondents to the survey for their age as we had monitored the debates about the decline in Facebook use amongst young people, a factor we also experienced when we tried to recruit participants (16–18 year olds) from a local Lewisham sixth form college to our project. Out of 500, only 2 said that they still used Facebook, preferring instead WhatsApp, Instagram and Snapchat, as a result of parents using Facebook to snoop on them. Most survey respondents declined to give their age, those who did were in their 30s and 40s.

In total, we received 154 responses, averaging 3 pages in length. We organized these into five connected themes that emerged from these. In future papers, we summarize the findings of the different elements of the research. Our third method used was to organize our participant's Facebook engagement into visualizations.

Using data visualizations

We used four different approaches to analyse our visualizations. Firstly, we employed traditional statistical analysis to identify patterns of engagement – when, where, how, whom – producing graphs and maps. We began with the interaction ratio, that is, the ratio of posts in the newsfeed which have been interacted with by the participant in relation to the total number gathered. We then mapped interaction rhythms to visualize the periods in the day in which the participant is most active on Facebook either writing or responding to posts and the gaps between these interactions. We measured average hourly activity over one day and average gaps between posts within a 24-hour period. We noted the most active posters and most active respondents – a listing of users who have responded to the most posts that the participant has interacted with. We measured the longest post durations – posts that have been acted upon over the longest periods. We then turned our attention to the newsfeed and mapped the length of longest newsfeed durations to identify posts that have been present in the newsfeed over the longest periods, the longest sponsored story durations – those that have recurred over the longest periods – and the longest Facebook advert durations – the Facebook sidebar adverts that have recurred over the longest periods. A more detailed description of the software and visualizations is available at: https://values.doc.gold.ac.uk/interactions/

Secondly, we mapped traditional network visualizations of our research participant's networks to make apparent the different networks of connectivity. Whereas such networks are normally egocentric, based

Figure 1. Network without user.

around the account user as a central node (such as in Hogan, 2008), we removed the account user from their own network in order to show the relations between other respondents that were based on communication activities such as posting, liking and commenting, rather than the participant's own friend network. This revealed the extent to which the participant was interacting with one cohesive group of people or several distinct groups and from this, we were able to infer some insights as to the influence of these networks on content in the newsfeed. It also provided empirical indicators relating both to issues of context collapse and to the survey question on how similar a participant considered their online correspondents to be. These visualizations are dynamic and can be manipulated by the user to see the density and form of the connection (Figure 1).

Thirdly, we inspected for rhythms and patterns, as per our methodology above. The visual graph (Figures 2) shows the flows over time of interactions between users on Facebook posts that participants have either created or responded to (through liking, commenting or sharing). We collected these data over a maximum six months and for some participants, over a shorter period of time as they struggled to stay connected (as per problems above). This is displayed against a visualization of the advertising data collected by Admonitor.

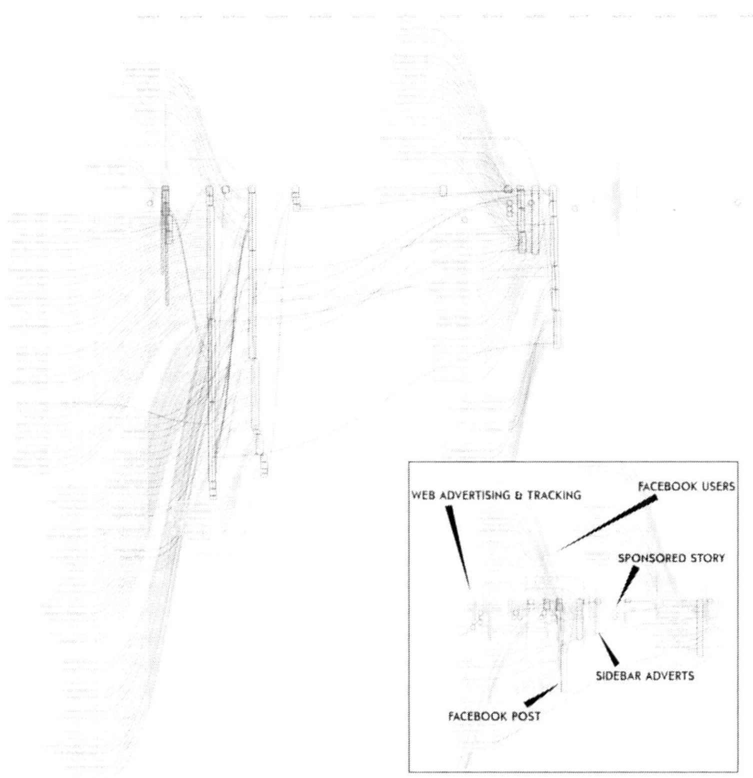

Figure 2. Interaction flows. The key to interaction flows is shown in the insert.

We read each method alongside each other; for instance, we read participant's survey responses to see if the engagement they described corresponded in any way to what they did and the networks they inhabited. We have always worked with a theory of the subject based on contradiction rather than coherence, for often, as subjects we may not know what we are doing and/or why. This is one of the big advantages of doing multi-method digital research that can connect different social relations over time and space.

Similarly, a great deal of traditional qualitative digital research rests on cognisant speech acts given via surveys (as above) or interviews, limiting understanding to one contrived encounter or discursive moment. This, as Savage and Burrows (2007) note, marks the 'old' sociology where accounts of action represented *the* sociological method. We are thus alert to these issues and are cautious when discussing visualizations with participants to see how they read and understand their rhythms, attention and 'lifeness'.

Fourthly, we analysed the adverts that appeared on participant's Facebook pages, their occurrence, if and when they were noticed and the actual content to see if they were indeed (algorithmically) effectively targeting their recipient as hyped by Facebook (Figure 3). The advertising data we gathered from Admonitor enabled us to relate patterns of user interaction with Facebook to possible patterns that might appear as participants browse the web. This enables us to learn more about how Facebook seeks to stimulate user interaction for the purposes of monetization.

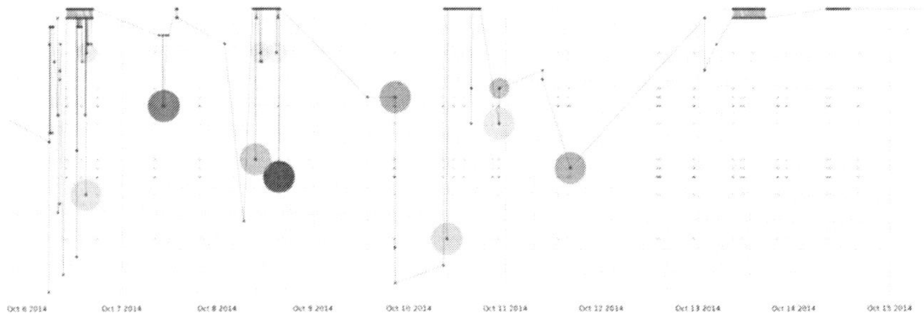

Figure 3. Advert tracking.

Our visualizations offer a picture not just of one person's rhythmic entanglements with others and the platform but also the ambient advertising that appears behind a person's Facebook page as they use it, and Facebook's algorithmic manipulation of the newsfeed (Figure 4).

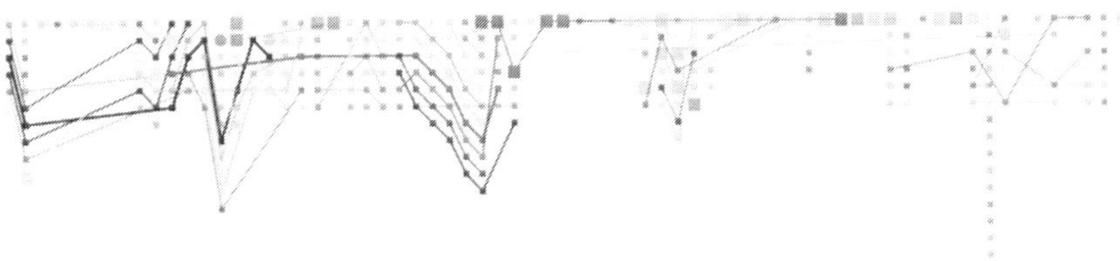

Figure 4. Newsfeed interaction.

The newsfeed data revealed how Facebook attempts to attract attention to what Facebook thinks it is important for the participant to see. We know that Facebook advertises itself to investors on the basis of its supposed capture of attention and here we can see how it attempts to do so. Napoli (2003) notes how advertisers struggle to secure attention by making the immaterial material: 'In selling audiences, media firms essentially deal in human attention, and human attention represents a much more abstract, elusive, and intangible product than, say, steel, insurance, or legal services' (p. 5). In a space where numerous platforms compete for our attention, for 'eyeballs', attention becomes a scarce resource, one that cannot easily be extracted and converted to value. An analysis of an 'attention economy' (Crogan & Kinsley, 2012) where values are directly connected to value thorough the idea of attention as a commodity, as well as debates about attention and governmentality (e.g. Bucher, 2012), will be developed from these visualizations. Here, our visualizations of newsfeed rhythms revealed how Facebook placed items for attention but also how our participants worked hard to avoid them, by scrolling down to find items they wanted to see.

We were also surprised to see how the advertising Facebook targeted, via stories on a user's newsfeed, was drawing not just on data collected on a participant's Facebook page but from all other internet sites they use with a Facebook connection (see the grey patterns in the background Figure 2 and the pathway in Figure 3). These are all the other sites from which Facebook is tracking and drawing information on use. At the start of our research, Facebook denied that it engaged in ambient tracking; however, since then, researchers commissioned by the Belgian data protection agency, from the Centre of Interdiscipinary Law and ICT and the Computer Security and Industrial Cryptography department (Cosic) at the University of Leuven, and the media, information and telecommunication department (Smit) at Vrije Universiteit Brussels, have drawn similar results to ours. Facebook tracks all usage on computers/devices via its social plugins which have been placed on more than 13 million sites including health and government sites (Gibbs, 2015). They are now being further investigated by the Belgian,

Dutch and EU Data Protection Commissioner, although as legal theorists (Mierzwinski & Chester, 2013) note, it is almost impossible for the legal profession to keep pace with global software companies who have far greater resources at their disposal.

Making sense of data

Kittler (1999) maintains that software is able to shape our experience and existence through a recursive or recombinant set of relations between code in space and time, in which infrastructures need code in order to function and people need infrastructures to make connections (Hayles, 2012). Yet, unlike Kittler, we do not think that the power of digital media is epistemological and computational rather than just ideological. We argue that algorithms are always/already ideological (Mager, 2012), in the way of all mediating technology (Postman, 1992) because they operate for particular interests (such as Facebook's capital expansion) and have been developed through different histories of power. If we reduced all aspects of life to computational capacity, we would lose the wider rhythms and the lifeness we were concerned to investigate. We understand how algorithms are essential to capitals' search for opening up and intervening in new markets (Poon, 2013) hence the proprietary nature of the major social media companies such as Facebook, Google and Amazon. Algorithms are the code that searches for value through capture and connectivity, maximizing their own potential through operations on themselves, as they model users and relationships. As Neyland (2014) notes, algorithms are also open to a world yet to be built but, as our newsfeed visualizations showed, unable to fully capture the attention they require.

Algorithmic data capture, although initially tied to an individual on Facebook, is made to work in ways independent of its source, independent of a person's production – hence why we argue that we need to understand the overdetermination involved in value production. The use of Facebook is not a straightforward exchange of an individual's labour for surplus value, rather data are partitioned, indexed, aggregated and disconnected from the labour of its producer before it is sold in a myriad of different ways (Nixon, 2014). It is also animated in many different ways as a commodity (Terranova, 2004). Facebook is a platform which mediates multiple features: to convert data into capital, to interest investors, to promote advertising to advertisers, to induce free development labour, to keep people connected, to encourage people to divulge, to experiment with users, to make users available to different forms of media. Facebook also operates to transmit many other media such a film, news and gossip; it mediates media.

CONCLUSION

Hopefully, we have established the terms for our inquiry into Facebook. This paper represents the start of the interrogation but we thought unpacking the methodology would be useful for other researchers. In this sense, our methodology is similar to what Jameson (1981) calls immanent critique, the idea that the description and criticism of a philosophical or cultural text (or in this case, a digital interface – Facebook) must be carried out in the same terms that the platform itself employs. We did 'get inside' one example of capital's new lines of flight. Our understandings of lifeness and rhythms, via software and methods that visualized networks, interfaces, entanglements, algorithms, encounters, person/a/s, conductivity, platformativity, time, tone, transduction and lifeness, hopefully open out ways of understanding some of the ways by which Facebook works. Our methodology enabled us to map a capta trail, of information captured as well as given (Kitchin & Dodge, 2011), recording lives over space and time. It is a small sample of big data, designed not to just describe patterns and networks, but also temporal rhythms, attention and activity beyond and through the interface. Through immanent research, we have advocated a practicable research ontology that goes beyond either an entrenched pessimism about algorithmic sociality or the empiricism which appears to characterize much of the study of ostensibly networked communication (Tufekci, 2014).

And we hope we address the concerns of Burrows (2012) who maintains that we need to rethink the descriptive power of the social sciences in order to reinvigorate a sociological imagination able to grasp the complexities of the data and to visualize, map and represent in ways that can claim back a distinctive jurisdiction over the social. We do feel we can say that our methodology enabled us to begin to examine our research questions: we were able to visualize how Facebook infra-structured connectivity, but also how people responded to connectivity differently, through diverse ratios of activity. We could trace and visualize how central advertising was to the infrastructure that lay behind our connections. These do shape our social relations on the interface but we provisionally suggest are less important once we leave. These questions will be investigated in more detail through our research findings. What we have outlined here is how we have developed the methods and mechanisms for further exploration. 'Lifeness' as a perspective enables us to understand the dialogic, complex and contradictory nature of technical forms that incite platformativity (people are limited by how they can engage). We agree with Terranova (2004) that sociality is folded into new forms of valorization where value resides in forms of life itself, within modes of expression, intensive relations, abstract knowledge, communications and affective interactions. Yet, we also think from our initial analysis that it is precisely the 'lifeness' that enables our engagement and entanglement, that sets our rhythms, such as energy, time and curiosity, that may establish limits to the way that valorization can encroach.

DISCLOSURE STATEMENT

No potential conflict of interest was reported by the authors.

FUNDING

This work was supported by the Economic and Social Research Council [grant number ES/K010786/1 "A Sociology of Values and Value"].

NOTES

1. When Facebook first launched as thefacebook.com, it described itself as 'an online directory that connects people through social networks' that provided 'a visualization of your social network.'. Screenshots of the original site are available at https://en.wikipedia.org/wiki/History_of_Facebook
2. Developed from Jenkins, Clinton, Purushotma, Robison, and Weigel (2009) which argued that 'new media literacies' comprise digital literacies plus media literacies.
3. API stands for Application Programming Interface; this is generally a surface layer provided by a software system to enable external programmes to interact with it or make use of its services.
4. Data from the Facebook API were updated on a daily basis whilst those coming from the browser plugin were updated as they arrived.
5. More detailed information on the software is available at the project website.
6. All Facebook apps intended for public use must first be evaluated and approved by Facebook. Facebook does allow developers free access to create private test versions of their apps before public release which do not require approval from Facebook. These test versions are restricted and can only be installed by users who are registered as part of the Facebook app developer scheme. Not all Facebook users can become developers, however, and we found in several cases, potential participants were refused developer status

due to, it appeared in some cases, them having quite tight privacy settings enabled on their accounts.

7. Scraping is a process through which data are extracted from a live webpage by software that looks for specific patterns, elements or phrases within a page and copies the content related to these.
8. The concept of transduction is adopted from Gilbert Simondon. It was originally applied to software and how algorithms process information in Mackenzie (2002).
9. See 'Monetary and moral value: What are the consequences of Capital Experimentation with Person Formation on Facebook', paper presented at the ICS Conference on 'Protest Participation in Variable Communication Ecologies', Alghero, Sardinia, June 2015.
10. Although clearly no longer. Facebook's demand for authentication was evidenced in the recent dragking/queen fiasco, September 2014, where Facebook tried to force people to reveal their state-legitimate names.

NOTES ON CONTRIBUTORS

Beverley Skeggs is Professor of Sociology at Goldsmiths, University of London. She is an ethnographer who researches the relationship between values and values through different aspects of social life, including class, space, violence, sexuality, reality TV and subject formation. [email: b.skeggs@gold.ac.uk]

Dr Simon Yuill is a software designer, postdoc researcher and artist. He has won many arts awards. His PhD was on software, legality and privacy. [email: sos01sy@gold.ac.uk]

The underlying research materials for this article can be accessed at https://values.doc.gold.ac.uk/

ORCID

Beverley Skeggs http://orcid.org/0000-0001-9354-5381

REFERENCES

1. Andrejevic, M. (2013). *Infoglut: How too much information is changing the way we think and know.* London: Routledge.
2. Back, L., & Puwar, N. (Eds.) (2013). *Live methods.* Malden, MA: Wiley-Blackwell.

3. Barad, K. (2007). *Meeting the universe halfway.* Durham, NC: Duke University Press.

4. Baym, N. K. (2007). The new shape of online community: The example of Swedish independent music fandom. *First Monday, 12*(8). Retrieved from http://firstmonday.org/ojs/index.php/fm/article/view/1978/1853

5. boyd, d. (2008). Why youth (heart) social network sites: The role of networked publics in teenage social life. In D. Buckingham (Eds.), *Youth identity and digital media* (119–142). Cambridge, MA: MIT Press.

6. Bucher, T. (2012). A technicity of attention: How software 'makes sense'. *Culture Machine, 13,* 1–13.

7. Bucher, T. (2013). Objects of intense feeling: The case of the twitter API. *Computational Culture, 3.* Retrieved from http://computationalculture.net/article/objects-of-intense-feeling-the-case-of-the-twitter-api

8. Burrows, R. (2012). Living with the h-index? Metric assemblages in the contemporary academy. *The Sociological Review 60*(2): 355–372. doi: 10.1111/j.1467-954X.2012.02077.x

9. Burrows, R., & Savage, M. (2014) After the crisis? Big data and the methodological challenges of empirical sociology. *Big Data and Society, 1,* 1–6. doi: 10.1177/2053951714540280

10. Crogan, P., & Kinsley, S. (2012). Paying attention: Towards a critique of the attention economy. *Culture Machine, 13,* 1–29.

11. DeLyser, D., & Sui, D. (2013). Crossing the qualitative-quantitative divide II: Inventive approaches to big data, mobile methods, and rhythmanalysis. *Progress in Human Geography, 37*(2), 293–305. doi: 10.1177/0309132512444063

12. Galloway, A. (2012). *The interface effect.* Cambridge: Polity.

13. Gibbs, S. (2015). Facebook tracks all visitors, breaching EU law. *The Guardian.* Retrieved from http://www.theguardian.com/technology/2015/mar/31/facebook-tracks-all-visitors-breaching-eu-law-report

14. Gillespie, T. (2010). The politics of platforms. *New Media and Society, 12*(3), 347–364. doi: 10.1177/1461444809342738

15. Hands, J. (2013). Introduction: Politics, power and 'platformativity'. *Culture Machine, 13*, 1–9.

16. Hayles, K. N. (2012). *How we think: Digital media and contemporary technogenesis.* Chicago, IL: Chicago University Press.

17. Hochman, N., & Manovich, L. (2013, June). Zooming into an instagram city: Reading the local through social media. *First Monday.* Retrieved from http://firstmonday.org/ojs/index.php/fm/article/view/4711/3698

18. Hogan, B. (2008). Analyzing social networks via the internet. In N. Fielding, R. Lee, & G. Blank (Eds.), *The handbook of online research methods* (pp. 141–161).Thousand Oaks, CA: Sage.

19. Ito, M. (2007). Introduction. In K. Varnelis (Ed.), *Networked publics* (pp. 1–14). London: The MIT Press.

20. Jameson, F. (1981). *Postmodernism, or, the cultural logic of late capitalism.* London: Verso.

21. Jenkins, H., Clinton, K., Purushotma, R., Robison, A. J., & Weigel, M. (2009). *Confronting the challenges of participatory culture: Media education for the 21st century* (The John D. and Catherine T. MacArthur Foundation Reports on Digital Media and Learning). Cambridge, MA: MIT Press. Retrieved from http://dmlcentral.net/resources/3756

22. Karpf, D. (2012). Social science research methods in internet time. *Information, Communication & Society, 15*(5), 639–661. doi: 10.1080/1369118X.2012.665468

23. Kember, S., & Zylinska, J. (2012). *Life after new media: Mediation as a vital process.* Cambridge, MA: MIT Press.

24. Kitchin, R., & Dodge, M. (2011). *Code/space: Software and everyday life.* Cambridge, MA: MIT Press.

25. Kittler, F. (1999). *Gramophone, Film, Typewriter.* Stanford: Stanford University Press.

26. Langlois, G., & Elmer, G. (2013). The research politics of social media platforms. *Culture Machine, 14*, 1–17.

27. Lefebvre, H. (2004). *Rhythmnanalysis: Space, time and everyday life.* London: Continuum.

28. Lievrouw, L. A. (2012). The next decade in internet time. *Information, Communication & Society, 15*(5), 616–638. doi: 10.1080/1369118X.2012.675691

29. Lievrouw, L. A., & Nguyen, L. U. (2007). Linking and the network imaginary. In *New Network Theory Conference*. Amsterdam Polytechnic, Hogeschool van Amsterdam: Amsterdam School for Cultural Analysis, the Institute of Network Cultures.

30. Light, B., & Cassidy, E. (2014). Strategies for the suspension and prevention of connection: Rendering disconnection as socioeconomic lubricant with Facebook. *New Media & Society, 16*(7), 1169–1184.

31. Mackenzie, A. (2002). *Transductions: Bodies and machines at speed*. London: Continuum.

32. Mackenzie, A. (2006). *Cutting code: Software and sociality*. New York, NY: Peter Lang Publishing.

33. Mager, A. (2012). Algorithmic ideology. *Information, Communication & Society, 15*(5), 769–787. doi: 10.1080/1369118X.2012.676056

34. Mair, M., Greiffenhagen, C., & Sharrock, W. W. (2015). Statistical practice: Putting society on display. *Theory, Culture & Society,* 1–27. doi:10.1177/0263276414559058.

35. Marres, N., & Weltevrede, E. (2013). Scraping the social? Issues in real-time social research. *Journal of Cultural Economy, 6*(3), 313–335. doi: 10.1080/17530350.2013.772070

36. Mierzwinski, E., & Chester, J. (2013). Selling consumers, not lists: The new world of digital decision-making and the role of the fair credit reporting act. *Suffolk University Law Review, XLVI*(3). Retrieved from http://suffolklawreview.org/selling-consumers-not-lists/

37. Miller, V. (2008). New media, networking and phatic culture. *Convergence: The International Journal of Research into New Media Technologies, 14*(4), 387–400.

38. Miyazaki, S. (2012). Algorhythmics: Understanding micro-temporality in computational cultures. *Computational Cultures, 3.* Retrieved from http://computationalculture.net/article/algorhythmics-understanding-micro-temporality-in-computational-cultures

39. Morozov, E. (2014). How much for your data? What you whistle in the shower. *LeMonde Diplomatique*. Kent, LeMonde 1408: 1, 10–11.

40. Munster, A. (2013). *An aesthesia of networks: Conjunctive experience in art and technology*. Cambridge, MA: MIT Press.

41. Murphie, A. (2007). The fallen present: Time in the mix. In R. Hassan & R. E. Purser (Eds.), *24/7: Time and temporality in the network society* (pp. 122–141). Stanford, CA: Stanford Business Books.

42. Napoli, P. (2003). *Audience economics: Media institutions and the audience marketplace*. New York: Columbia University Press.

43. Neyland, D. (2014). On organizing algorithms. *Theory, Culture & Society*. doi:10.1177/0263276414530477

44. Nixon, B. (2014). Toward a political economy of 'audience labour' in the digital era. *tripleC, 12*(2), 713–734.

45. Peters, J. D. (2013). Calendar, clock, tower. In J. Stolow (Ed.), *Tower in Deus in Machina: Religion, technology and the things in between* (pp. 25–42). New York: Fordham University Press.

46. Poon, M. (2013). Response to tarleton Gillespie's 'the relevance of algorithms'. In *Governing Algorithms Conference*. New York University, US. Retrieved from http://governingalgorithms.org/

47. Postman, N. (1992). *Technopoly: The surrender of culture to technology*. New York, NY: Vintage Books.

48. Rains, S. A., & Brunner, S. R. (2014). What can we learn about social network sites by studying Facebook? A call and recommendations for research on social network sites. *New Media & Society*. doi:10.1177/1461444814546481

49. Rogers, R. (2013). *Digital methods*. Cambridge: Cambridge University Press.

50. Savage, M., & Burrows, R. (2007). The coming crisis of empirical sociology. *Sociology, 41*(5), 885–899. doi: 10.1177/0038038507080443

51. Skeggs, B. (2004). *Class, self, culture*. London: Routledge.

52. Skeggs, B. (2009). The moral economy of person production: The class relations of self-performance on 'reality' television. *Sociological Review, 57*(4), 626–644. doi: 10.1111/j.1467-954X.2009.01865.x

53. Skeggs, B. (2014a). Values beyond value? Is anything beyond the logic of capital? (The 2013 BJS annual lecture). *The British Journal of Sociology, 65*(1), 1–20. doi: 10.1111/1468-4446.12072

54. Skeggs, B. (2014b). A global sociology of values and value: How to move beyond the analytic of capital. International Sociology Association, Yokohama, Japan.

55. Skeggs, B., & Wood, H. (2012). *Reacting to reality television: Performance, audience, value.* London: Routledge.

56. Terranova, T. (2004). *Network culture: Politics for the information age.* London: Pluto.

57. Tufekci, Z. (2014). Big questions for social media big data: Representativeness, validity and other methodological pitfalls. In ICWSM '14: Proceedings of the 8th International AAAI Conference on Weblogs and Social Media, pp. 505–514.

58. Turkle, S. (2011). *Life on the screen.* New York, NY: Simon and Schuster.

59. Van Dijck, J. (2012). Facebook as a tool for producing sociality and connectivity. *Television and New Media, 13,* 160–176. doi: 10.1177/1527476411415291

60. Wellman, B., & Gulia, M. (1999). Virtual communities as communities: Net surfers don't ride alone. In P. Kollock & M. Smith (Eds.), *Communities in cyberspace* (pp. 167–194). London: Routledge.

61. Weltevrede, E., Helmond, A., & Gerlitz, C. (2014). The politics of real-time: A device perspective on social media platforms and search engines. *Theory, Culture & Society,31*(6), 125–150. doi: 10.1177/0263276414537318

62. Wittel, A. (2001). Toward a network sociality. *Theory, Culture & Society, 18*(6), 51–76. doi: 10.1177/02632760122052048

2

CHAPTER

The Google Voter: Search Engines and Elections in The New Media Ecology

Filippo Trevisan[a], Andrew Hoskins[b], Sarah Oates[c] and Dounia Mahlouly[d]

[a]*School of Communication, American University, Washington, DC,USA;*

[b]*College of Social Sciences, University of Glasgow, Glasgow, UK;*

[c]*Philip Merrill College of Journalism, University of Maryland, College Park, College Park, MD, USA;*

[d]*Department of War Studies, King's College, London, UK*

ABSTRACT

This paper explores key Internet search trends for electoral information vis-à-vis the broader media ecology in the UK and the US. An innovative methodology is introduced that maps the informational trajectories of key election events by combining Google Trends data linked to significant news events during the campaigns. While the research found spikes of search that suggested a seemingly trivial event in the US could drive voters to search out deeper information on related policy issues, the study did not find the same phenomenon in the UK. This invites a reflection on the opportunities available to Internet users/voters in the changing informational landscape and how scholars can leverage

Google Trends records to better understand how voters seek information in new and evolving media ecologies.

Keywords: Google Trends, elections, Internet search, media ecology, UK, US

The mass media landscape in which elections take place has changed dramatically in recent years. Although established news media outlets continue to be major sources of information about elections, their ability to act as gatekeepers and frame campaigns, issues, and events in specific ways has been mitigated by the rise of new media technologies as direct communication channels between politicians and citizens. In light of these changes, some scholars have expressed concerns about personalized information feeds and tailored messages, which could create information cocoons that restrict pluralism. As Howard (2006) put it, by 'redlining some constituents and communities and then narrowcasting political content, hypermedia campaigns diminish the amount of shared text in the public sphere' (p. 183). Many others expanded on this point, explaining that 'campaigns simply are not designed to be the training grounds of radical democratic participation that many desire' (Kreiss, 2012, p. 183) and that nowadays 'the decidedly undemocratic view of *controlled interactivity* is how most campaigns operate' (Stromer-Galley, 2014, p. 2).

Despite these concerns, however, the vast majority of the research carried out to date to investigate the diffusion of electoral information in this changing media landscape has focused primarily on the 'supply-side' (Lilleker & Vedel, 2013), that is, the online activities of candidates and the content of official campaign messages. Therefore, it is essential to look more broadly at how information about elections is accessed, shared, and re-mediated in today's hyper-connected public sphere. Crucially, neither 'wired' voters nor politicians operate in an informational vacuum. Instead, they are flanked by established print and broadcast outlets, which continue to perform as influential news providers. Researchers need new methods and types of data that can map the trajectories followed by electoral information in such shifting media ecologies.

In this paper, we contribute to methodological development by presenting a new approach that combines the analysis of open access Google Trends data (www.google.com/trends) with the exploration of relevant news coverage. This strategy is applied to the investigation of electoral information flows in two democratic countries with variant political and media structures: the United Kingdom and the United States. Google Trends is employed to explore how users/voters seek information in the new media ecology. This work contributes to the development of new methods for political and public communication research.

1. ELECTIONS IN THE 'NEW MEDIA ECOLOGY'

There have been some useful attempts to re-conceptualize contemporary news production and event mediatization processes. In particular, the idea of 'media ecology' – that we define as not only the media environment, but also the media 'imaginary' – of the day that makes possible and shapes a set of relationships between individuals, groups, and communication technologies has gained renewed popularity. It is a useful way of setting out how the avalanche of open access data and Internet search opens up new media imaginaries both for citizen publics and for scholars.

'Media ecology' is a term that has been variously applied to highlight 'the complex symbiotic relationship between the media and, on another level, between media and the various forces in society' (Lum, 2000, p. 1). Many associate the term with the work of Marshall McLuhan, although it has a much longer history.[1] Media ecology is often traced to its naming by Neil Postman in his early 1970s work that attempted to imagine the future of American secondary education. Postman championed the teaching of media ecology as an alternative to English, namely as 'the study of media as environments' (Postman, 1970, p. 161), but also the 'study [of] the interaction between people and their communications technology' (Postman, 1970, p. 161).

Building on this tradition, our study draws upon a more recent development of the idea of a 'new media ecology' (Awan, Hoskins, & O'Loughlin, 2011; Hoskins & O'Loughlin, 2010). This we define as the global hyper-connected media environment that gives rise to distinct media ecologies that are shaped by more localized (and particularly national) factors such as media regulation, journalistic culture, and the dynamic relationship between established and emergent media (Hoskins & Tulloch, 2016). As Kluver (2002) noted, Internet-based media operate on the basis of database logic, which is different from the narrative logic of established news media. This affords Internet users opportunities to gather unprecedented amounts of content and, at least in theory, escape the information 'hegemony' of legacy media. Yet, 'because the [Internet] data must be arranged, the power of arrangement, and thus the power of authority, typically rests with the database operator or designer' (Kluver, 2002, p. 503). This important role is performed largely by search engines and social networking sites. In addition, the use of new media is influenced by local culture, habits, and user expectations, as well as interactions with established media outlets (Oates, 2011). In light of this, it seems reasonable to assume that, despite the global reach of online media, the ways in which citizens in different countries acquire electoral information are guided by and through distinct media imaginaries.

From this perspective we can investigate empirically how new media technologies challenge the traditional media balance in elections. So far, cross-media research on political information has focused mostly on the relationship between Twitter and television (see, e.g., Anstead & O'Loughlin, 2011; Elmer, 2013). However, Twitter is only one platform in a wide range of new technologies. In addition, Twitter users are a relatively small minority of Internet users, albeit vocal and influential. Search engines are important players in the 'new media ecology,' but so far their use has not been studied in conjunction with traditional mass media. Focusing on key search trends and comparing them with other sources of information about elections, particularly established news media outlets, is an important way for cross-media research dig deeper into the 'demand' side of the political information 'market.'

2. SEARCH ENGINES, INFORMATION, AND POLITICS

Despite the rise of social media as information providers, search engines remain the primary channel through which Internet users access online information in democratic countries such as the UK (Dutton & Blank, 2013) and the US (Purcell, Brenner, & Rainie, 2012). Internet users appear to trust search engines – especially Google – almost blindly, to the point that they question their own ability to search properly before doubting the effectiveness of Google's algorithm (Hillis, Petit, & Jarrett, 2013). This is especially true for young people, who have no experience of the Internet prior to Google (Gunter, Rowlands, & Nicholas, 2009). In light of this, social scientists have spoken of a 'Googlization' phenomenon capable of affecting multiple economic, social, and political aspects of life (Lovink, 2009) and asked whether 'anything (or anyone) matter[s] if it (or she) does not show up on the first page of Google results' (Vaidhyanathan, 2011, p. 7).

These trends have implications for democratic elections. Recent experimental work in this area hypothesized the existence of a 'search engine manipulation effect' for which biased search rankings can directly influence voter choices, affecting election results (Epstein & Robertson, 2015). These claims expanded more established concerns that search engines perpetuate the status quo by favoring dominant information sources (e.g., established news outlets, major political parties, corporations, academic institutions, etc.) while at the same time marginalizing minorities and alternative narratives (Hindman, 2009). This goes back to the fact that search engines essentially operate a loop model for which they 'both contribute to the selection of more prominent sites, and in turn are more influenced by them' (Halavais, 2009, p. 59). This model is set to continue unless a radical shift in consumer preferences or regulative intervention in the construction of search engines takes power away from commercial players such as Google (Mager, 2012).

Concerns about the implications of search for democracy are justified. However, focusing exclusively on search results generates a partial understanding of this phenomenon. Experimental methods rely on artificial settings that privilege search outcomes and therefore cannot clarify when and why people search for information in natural contexts. Instead, fully exposing patterns of search engine use in pivotal political moments such as elections requires a focus on search 'inputs' as well as 'outputs.' Do users/voters tend to search the Internet at particular moments in time? How do search patterns compare with the election information provided by other sources, including established news media outlets? Long before the Internet became commercially available, Molotch and Lester (1974) described news consumption as a 'purposive behavior' through which audiences filter and interpret information obtained from the mass media according to their own priorities and interests. Internet search, one could argue, expands the purposive nature of information gathering by providing citizens with tools that they trust to sort through virtually limitless amounts of content to fulfill specific informational needs and interests.

Online information-gathering in the new media ecology occurs alongside and in combination with, not in opposition to, the continued contribution of legacy news media to electoral contests (Vergeer, 2013). Therefore, investigating search engine use requires us to track the informational trajectories of key electoral issues across both search engines and established forms of media. Researchers must be able to identify when new issues gain prominence on different forms of mass media, which can provide useful cues as to who (or what) drives election information flows in the new media ecology. In an effort to better understand this dynamic informational space, this paper proposes to compare search trends over time with relevant news media trends. Our use of Google Trends is a key element in transforming the understanding of electoral information flows in this complex emergent environment.

3. GOOGLE TRENDS: OPPORTUNITIES AND DRAWBACKS

Google Trends is a freely available tool that elaborates archived search records to show fluctuations in the popularity of any given keywords on Google, provided that a minimum number of relevant searches were carried out over a set period of time. Google Trends has become popular with researchers in a number of fields, from economics (Kaeserbauer, Hohenstatt, & Reed, 2012) to environmental studies (Chai & Sasaki, 2011). In particular, the number of scientific publications including Google Trends data has grown steadily in the areas of epidemiology and public health (Nuti et al., 2014). Some public opinion scholars have pioneered the use of Google Trends as a 'salience' barometer (Scharkow & Vogelsgang, 2011; Scheitle, 2011). Yet, so far the application of this tool in political and public communication research has been much more limited than in other fields. To understand its relevance for this type of research, it is useful to briefly review how it works.

Google Trends output consists of normalized data measured against the total number of searches registered during the period under scrutiny. Individual Google Trends scores are derived from actual search records normalized on a 0–100 scale, where 100 represents the day, week, or month – depending on the amount of time under examination – in which the given keywords were searched most frequently. All the other values are calculated in function of their distance from the top score. Thus, Google Trends identifies swings in user-interest for specific topics, people, or events by carrying out longitudinal comparisons of the search frequency for inter-related terms within one or multiple countries, as well as on a global scale. The output can be downloaded in CSV format and used to build additional visualizations. Google Trends also automatically identifies the terms that users searched most frequently in conjunction with those suggested by the researcher. This may offer insights into how users move between topics or different aspects of the same topic. Superimposing this type of information on a timeline of key campaign events and relevant news media coverage could help to clarify information flows in contemporary elections.

Nevertheless, the actual number of relevant search queries carried out for an individual keyword or a set of keywords remains undisclosed. In addition, Google Trends tells us how likely users were to run a certain search query at any one time, but it does not reveal the results of these searches or which URLs users clicked on. The lack of even a rough estimate of search volume makes the use of Google Trends output in lieu of public opinion survey data impractical and inappropriate. Thus, recent studies found that Google Trends is a poor predictor of election results (Lui, Metaxas, & Mustafaraj, 2011; Yasseri & Bright, 2013). Although Google is the preferred search engine for a large majority of Internet users in the UK and the US, this is still not the case for every single user. Similarly to Twitter-based research, studies using Google Trends should clarify that findings relate only to Google users and not all Internet users or, worse, an entire country population (Trevisan, 2014). As with other digital data, it would be misleading to assume that search engine records provide a cheaper alternative for significant population indicators or more expensive methods such as opinion polls and surveys.

Given that Google Trends' results depend on the specific keyword(s) and timeframe selected for analysis, its output can be interpreted correctly only in conjunction with the surrounding political and media context. Search analysis should not be developed in isolation. Instead, it is most useful in combination with other methods to study increasingly complex media ecologies. One way of doing this is by creating a measure of 'news popularity' for the issues, people, or events of interest that can be compared directly with the measure of 'search popularity' obtained from Google Trends. In this study, this was achieved by retrieving all the news items that focused specifically on the issues under scrutiny through Lexis Nexis and subsequently normalizing frequency scores obtained for each day on a scale from 0 to 100, where 100 represents the day with the biggest number of relevant news items and all the other values were derived from it.[2] Results were then plotted on the same graph as Google Trends scores to establish whether Internet search patterns mirrored trends in news coverage, or Google users had pursued a different informational agenda. Given the exploratory nature of this paper, major search peaks were identified by visually inspecting Google Trends charts and reviewing the associated CSV records. Although they constitute exceptional events, major search peaks provide a useful starting point in exploring the use of search engines as they involve an especially large number of users, with numbers being the key determinant of election success. Future research could test also lagged time series analysis to explore the relationships between smaller search peaks, which visual analysis may miss, and relevant news coverage.

3.1. Case studies: search in different media ecologies

This paper explores two case studies: the UK and the US. There are differences between these democracies that could affect the use of search engines and their relationship with traditional mass media. The UK features a parliamentary system. In addition, commercial players dominate the British print media sector while public-service broadcasters are key producers of TV, radio, and online news that is arguably perceived as 'objective.' In contrast, the US is characterized by a presidential system and commercial ownership across both print and broadcast media, which is also more polarized.

Both UK and US publics are familiar with online politics, in the context of digital electioneering and through other activities such as e-petitions (Chadwick, 2012). However, these countries are approached as having media ecologies at different stages of development, with different dynamics fostered between 'traditional' and 'emergent' media. Recent studies highlight an interesting difference between these two countries, with American Internet users especially keen on engaging with digital native news outlets such as Yahoo News, the Huffington Post, and Buzzfeed, while the British online news market is dominated by legacy brands, especially the BBC (Newman, Fletcher, Levy, & Nielsen, 2016, pp. 32–35). These findings seem to suggest that US publics are less reliant on traditional news sources and more inclined to experiment with alternative online information providers. Conversely, UK audiences continue to rely more frequently on established news providers.

As we consider these two country case studies, we also need to remember that media ecologies are inherently dynamic (Strate & Lum, 2000). Several factors, both internal and external, can affect the perspective of citizens on media, as well as relationships between different forms of media. For example, in recent years the BBC's reputation for integrity has been challenged by scandals involving former presenters and executives. Regional differences have also emerged, with doubts raised about the impartiality of the BBC's coverage of political events such as the 2014 Scottish independence referendum (Robertson, 2016). This could affect the levels of trust that people have in mass media, with likely implications for how publics use Internet-based tools, including search engines, to access alternative sources of information. Therefore, research that takes an ecological approach should always contextualize its results and remain alert to the limitations involved in trying to assess such a moving target.

For the purpose of this research, Google Trends was used to investigate the popularity of searches involving the names of the main candidates in the 2012 US presidential election (Barack Obama and Mitt Romney) and those of the three main UK party leaders – and therefore candidates to the role of Prime Minister – in the 2010 UK general election (David Cameron, Gordon Brown and Nick Clegg). Geographical parameters were set in Google Trends so that the datasets generated for this study included only domestic searches and expressed local dates and times in each country. Findings from this initial part of the study informed a deeper investigation of country-specific trends.

Party leaders and presidential candidates were useful comparative cases because, despite differences in electoral processes, they tend to catalyze the attention of both traditional news providers and voters in each country. Although the politics of personalization is significantly more important in the US system, personal traits of party leaders, history, and behavior are also widely reported in British campaigns and news coverage (Langer, 2012). In addition, the centrality of party leaders continues to increase as TV debates, which originated in the US, have now become a mainstay of elections in multiple parliamentary democracies, including the UK (Anstead, 2016). That said, future research should look into a broader set of election-related factors and topics as comparative search terms, including political parties and flagship policies.

Having identified key search moments in elections and the potential drivers behind them, we retrieved the news coverage associated with these events from Lexis Nexis's 'US publications' and 'UK publications' collections. Given our media ecology approach, these collections were chosen because they are the largest and most representative ones in Lexis Nexis, including all the traditional news sources – both print and broadcast transcripts – that this database carries for each country. Search parameters were set to draw all the news items that mentioned the issues associated with Internet search peaks in their respective headline or lead paragraph, leaving out those that referred to them only in passing. Highly similar items were eliminated automatically by Lexis Nexis. Results were also screened manually for accuracy. This led to the elimination of a small number of news items such as letters to the editor and others that were not directly relevant to the case studies examined here, most notably reports on policy issues in other countries.

4. FINDINGS AND DISCUSSION

Overall, search patterns identified some broad similarities. However, a more detailed exploration of seemingly equivalent 'search events' – that is, particularly intense search traffic surges – in both countries revealed some important differences. Except for those on Election Day (6 May 2010 in the UK; 6 November 2012 in the US) and the day on which David Cameron was appointed new British Prime Minister (11 May 2010), the most substantial search peaks in the three-month periods examined for this study were associated with TV appearances by presidential candidates in the US (Figure 1) and party leaders in the UK (Figure 2). In particular, these surges in search popularity coincided with the televised election debates in both countries. This suggests that major scheduled media events such as the

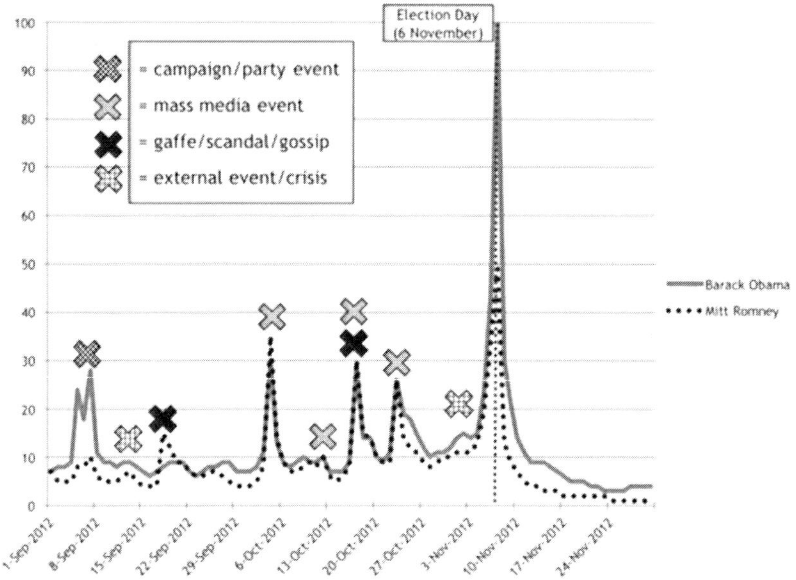

Figure 1. Search popularity for 'Barack Obama' and 'Mitt Romney' in the US, Sept.-Nov. 2012 (Google Trends scores on Y axis, dates on X axis).

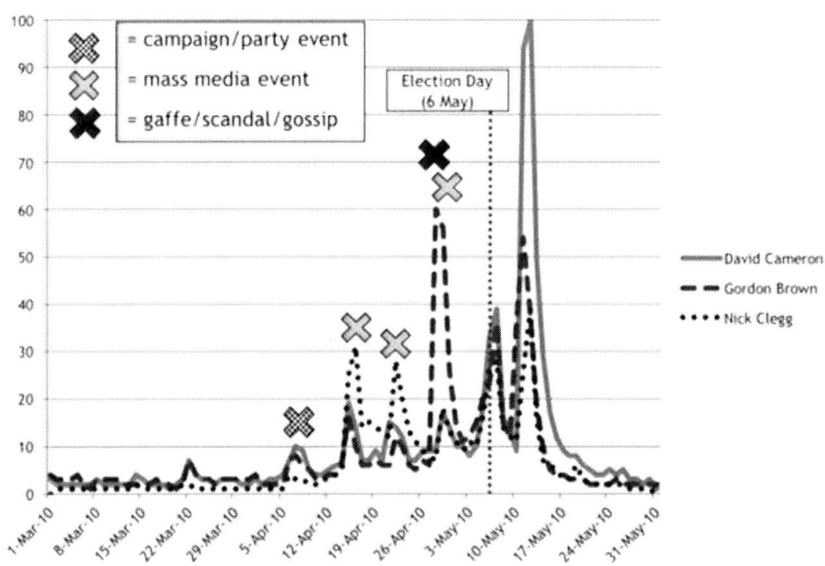

Figure 2. Search popularity for 'Gordon Brown,' 'David Cameron,' and 'Nick Clegg' in the UK, March–May 2010 (Google Trends scores on Y axis, dates on X axis).

debates continue to be key drivers of citizen interest in politics even in times of unprecedented digital connectivity and hypermedia campaigns.

It was especially interesting to note that, in conjunction with the first two UK TV debates (15 and 22 April), the Liberal-Democrat leader Nick Clegg was 'Googled' considerably more than both the incumbent Prime Minister (Gordon Brown) and the leader of the opposition (David Cameron). Prior to the first debate, 'journalists and commentators had largely ignored Clegg' (Parry & Richardson, 2011, p. 479) so that for the majority of the British electorate 'he did not really exist' (Parry & Richardson, 2011, p. 479). As Clegg performed particularly well in that debate, it seems reasonable to hypothesize that this successful TV appearance spurred a flurry of Google searches by citizens who wanted to know more about him. This corroborates the impression that high-profile TV events tend to act as powerful catalysts of online behavior in elections. That said, the fact that the Liberal-Democrats actually lost five seats in the House of Commons at the 2010 election suggests that increased media visibility, including on search engines, does not necessarily translate into more votes.

Although TV debates appeared to drive up Internet search in both countries, it is worth noting that search peaks occurred at slightly different moments in the UK compared to the US. British Google users were especially likely to search for information on party leaders on the same days as the televised debates (15, 22, and 29 April). On the other hand, Americans were more likely to query Google for information about Barack Obama and Mitt Romney on the days that immediately followed each of the presidential TV debates, which took place on 3, 16, and 22 October. These results raise the issue of whether temporal differences in otherwise similar search patterns signal different approaches to search engines on the part of the users.

For example, one could hypothesize that British voters used search engines both to learn about party leaders before each debate and check the issues mentioned during the debates in 'real time.' Conversely, US voters might have turned to Google in order to review and reflect on what candidates said during the TV debates, possibly looking for video clips, transcripts, or commentary. While testing these hypotheses goes beyond the scope of this paper and is likely to require in-depth research 'to understand the psychology of online political participation' (Lilleker & Vedel, 2013, p. 415), this difference provided the first clue that platforms such as Google may indeed support different types of voter activity in different democratic contexts.

Other notable 'search catalysts' in both countries included planned campaign events such as party conventions in the US (Figure 1)[3] and the day on which the election was officially called – 6 April 2010 – in the UK (Figure 2), as well as major events unrelated to the campaigns that granted candidates intense media exposure. For example, there was a substantial surge in searches for Barack Obama in conjunction with Hurricane Sandy – especially 31 October – and the president's visit to disaster-stricken areas. Another external event connected to an increase in searches for the two main presidential candidates in the US was the attack against the American consulate in Benghazi between 11 and 12 September 2012 in which four Americans, including the ambassador, lost their lives. However, other important search peaks did not appear to be linked to any immediately identifiable event. In particular, two of these search peaks stood out: the one registered for US Republican candidate Mitt Romney on 18 September 2012; and the one registered for then British Prime Minister Gordon Brown on 28 April 2010. No key campaign events or planned candidate media appearances took place on those days. Thus, the source of such strengthened user interest was bound to be something less conventional and more spontaneous, outside the direct control of campaign managers.

4.1. Just a laughing matter? Gaffes as search triggers

A look at news reports for those days revealed that both Mitt Romney and Gordon Brown found themselves at the center of rather embarrassing and widely publicized gaffes. On 28 April 2010, a Sky News microphone left inadvertently on Gordon Brown's jacket caught him describing a Labour party supporter whom he met during a campaign event in Rochdale as a 'bigoted woman.' This episode, which British news media immediately labeled as 'bigot-gate,' generated huge media attention.

In another embarrassing incident, a secretly recorded video was posted to the website of magazine Mother Jones on 17 September 2012, showing Mitt Romney talking with contempt about the '47 per cent of the people who will vote for the president [Obama] no matter what' during a fundraising event. Furthermore, a look through the other search peaks identified with Google Trends revealed that at least one more of these was connected to a further gaffe by Romney. When he addressed a question about pay equity during the second presidential debate, Romney said that, after he was elected governor of Massachusetts, he asked women's organizations to recommend talented female candidates to appoint to his cabinet. In response, Romney said during the debate, he was given 'binders full of women.' This comment immediately sparked a discussion about Romney's presumed sexist attitudes and popular critique through a considerable number of online memes (Rentschler & Thrift, 2015).

At face value, the emergence of politicians' gaffes as drivers of Google searches could seem part of a tendency for voters to be more interested in trivial, almost comedic episodes than policy issues. Indeed, this argument resonates with work that shows how citizens have become especially interested in politicians' personal lives (Stanyer, 2013), which digital media help to expose by 'overhearing' conversations that were not meant for a public audience (Dayan, 2013). However, one has to ask also whether there is more to these search peaks than the mere 'tabloidization' of politics. Could it be that these seemingly frivolous episodes became springboards for Google users to search for additional information about policy issues, making a 'thematic leap' in their searches? If so, how did these searches compare to news media coverage? We addressed these questions by querying Google Trends for searches about policy issues that voters might have associated with these episodes, namely: women's rights and gender equity for 'binders full of women'; and immigration for 'bigot-gate.' Thereafter, we compared search results with the popularity of the same issues in the news media in each country.

4.2. 'Binders full of women' and gender equity

To explore the interaction between different forms of media, it was useful to analyze in depth the 'binders full of women' gaffe, which – like 'bigot-gate' in the UK – occurred on TV. For this reason, we queried Google Trends for an aggregate set of keywords closely connected to women's rights and gender equity issues, including: gender equity; sex equity; women's rights; women's pay; pay equity; pay gap; gender gap; and glass ceiling. First, it was apparent that searches for information on women's rights and gender equity peaked to their maximum level immediately after the 'binders full of women' gaffe (17 October, see Figure 3). In addition, on average US Google users were markedly more likely to search for information on women's rights and gender equity in the period after Romney's ill-fated remark compared to the six weeks of campaigning that preceded it (Table 1).

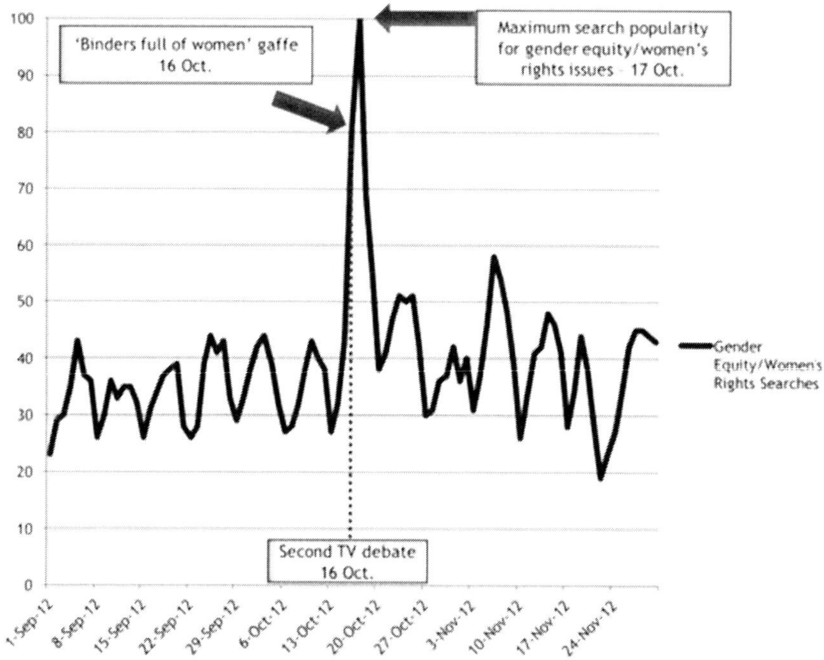

Figure 3. Search popularity for gender equity and women's rights issues in the US, Sept.-Nov. 2012 (Google Trends scores on Y axis, dates on X axis).

Table 1. Average Google Trends daily score for gender equity and women's rights issues before/after 'binders full of women.'

Dates	Average Google Trends score	
	Mean	**Median**
Before 'Binders full of women' (Sept. 1st–Oct. 15th)	34.48	34.5
After 'Binders full of women'(Oct. 18th–Nov. 30th)	40.5	41

One key hypothesis to explain this increase in interest for women's rights issues among American Google users would be that the way in which the news media covered Romney's gaffe prompted them to look for more information on this topic. In order to verify this, we searched Lexis Nexis's 'US publications' collection for news items that mentioned the same keywords we used to query Google Trends in their headline and/or first paragraph between the start of October and Election Day (6 November). The daily frequencies obtained through this process were then normalized to create a 'news popularity index' based on the same 0–100 scale used for Google Trends scores as discussed above.

Lexis Nexis searches revealed that news coverage of women's rights and gender equity increased considerably after the second presidential debate with 205 news items focused primarily on these issues published between October 16 and Election Day, compared to just 99 in the first half of October. That said, looking more closely at the temporal distribution of these news items and comparing it to Google Trends scores showed that, in the immediate aftermath of 'binders full of women,' the popularity of women's rights and gender equity issues among Google users greatly outpaced their popularity on US news outlets (Figure 4). While Google searches for information on these issues peaked to maximum levels on the day after Romney's gaffe on live TV (17 October), as was discussed above, the highest concentration of news coverage on these issues occurred only several days later (22–26 October).

One possible explanation for this would be that journalists needed some time to interpret the significance of the gaffe and incorporate these issues in their coverage, while Google users were able to search for information on related policy topics much more rapidly. However, it is also worth noting that the 'binders full of women' gaffe generated a great amount of coverage and was at the center of 117 news items up to Election Day, 35 of which just on 17 October. While this is not the place for a detailed framing analysis of this coverage, it is interesting to note that only 18 of the news items that discussed the 'binders full of women' episode in this entire period explicitly mentioned women's rights and gender equity. This suggests that, while American news media were eager to cover Romney's gaffe, they also did not connect it to broader policy issues, at least not immediately, while they may have associated it with the candidate's personality and attitude toward women. In contrast, users/citizens were the first ones to show increased levels of interest in women's rights and gender equity, turning to Google before established media outlets grasped the newly augmented salience of these issues in the wake of Romney's words (Figure 4).

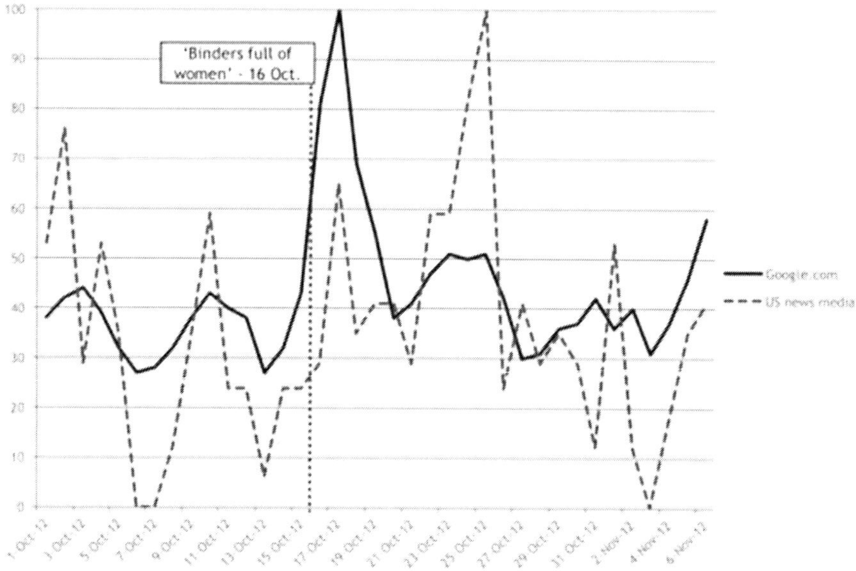

Figure 4. Popularity of women's rights and gender equity issues on Google.com vs. US news media.

It is possible to characterize these results as showing citizens as 'entrepreneurial' Internet users in the electoral information space. We acknowledge that it is impossible to know from the search data and news content analysis whether citizen and journalistic interest in women's rights and gender equity arose spontaneously. What can be interpolated from the data, though, is that a large number of Google users in the US – and therefore a considerable snapshot of the American electorate at that time – showed a very high interest in both the 'binders full of women' gaffe and the policy issues surrounding it. While these data are not sufficient to claim that Romney's 'binders full of women' gaffe directly caused a surge in interest for women's rights issues, it nevertheless shows that, following this episode, users/voters became interested in information that traditional news outlets did not provide immediately and one way for citizens to remedy this gap was to search for information on the Internet. Further work is needed to better understand the influence that the campaigns' responses to this episode – in particular Obama's – had on the way in which voters interpreted it before it can be confirmed whether Google users truly acted as 'emancipated' information scouts on this occasion. However, what these data do highlight is potential to understand voter interest and engagement with campaign events at a more granular level by using Google Trends.

4.3. 'Bigot-gate' and immigration policy

Results for the relevance of 'bigot-gate' in UK searches told a very different story. In addition to being associated with the highest search peak for information about Gordon Brown in the entire election period (Figure 2), 'bigot-gate' was at the center of considerable media attention in the UK. Searching Lexis Nexis's 'UK publications' collections for news items that mentioned this incident in their headline or lead paragraph up to one week after the election provided a sense for the amount of coverage dedicated to it. Direct coverage and commentary spiked immediately after the incident, with over 140 news items published on the day after 'bigot-gate' alone (29 April). However, this trend also dissipated very rapidly (Figure 5).

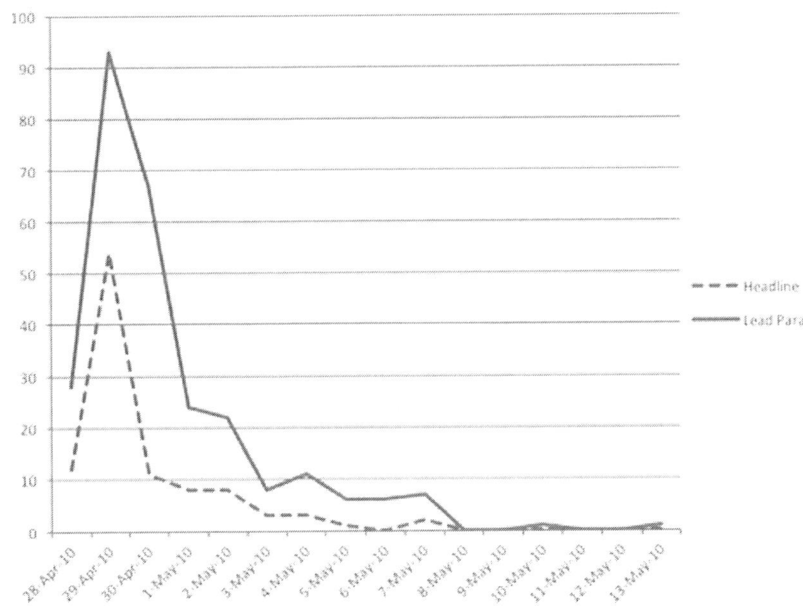

Figure 5. Number of UK news items in which 'bigot-gate' appeared in the headline or lead paragraph.

In this context, it is important to ask whether this gaffe was accompanied by a surge in Google searches for information on policy-relevant issues in the same way as Romney's 'binders full of women' in the US. Given that Gordon

Brown used the word 'bigoted' in response to a negative comment about migrants made by the Labour supporter he met in Rochdale, some authors have argued that this incident strengthened the centrality of immigration to voter concerns (Bates, 2010; Flynn, Ford, & Somerville, 2010). Thus, Google Trends was queried for keyword aggregates focusing on immigration policy plans (immigration policy, immigration plan(s), immigration cap).

In contrast with expectations, Google Trends results showed that there had not been a noticeable change in the likelihood that British users would search the Internet for information on immigration-related issues following Gordon Brown's gaffe (Figure 6). In fact, a dip in the number of Google searches for specific information on immigration policy was registered during the week of 'bigot-gate' compared to the previous one. The very fact that Google Trends only provides weekly scores for these searches signals that the daily volume of relevant searches carried out by British users in this period was relatively low, contradicting the assumption that 'bigot-gate' had increased voter concerns around immigration and heightened the interest in relevant policies. Instead, it seems that Gordon Brown's gaffe, unlike Mitt Romney's 'binders full of women,' fell short of fueling a demand for 'thematic' information among British Google users.

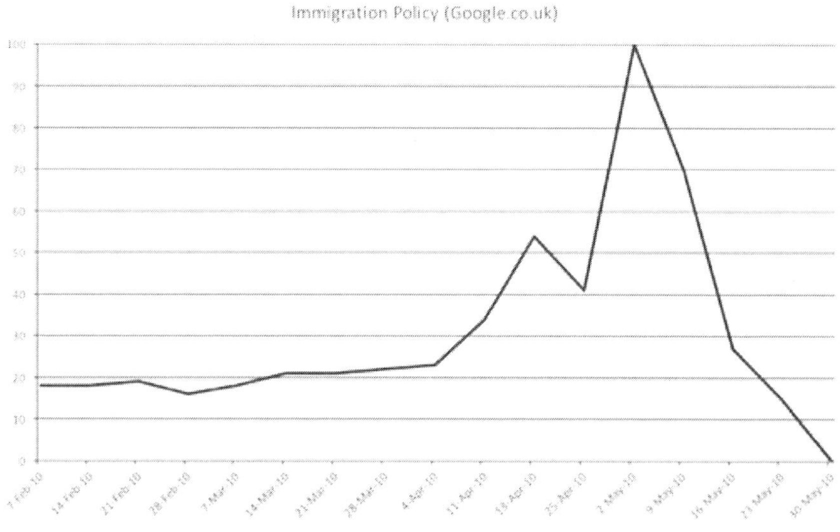

Figure 6. Search popularity for immigration policy issues on Google.co.uk, Feb.-May 2010 (Google Trends scores on Y axis, dates on X axis).

These search trends were in line with the way in which the British press covered 'bigot-gate,' given that only a really small number ($n = 16$) of the hundreds of news items published about this episode explicitly mentioned immigration. British voters did not remedy this lack of coverage about an important policy area by turning to Internet search. Instead, these data corroborate the hypothesis that the news narrative around 'bigot-gate' encouraged citizens to associate Gordon Brown's words with his 'dysfunctional' personality and perceived inability to connect with ordinary people (Cantijoch, Cutts, & Gibson, 2013; Higgins & Smith, 2013; Wring & Ward, 2010). This, in turn, promoted an 'episodic' interpretation of this incident over one connected to broader policy themes, suggesting that British Google users approached Internet search as a way to essentially confirm their pre-existing views of a candidate.

5. CONCLUSIONS

For this research, we used Google Trends to highlight the distinctive roles played by Internet search in two different media ecologies. We also explored the contribution of Google Trends to the information tool-kit in analyzing voters in the new media ecology. The evidence collected for this study shows that high-profile mediated events such as TV debates between candidates constitute key drivers of Internet searches in elections. Among televised moments, politicians' gaffes emerged as particularly powerful search catalysts. Furthermore, major 'staged' campaign events and external crises – for example, the Benghazi attack and hurricane Sandy – can act also as search boosters. While these general trends applied to both the US and UK case studies examined here, there were also some important differences in how Google users in each country followed up on these search 'triggers' to find more information online.

In particular, the analysis of key search trends following Mitt Romney's 'binders full of women' gaffe in 2012 showed that US Google users were able to use an arguably trivial episode as a springboard to complete a 'thematic leap' in their Internet searches, looking for information about policy issues related to that episode (gender equity and women's rights). Crucially, this spike in 'thematic' searches occurred before traditional news media intercepted and responded to the demand for information in that area. In contrast, British Google users did not complete a comparable 'thematic leap' following Brown's 'bigot-gate' gaffe in 2010. Indeed, more research is needed to clarify whether this episode signaled a broader tendency for UK voters to be less entrepreneurial in seeking out campaign information in Internet searches. There could be varying hypotheses for this, linked to the specific features of British political communication and the local media ecology.

Notably, unlike US voters, British voters have strong, programmatic political parties that communicate their issues more systematically and effectively than US parties the vast majority of the time. Another intriguing hypothesis to explore would be whether British publics are more likely to align their online searches with the topics and personalities promoted on legacy media outlets, using the Internet as a complementary rather than alternative source of information. The fact that Nick Clegg, who was relatively unknown to most British voters before the 2010 televised debates, became the most searched party leader following those pivotal media events corroborates the impression that TV in particular can strongly influence search trends in the UK.

Indeed, the latest data on news consumption show that British audiences have a relatively high level of trust in established news organizations compared to other countries, particularly the US (Newman et al., 2016, p. 25). Nevertheless, this is always subject to change. In particular, since the 2010 election the BBC has been mired in a series of scandals and the objectivity of its reporting has been questioned over coverage of events such as the 2014 Scottish independence referendum (Robertson, 2016). These developments could have important implications for search engine use and highlight one of the main limitations associated with case study research, which analyzes in detail a specific set of events while media ecologies are constantly evolving. Finally, the distinct reactions of British and American Google users may be the result of differences between gaffes too, with 'bigot-gate' more readily reflecting Gordon Brown's difficulties in connecting with voters and Mitt Romney's 'binders full of women' interpreted as a slip of the tongue that nevertheless served to catalyze the public's attention on policy issues.

Aside from the specific findings outlined above, how can publicly available data from Google Trends more broadly inform our understanding of electoral information flows? The research presented here suggests there are two key contributions from a study of Google Trends data. First, it contributes to the idea of information trajectories within the new media ecology; thinking about and studying search results allows us to consider theoretically how individual information-seeking agency through search may shift informational boundaries, as well as actually observe this process in action. As acknowledged above, there are limitations to this process. As search patterns are aggregated and standardized on Google Trends, it is impossible to examine individual behavior. However, by situating Google Trends data within a media ecology framework – which is concerned with flow and cross-fertilization between traditional and online forms of news –

we can better see how the affordances of search affect information-seeking behavior. We can see when search results as expressed in Google Trends give compelling evidence that voters are being inspired by events to become more informed on broader points of policy. This could contribute to ideas about how searches specifically and online information seeking more generally may challenge agendas and frames set by traditional news media.

We argue also that the evidence from this study suggests that search engines expand the purposeful nature of information gathering (Molotch & Lester, 1974). While search engines are broadly used as a convenient way of retrieving data, they are criticized for filtering and limiting the amount of information that is visible to users. This has led some to argue that search engines reinforce the informational status quo and can become significant barriers to democratization and political literacy. While we are not directly challenging those concepts, we feel it important to highlight that Google Trends can demonstrate how Internet users, including potential voters, have greater agency as they can use search to purposively seek out highly specific information in campaigns. Indeed, the scope of the findings is limited by the nature of the publicly available Google Trends data. Yet, we can show how Google Trends both gives us specific insight into voters in key national elections, as well as serves as a useful part of a social-science tool-kit, particularly valuable in analyzing audiences and media interaction in the new media ecology.

NOTES

1. For a comprehensive history of competing versions of 'media ecology' and its theoretical influences and uses, see Strate (2006).
2. This required a simple equation: N : 100 = n : X.
3. The graph shows a search peak in conjunction with the 2012 Democratic Convention held on 4–6 September. A separate Google Trends query returned a similar search peak for Mitt Romney during the 2012 Republican Convention, which took place between 27–30 August.

DISCLOSURE STATEMENT

No potential conflict of interest was reported by the authors.

FUNDING

This work was supported by the Economic and Social Research Council [grant number ES/K007890/1].

NOTES ON CONTRIBUTORS

Filippo Trevisan is Assistant Professor in the School of Communication at American University in Washington, DC, USA [email: trevisan@american.edu].

Andrew Hoskins is Interdisciplinary Research Professor in the College of Social Sciences at the University of Glasgow, UK [email: andrew.hoskins@glasgow.ac.uk].

Sarah Oates is Professor and Senior Scholar at the Philip Merrill College of Journalism at the University of Maryland, College Park, USA [email: soates@umd.edu].

Dounia Mahlouly is a Research Associate in the International Centre for the Study of Radicalisation in the Department of War Studies at King's College, London, UK [email: dounia.mahlouly@kcl.ac.uk].

ORCID

Filippo Trevisan http://orcid.org/0000-0003-2463-4596

Sarah Oates http://orcid.org/0000-0003-2154-0121

REFERENCES

1. Anstead, N. (2016). A different beast? Televised election debates in parliamentary democracies. *The International Journal of Press/Politics*, *21*, 508–526.
2. Anstead, N., & O'Loughlin, B. (2011). The emerging viewtariat and *BBC question time*: Television debate and real-time commenting online. *The International Journal of Press/Politics*, *16*, 440–462.
3. Awan, A., Hoskins, A., & O'Loughlin, B. (2011). *Media and radicalisation: Connectivity and terrorism in the new media ecology*. London: Routledge.

4. Bates, D. (2010). Making the white folk angry: The media, 'race' and electoral politics in the United Kingdom in 2010. *Platform: Journal of Media and Communication, 3,* 8–25. Retrieved from http://platformjmc.com/

5. Cantijoch, M., Cutts, D., & Gibson, R. (2013). Internet use and political engagement: The role of e-campaigning as a pathway to online political participation. UC Irvine Center for the Study of Democracy.

6. Chadwick, A. (2012). Recent shifts in the relationship between the Internet and democratic engagement in Britain and the United States: Granularity, informational exuberance, and political learning. In E. Anduiza, M. Jensen, & L. Jorba (Eds.), *Digital media and political engagement worldwide* (pp. 39–55). Cambridge: Cambridge University Press.

7. Chai, S., & Sasaki, N. (2011). Using online tools to assess public responses to climate change mitigation policies in Japan. *Future Internet, 3,* 117–129. doi:10.3390/fi3020117

8. Dayan, D. (2013). *Overhearing in the public sphere.* Retrieved from http://www.deliberatelyconsidered.com/2013/02/overhearing-in-the-public-sphere/

9. Dutton, W. H., & Blank, G. (2013). *Cultures of the Internet: The Internet in Britain.* Oxford: Oxford Internet Institute. Retrieved from http://oxis.oii.ox.ac.uk/reports/

10. Elmer, G. (2013). Live research: Twittering an election debate. *New Media & Society, 15,* 18–30. doi:10.1177/1461444812457328

11. Epstein, R., & Robertson, R. (2015). The search engine manipulation effect (SEME) and its possible impact on the outcomes of elections. *Proceedings of the National Academy of Sciences of the United States of America, 112,* E4512–E4521. doi:10.1073/pnas.141828112

12. Flynn, D., Ford, R., & Somerville, W. (2010). Immigration and the election. *Renewal, 18,* 102–114.

13. Gunter, B., Rowlands, I., & Nicholas, D. (2009). *The Google generation: Are ICT innovations changing information-seeking behaviour?* Oxford: Chandos.

14. Halavais, A. (2009). *Search engine society.* Cambridge: Polity Press.

15. Higgins, M., & Smith, A. (2013). 'My husband, my hero': Selling the political spouses in the 2010 general election. *Journal of Political Marketing, 12*, 197–210.

16. Hillis, K., Petit, M., & Jarrett, K. (2013). *Google and the culture of search*. New York, NY: Routledge.

17. Hindman, M. (2009). *The myth of digital democracy*. Princeton, NJ: Princeton University Press.

18. Hoskins, A., & O'Loughlin, B. (2010). *War and media: The emergence of diffused war*. Cambridge: Polity Press.

19. Hoskins, A., & Tulloch, J. (2016). *Risk and hyperconnectivity: Media and memories of neoliberalism*. Oxford: Oxford University Press.

20. Howard, P. (2006). *New media campaigns and the managed citizen*. Cambridge: Cambridge University Press.

21. Kaeserbauer, M., Hohenstatt, R., & Reed, R. (2012). Direct versus search engine traffic: An innovative approach to demand analysis in the property market. *International Journal of Housing Markets and Analysis, 5*, 392–413.

22. Kluver, A. R. (2002). The logic of new media in international affairs. *New Media & Society, 4*, 499–517.

23. Kreiss, D. (2012). *Taking our country back: The crafting of networked politics from Howard Dean to Barack Obama*. New York: Oxford University Press.

24. Langer, A. (2012). *The personalisation of politics in the UK: Mediated leadership from Attlee to Cameron*. Manchester: Manchester University Press.

25. Lilleker, D., & Vedel, T. (2013). The Internet in campaigns and elections. In W. Dutton (Ed.), *The Oxford handbook of Internet studies* (pp. 401–420). Oxford: Oxford University Press.

26. Lovink, G. (2009). Society of the query: The Googlization of our lives. In K. Becker & F. Stalder (Eds.), *Deep search: The politics of search beyond Google* (pp. 45–53). Innsbruck: Studien Verlag.

27. Lui, C., Metaxas, P. T., & Mustafaraj, E. (2011). On the pre-dictability of the US elections through search volume activity. Retrieved from http://bit.ly/gJ6t8j

28. Lum, C. M. K. (2000). Introduction: The intellectual roots of media ecology. *The New Jersey Journal of Communication, 8*, 1–7.

29. Mager, A. (2012). Algorithmic ideology: How capitalist society shapes search engines. *Information, Communication & Society, 15*, 769–787.

30. Molotch, H., & Lester, M. (1974). News as purposive behavior: On the strategic use of routine events, accidents, and scandals. *American Sociological Review, 39*, 101–112. Retrieved from http://www.jstor.org/journal/amersocirevi

31. Newman, N., Fletcher, R., Levy, D. A., & Nielsen, R. K. (2016). *Reuters institute digital news report 2016*. Oxford: Reuters Institute for the Study of Journalism.

32. Nuti, S. V., Wayda, B., Runasinghe, I., Wang, S., Dryer, R. P., Chen, S. I., & Murugiah, K. (2014). The use of Google Trends in health care research: A systematic review. *PLOS One, 9*,

33. Oates, S. (2011). Going native: The value in reconceptualizing international internet service providers as domestic media outlets. *Philosophy and Technology, 24*, 391–409.

34. Parry, K., & Richardson, K. (2011). Political imagery in the British general election of 2010: The curious case of 'Nick Clegg'. *British Journal of Politics and International Relations, 13*, 474–489.

35. Postman, N. (1970). The reformed English curriculum. In A. C. Eurich (Ed.), *The shape of the future in American secondary education* (pp. 160–168). New York, NY: Pitman.

36. Purcell, K., Brenner, J., & Rainie, L. (2012). *Search engine use 2012*. Washington, DC: Pew Internet and American Life Project.

37. Rentschler, C. A., & Thrift, S. C. (2015). Doing feminism in the network: Networked laughter and the "Binders full of Women" meme. *Feminist Theory, 16*, 329–359.

38. Robertson, J. (2016). Scottish TV coverage of the referendum campaign from September 2012 to September 2014. In N. Blain, D. Hutchison, & G. Hassan (Eds.), *Scotland's referendum and the media: National and international perspectives* (pp. 59–69). Edinburgh: Edinburgh University Press.

39. Scharkow, M., & Vogelsgang, J. (2011). Measuring the public agenda using search engine queries. *International Journal of Public Opinion Research, 23*, 104–113.

40. Scheitle, C. (2011). Google's insights for search: A note evaluating the use of search engine data in social research. *Social Science Quarterly, 92*, 285–295.

41. Stanyer, J. (2013). *Intimate politics: Publicity, privacy and the personal lives of politicians in media-saturated democracies*. Cambridge: Polity Press.

42. Strate, L. (2006). *Echoes and reflections: On media ecology as a field of study*. Cresskill, NJ: Hampton Press.

43. Strate, L., & Lum, C. M. K. (2000). Lewis Mumford and the ecology of technics. *The New Jersey Journal of Communication, 8*, 56–78.

44. Stromer-Galley, J. (2014). *Presidential campaigning in the Internet age*. New York: Oxford University Press.

45. Trevisan, F. (2014). Search engines: From social science objects to academic inquiry tools. *First Monday, 19*(11).

46. Vaidhyanathan, S. (2011). *The Googlization of everything (and why we should worry)*. Berkeley, CA: University of California Press.

47. Vergeer, M. (2013). Politics, elections and online campaigning: Past, present ... and a peek into the future. *New Media & Society, 15*, 9–17.

48. Wring, D., & Ward, S. (2010). The media and the 2010 campaign: The television election?*Parliamentary Affairs, 63*, 802–817.

49. Yasseri, T., & Bright, J. (2013). Can electoral popularity be predicted using socially generated big data? Retrieved from http://arxiv.org/pdf/1312.2818v1.pdf

3
CHAPTER

Opportunities and Challenges of The Digital Lifespan: Views of Service Providers and Citizens in The UK

Kathryn M. Orzech[a], Wendy Moncur[a,b], Abigail Durrant[c] & Diego Trujillo-Pisanty[c]

[a]*Duncan of Jordanstone College of Art & Design, Dundee University, Dundee, UK;*

[b]*Design, Architecture and Building, University of Technology Sydney, Ultimo, Australia;*

[c]*Open Lab, Newcastle University, Newcastle upon Tyne, UK*

ABSTRACT

Information about UK citizens' use of digital technologies is often expressed in statistics – x% lack Internet access; y% get online to engage in online banking, update social media sites, or participate in online auctions. There are many social implications to digital technology use, however – individuals may communicate online as a major way to stay in touch with friends and family, and as Internet access rises and government and public sector budgets shrink, online services become an increasingly attractive way for government and public sector service providers to communicate with citizens. This paper presents selected results of an exploratory study designed to investigate the digital personhood of UK citizens through interviews with participants at three life transitions: leaving secondary school, becoming a parent, and retiring from work. Digital personhood in this paper implies identity information online, and some interaction with others around that information. We then report on our presentation of a selection of these results to thirteen stakeholders who represented UK government departments, public sector organisations, and industry. We found that citizen and stakeholder concerns were quite different, especially at the new parent life transition, and that stakeholders tended to *underestimate* the willingness and ability of citizens to become involved online with the government and public sector, and *overestimate* citizens' vulnerability online. Future research should investigate practical strategies for increasing communication between stakeholders and citizens, and also how to encourage stakeholders to work together to benefit their common clientele – the citizens.

Keywords: E-government, digital literacy, civil servants, public sector, online privacy, trust, impact

INTRODUCTION

Old and young alike, UK citizens' engagement with digital technologies and services has skyrocketed in the past 10 years. In all, 86% of UK households currently have Internet access, and 78% of the British population spend time online every day or almost every day. This engagement presents opportunities and challenges to government and public sector agencies. Whilst engagement happens across all age groups, the needs of people at *different* points in their lives are underexplored. To help remedy the imbalance, this paper presents findings from our research project that involved (i) citizens and long-term residents of the UK (henceforth, citizens) who had recently lived through one of three life transitions – leaving secondary school, becoming a parent, or retiring from work; and (ii) representatives of the UK government, public sector, and industry (henceforth, referred to as our stakeholders). Our project, *Charting the Digital Lifespan (CDL)* is one of the first to incorporate both citizen and stakeholder perspectives on opportunities and challenges associated with the digital lifespan.

The two-year CDL project investigated how UK citizens create and manage their digital identities across the human lifespan, focusing on the three significant life transitions named above. Combining expertise in anthropology, design, cultural studies, and computer science, the aim was to develop a concrete understanding about how self-representation in a digital context (digital personhood) was experienced at the current time (2013–2015) by different generations, and how it was envisioned in the near future.

A central project goal was to generate social, cultural, and technical insights to inform UK policy-making and service innovation for enhancing digital literacy and enabling self-representation online. A Stakeholders' Workshop ('the Workshop') was held to support this goal, focusing on the project's qualitative insights. At the Workshop, we invited reflection on how stakeholders perceive citizens' understanding of and engagement with digital tools and media – both currently available, and envisioned. A bespoke 'Picture Book' with illustrated scenarios was designed for the Workshop as a resource to prompt questions from stakeholders about how services and systems can be designed to respond to and scaffold expressions of digital personhood made by citizens as they live through changing life circumstances.

Digital identity and digital personhood are interrelated terms; whilst digital identity refers to data or information about people created, captured, verified and used in life and death (Engineering and Physical Sciences Research Council, 2015), digital personhood assumes digital identity (both in traditional forms such as name and address, and specifically in digital forms like email addresses and online traces like photos or status updates) to feature in citizens' lives, and encourages individuals' interaction with others in digital realms – for example, by creating a profile, seeking friends online, and participating in digital friendship by validating the digital personhood of others (Lee, Goede, & Shryock, 2010).

In this paper, we provide a background to the central topics that emerged out of our qualitative studies and the Workshop, which fall under the broad heading of social implications of technology use:

- Living as digital citizens
- Privacy concerns
- Online danger and being vulnerable online
- Challenges and opportunities of connecting with citizens

We then give an overview of the qualitative studies conducted, and the subset of interdisciplinary findings generated that were subsequently presented to stakeholders. We report findings from the Workshop, and examine topics of concern to citizens and stakeholders. We suggest ways forward that encourage communication and an integration of perspectives between these two groups, particularly highlighting research that has incorporated both citizen and stakeholder perspectives.

BACKGROUND: SOCIAL IMPLICATIONS OF TECHNOLOGY USE

Living as digital citizens

We focused on the daily routines enacted via digital technologies of UK citizens who had recently undergone one of three life transitions. We took a 'snapshot' approach to citizens' current online practices, whilst also encouraging comparison with practices enacted prior to the transition recently undergone. The transitions that we focused on were (i) leaving secondary school, (ii) becoming a parent, and (iii) retiring from work. Such life transitions are characterised by a period of instability and change, as the central actor typically makes major adjustments, learning to cope with new experiences and developing new skills (Hulme, 2014). Anderson and Tracey (2001) identify that such life transitions can have a significant effect on how people engage with and use the Internet.

For those leaving secondary school (who we also refer to as 'emerging adults'), this period of change marks the departure from adolescence. It is distinguished by 'relative independence from social roles and normative expectations' (Arnett, 2000, p. 469) and is a time of exploration. Emerging adults may exhibit varying degrees of maturity and dependence (Hulme, 2014), governed in part by the age at which they begin to live independently (Furlong & Cartmel, 1997). Evidence of increasing maturity is seen through factors including leaving school, moving away from home, getting a job, or going to University, and adoption of economic responsibilities (e.g., controlling a budget) (Lenz, 2001) – factors that lead to a growing level of citizenship, and that are extensively intertwined with the use of online services.

The transition into parenthood involves a huge emotional and physical impact, and may well come at the same time as many other changes – for example, in relationships, careers, finances, and general life balance (Hulme, 2014). Gibson and Hanson (2013) identify the Internet as a helpful lifeline during this period, with new mothers using online forums and resources to assist them in developing confidence as new parents. Further, Bartholomew, Schoppe-Sullivan, Glassman, Kamp Dush, and Sullivan highlight that 'Facebook may be a particularly important means for mothers … to build and maintain social capital for themselves, their children, and their families' (2012, p. 465). As well as interacting online on their own behalf, parents also manage their children's nascent digital footprints, which begin at (or even before) birth. This management will usually tail off over time as children take over responsibility for crafting their own online personas as they get older (e.g., Ammari, Kumar, Lampe, & Schoenebeck, 2015).

Retiring from work can bring changes in health, housing, social interaction, work life, and personal finance. This transition may force individuals to reformulate their social identity, often through 'project-like' activities that act as stepping stones to a new identity as a retired yet still active person (Antti, Asko, Leif, Paula, & Jaana, 2010). Digital devices have been found to aid:

in coping with the transitions experienced, by making it easier to retrieve information (e.g., via the Internet), to maintain and extend social networks (e.g., through cell phones and e-mail), and to maintain a feeling of actually being 'connected to the world'. (Antti et al., 2010, p. 814)

Such devices have also served to sustain communication between older adults and their remotely located family (Lindley, Harper, & Sellen, 2009; Rodríguez, Gonzalez, Favela, & Santana, 2009).

Honesty and privacy

UK citizens experiencing any of the life transitions described above increasingly engage online with government agencies, as well as with corporations offering products and services (Dutton & Blank, 2013). These citizens desire efficiency and ease of use from the online systems that they engage with – especially when the systems are delivered by government agencies (Connolly, Bannister, & Kearney, 2010). In turn, government agencies hope not only for citizens who are competent users, but also for citizens' trust and honesty during online engagement. The difficulty for government agencies is that trust and honesty are typically associated with interpersonal relationships, not online efficiency and ease of use. For government agencies to inspire trust, they may need to reassure citizens that their data will not be mismanaged (Connolly, Bannister, & Kearney, 2010), and ensure that citizens feel well-informed, in control of their personal information, and have a sense of influence rather than powerlessness (Grimsley & Meehan, 2007; Smith, 2011).

Part of citizens' concerns about trust and honesty online may be related to online privacy, which has been the subject of increasing interest amongst UK citizens over the past 5–10 years – an interest reflected by the popular press (Bartlett, 2015). Online privacy has also been the subject of legislative changes in the European Union that affect UK citizens. For example, the 'Right To Be Forgotten' legislation mandates that individuals can request that search engines (e.g., Google) do not return certain content about them in online search results (Rosen, 2012). Online privacy is a somewhat plastic concept. Burkell, Fortier, Wong, and Simpson suggest that even though users of social media expect their privately posted content to leak into the public domain, privacy may be perceived to be violated 'if publicly revealed information is used for unintended purposes by unanticipated parties' (2014, p. 983). Ahern et al. further observe that users may lack awareness of the privacy implications associated with the aggregation of information disclosed online by either 'the system ... or ... people examining the user's online activity' (2007, p. 365).

There are differing views on whether older or younger adults are most aware – and protective – of their online privacy. In a small study of eight younger adults (aged 16–33 years) and eight older adults (aged 40–64 years) in Norway, Brandtzæg, Lüders, and Skjetne found that younger adults were likely to be more aware of privacy issues than older users, and more likely to use privacy settings (2010). In contrast, in a nationally representative telephone survey of 1000 adults in the U.S.A, Hoofnagle, King, Li, and Turow found that younger and older adults are 'more alike on many privacy topics than they are different', but that 'young-adult Americans ... aspiration for increased privacy' is tempered by their desire to 'participate in an online reality that is optimized to increase their revelation of personal data' (2010, p. 20). Amongst new parents, Ammari et al. suggest that gender is a factor in decisions around online privacy, with mothers more likely to take the lead in decisions around sharing content about their baby than fathers do (2015), although a larger study by Bartholomew et al. finds no significant difference in the frequency with which mothers and fathers shared photos of their baby, amongst those who engaged in photo-sharing online (2012).

Online danger and being vulnerable online

To defend their online privacy and to keep their data secure, UK citizens use antivirus software and firewalls to protect themselves, check for indicators of website safety (particularly for adults aged over 35 years), and report awareness of ways to flag inappropriate or offensive content (Ofcom, 2015). However, nearly two-thirds (64%) of surveyed UK Internet users reported using the same passwords for most or all websites (Ofcom, 2015). UK citizens' attitudes to privacy are inconsistent – whilst they are cautious about revealing personal information online, 26% report not reading website terms and conditions, and 68% balance the potential risks involved in providing personal information to companies with the convenience of getting what they want (Ofcom, 2015). This may lead to unintended and unforeseen consequences in the digital technology realm (Debatin, Lovejoy, Horn, & Hughes, 2009), such as personal data being sold for use by third parties in targetted marketing.

Challenges and opportunities of connecting with citizens

One of the challenges of using digital technologies to connect government and the public sector with citizens lies in understanding the needs of both service providers and service users. Few projects speak to both stakeholders and citizens: researchers more commonly seek the perspective of *either* government employees (Clarke & Margetts, 2014; Mergel, 2014; Shackleton, Fisher, & Dawson, 2004) *or* citizens (e.g., Daou, Karuranga, Thiam, Mellouli, & Poulin, 2012; Kumar, Ureel, King, & Wallace, 2013). In a case study of the Singapore tax authority, Tan, Pan, and Lim (2005) identified stakeholders and their interests , but only spoke to government employees, relying on work from the early 1990s to provide taxpayer perspectives, thus leaving a gap in the literature for the exploration of the concerns of *both* stakeholders and citizens in an increasingly digitally mediated society.

METHODS

Two qualitative studies

Empirical findings presented at the Workshop were drawn from two linked qualitative studies that used the complementary methods of ethnography and experience-centred design (McCarthy & Wright, 2004; Wright & McCarthy, 2010). We first detail these qualitative studies, and insights generated which were subsequently presented to stakeholders, before describing the Workshop that made use of these insights.

Study 1: ethnography to understand today's digital life across the lifespan

The first author conducted an ethnographic inquiry (Schensul, Schensul, & LeCompte, 1999, p. 1) into participants' online lives *as they are lived now*. This involved individual, semi-structured interviews (1 ½ to 2 hours), with 46 research participants (Sample 1). Interviews were augmented with participant observation, including informal conversations with individuals about technology use and observing technology use in public locations within the city. Interviews and observations all took place in the same mid-sized city (~150,000 residents) in the UK between December 2013 and December 2014. Participants were recruited through community contacts, maximising diversity in terms of occupation within each group. The University of Dundee granted ethical approval for this project.

Study 2: experience-centred design to inspire new strategic design directions

The third and fourth authors adopted an experience-centred design approach to study how another set of 18 participants (Sample 2) made sense of the technology they use in their daily lives, with the aim of *inspiring new design directions*. Participants took part in an initial interview about their everyday experiences with technology through the life transition being studied, and their perspectives on supportive technologies for their future online life. Design prototypes were designed and then 'deployed' for these participants to live with for up to a month between January 2014 and May 2015 (Trujillo-Pisanty, Durrant, Martindale, James, & Collomosse, 2014). These designs were interventions that provided new interfaces to the participant's online representations of self, inviting reflection and ideation on the research topic. Each deployment was followed by an exit interview about the participant's experiences and perspectives on future technology opportunities. This study took place in a UK city of ~300,000 residents. Newcastle University granted ethical approval for this project. Demographic characteristics of both the ethnography and experience-centred design participants are summarised in Table 1.

Table 1. Participant demographics.

Group	**Young adults**	**New parents**	**Older adults**	**Total participants**
Sample 1: Ethnography Participants				
Male	8	6	8	22
Female	7	10	7	24
Age range	18–23 years	17–50 years	59–70 years	
Sample 2: Experience-centred Design Participants				

Group	Young adults	New parents	Older adults	Total participants
Male	2	3	3	8
Female	2	5	3	10
Age range	19–21 years	25–35 years	57–73 years	

Data analysis

The ethnographic studies were analysed using a grounded theory (GT) approach (Charmaz, 2011; Strauss & Corbin, 1990), which involved letting theory develop out of the collected data. This approach allowed us to identify individual perceptions and actions in the context of everyday life experiences, without preconceptions. The experience-centred design studies adopted an interpretative phenomenological analytical framework (Smith, Flowers, & Larkin, 2009). This involved the researcher coding transcribed interview data to understand participant 'sense making' at interview and about the design deployment. Further details about our data analysis can be found online in Supplementary Table A.

Findings

Distilling findings from qualitative studies

After data collection, we selected those of our research findings that were relevant and useful for stakeholders. Research insights to focus on in the Workshop had to:

1. Be related to stakeholders' known interests.
2. Be common across the ethnographic and experience-centred design research strands.
3. Open up discussions on service design.

Based on these criteria, we chose a subset of concerns raised by participants to focus on. In the qualitative data presented below, participants are identified using a pseudonym, followed by their group

(Young Adult – YA, New Parent – NP, and Recent Retiree – RR), and age at the time of interview.

Young adults' concerns

One of the major concerns of young adults was *building and adding to an online persona* – including sharing attitudes and beliefs online, resharing and remixing online content, and having their digital identity shaped by others. Lauren talks about this concern in discussing why her use of online services has increased since she left school:

... when I was in sixth year, I think because I was still at school with everyone I didn't feel like I needed to ... really catch up or look at what everybody was doing, because I saw them every day. I was constantly in contact with them ... at the weekends. We'd even go out and things like that. Whereas now we're in different jobs, different cities, doing different courses, so there's more of a need to be online .... (YA19)

A second major concern was *getting things done online* – chiefly not only through communication with friends, but also by engaging in administrative activities online such as using banking apps or University course management systems.

New parents' concerns

New Parents had deep *privacy concerns*. These included a focus on the appropriateness and security concerns associated with sharing photos of babies, concerns about a hypothetical near future with pervasive monitoring, and concerns about realities of visibility on Facebook that meant their past could not ever really be left behind as long as they stayed on social media (e.g., Fox & Tokunaga, 2015).

I'm not paranoid but I just don't feel the need ... I mean you hear these things about once stuff is online, basically ... you've got no control of it ... I've got ... an album of pictures of her month-by-month, and every month I use my cheap camera to shoot a little video snippet that demonstrates some sort of a milestone, but again this is not online. This is purely for us, and in private. (Laura – NP34)

New parents also focused on *developing their new role as parents*, seeking advice online. As they were often holding their baby, some parents adopted devices that allowed them to access the Internet one-handed, for example, preferring phones and tablets to laptop computers for most online tasks. Parents *maintained their own digital identities as they raised children*.

... although Facebook sounds like one of things that should [be the first to] go it tends to be one of the first things to keep because that is my contact with friends, family, where I get all my advice that is baby related ... through my groups, that is kind of why [I keep it]. (Alison – NP21)

Retirees' concerns

Recent retirees' concerns included *online privacy*, ideas of what information was appropriate to share online, and knowing whom to trust.

I think there's always a risk that whatever you think is private may suddenly become non-private, because you can be hacked. We all know people who've had their mailboxes hacked, so the best thing is not to have anything on there that they don't want people to see. I mean you wouldn't put your bank statement on for example, or your medical records .... (Stuart – RR66)

Retirees also wanted to *maintain and strengthen social ties with friends and family at a distance*, which led to discussions with retirees of what would empower them to stay in touch.

I put my iPad on, that sits on the kitchen table so I just pop it on. It's a habit I've got into from my working days just [to] check emails. [Then] probably just catch up with the news there, watch television. And then probably about tea-time, I [check] my emails again. See if there's any news I want to catch up with, if I want to have a look at on my iPad or Mac to get further information. Mostly sports and news that I want to catch up with. (Tom – RR69)

Retirees engaged in daily routines with digital technology, and tended to browse and exchange information.

Organising the Stakeholders' Workshop

Our intent was to create a highly interactive, discursive workshop, with a format that our stakeholders could relate to and identify value from. To achieve this, we drew upon the 'Seven Questions' technique in the 'Futures Toolkit for Policymakers and Analysts' (Cabinet Office & GO-Science, 2014) developed by the UK Government as part of its approach to Futures work, focusing on three of the seven questions:

1. *What would you identify as critical issues for the future related to this scenario?*
2. *If things went wrong, what factors would you worry about – what are the risks you identify for your organisation?*
3. *Looking at your organisation, how might processes need to be changed to bring about desired outcome(s)?*

To enable stakeholders to respond to these questions required an engaging and effective distillation of our research insights. To achieve this, we developed a bespoke 'Picture Book' that presented three narratives, each representing one of the three life transitions. Each narrative consisted of a character-driven scenario alongside selected empirical research insights (as described above) and service design implications. These were used to prompt stakeholder dialogue. The design and use of the Picture Book are the subjects of a separate publication (Durrant, Moncur, Kirk, Trujillo-Pisanty, & Orzech, 2016).

Workshop planning and execution

The planning and delivery of the Stakeholders' Workshop was a multi-stage process that involved iteratively developing and testing resources; logistical planning; inviting the right mix of stakeholders; and careful scripting, facilitation, and evaluation of the event. This process is summarised online in Supplementary Table B. Details of participating stakeholders are presented in Table 2.

Table 2. Stakeholder (service provider) demographics.

Group	**Government civil servant**	**Employee of public sector organisation**	**Employee of an industrial enterprise**	**Total participants**
Male	3	4	1	8
Female	3	1	1	5

On the afternoon of the Workshop, we welcomed Stakeholders and introduced the project. Each stakeholder was directed to one of three tables. At each table, a facilitator focused on a specific life transition and associated narrative from the Picture Book. Stakeholders first read the Picture Book narrative to contextualise, ground, and open up the facilitated discussion. They were then invited to respond to Futures Toolkit questions (described previously). Responses were captured on flip-chart paper by the table's facilitator. Stakeholders rotated around each of the three tables in turn. After engaging with all three life transitions, the stakeholders regrouped, and responses from the three tables were shared and further discussed. This final discussion was audio-recorded for later analysis. Data collected from the Workshop included the flip-chart notes and audio recordings. We analysed data from the Workshop using a GT approach, as described above for the ethnographic data. Further information about data analysis for the Workshop may be found in Supplementary Table A.

Findings from the Stakeholders' Workshop

Below, we present the major concerns of the stakeholders voiced at the Workshop, relating to the use of digital technologies by young adults, new parents, and recent retirees. For each life transition, stakeholders touched on some, but not all, of the themes uncovered through our ethnography and experience-centred design work. Some stakeholder concerns spanned multiple life transitions; we present these at the conclusion of this section.

Young adults

Young adults were perceived by stakeholders as true *digital citizens*, and so our workshop participants were generally concerned with veracity online amongst young adults. This included a perception that young adults were creating a 'synthetic, artificial' version of self, which could embellish the truth about their lives for an online audience. Two reasons that stakeholders suggested for this perceived lack of veracity were (1) that online was perceived as a playful rather than a truthful space and (2) the view that others [online] have 'perfect' lives, leading to aspirations amongst young adults to make their own lives seem more perfect. Collectively, stakeholders perceived young adults as having their digital identities shaped by others, and expressed concern that young adults could be sharing attitudes and beliefs related to radicalisation or crime online. Following on from this, stakeholders worried that young adults might not trust the government, and might fear that engagement with online services could open a window for government surveillance on their lives.

Stakeholders also expressed concern that familiarity with the online environment and doing so much remotely (rather than face to face) could *increase danger* for young adults. Young adults were seen as especially vulnerable to online risk, with our stakeholders concerned by their perceived inability to notice when they crossed a line into risky online behaviour. There were also concerns raised around vulnerability to crimes such as rape that resulted from meeting online acquaintances face to face without taking steps to ensure personal safety or verify that the acquaintance was who they claimed to be. A further perceived risk lay in citizens (especially but not exclusively young adults) not knowing who they are really interacting with online, as the contextual clues provided by offline interactions, including non-verbal signals and a network of trust based on personal relationships and local knowledge, are largely absent online.

This overall negative view of young adults online amongst stakeholders was mediated by a few *opportunities presented by connecting with citizens* – namely, how the perceived 'always online' status of young adults could be harnessed for civic good, through mentoring, advocacy, or as a communication network in an emergency.

New parents

Stakeholders mentioned only one concern particular to new parents – *privacy*. Stakeholders briefly discussed how individuals could give consent for data sharing, especially in the case where the individual was a young child. They did not single out this life transition group as particularly *vulnerable to online danger*, or as associated with particular *challenges in connecting with citizens*. New parents were seen as young enough to be familiar with the online environment, but old enough to be wary of the danger they could find there.

Retirees

Stakeholders were particularly concerned about retirees' *vulnerability online*, buying into the stereotype of older computer users as unfamiliar with the online world and more likely than younger individuals to *reveal private information online* to individuals with malign intent. Several stakeholders noted that scams may target citizens at crisis moments, suggesting that even cautious older adults could fall prey to well-designed scams.

Stakeholders were worried about *connecting with retired citizens online* and concerned that retirees would not want to engage with online services, although they acknowledged that the digital literacy and connectivity of some retirees may be positively affected by their housing situation (multigenerational housing in London) or by relationships with children or other digitally literate friends.

Based on our interactions with digitally engaged retirees, who sometimes overtly asked to be involved in helping to develop services targeted at them, we asked stakeholders 'How could government harness the civic participation (and time) of retirees?' Stakeholders responded by brainstorming *opportunities presented by connecting with retired citizens*, ways that older UK citizens might help each other and their communities. For example, digital channels could offer opportunities for recent retirees to make use of their skills, perhaps enabling them to act like 'a Civil Society Mechanical Turk', carrying out socially valuable tasks on an ad hoc basis and volunteering online – subject to careful vetting. Stakeholders suggested the provision of an 'Outduction' (as opposed to an Induction) when citizens retired, helping them to (continue to) be digitally engaged. This could contain advice on the digital aspects of life from getting a pension to avoiding online fraudsters.

Across the lifespan

Living as digital citizens: trust

Stakeholders raised questions about *trustworthiness online*, with the discussion touching on questions such as 'Can citizens trust government online?' 'Can government trust citizens?' and 'How does the government know citizens are who they say they are?' The provision of dishonest information by citizens to government was viewed as generating negative social and economic consequences. Worries also arose about services such as peer-to-peer mentoring, which might be seen as Government 'just trying to deliver services cheaply'.

Privacy concerns

The topic of privacy was intertwined with discussions on digital literacy, online vulnerability, and persistence of online content. Stakeholders perceived efforts at policing online content as potentially problematic. However, the European Right to Forget legislation,[1] which excludes content from online search results, was seen as potentially beneficial in helping individuals to protect themselves from unwelcome reputational damage. Stakeholders also raised the issue of Government working with the 'really private' data of individuals, debating whether citizens will be less honest in the provision of private personal data if they are concerned that their privacy will be compromised.

Online danger and vulnerability online

Many aspects of online life were seen as beneficial, including instant access to information and facilitation of social interactions that are linked to better mental health (Kawachi & Berkman, 2001). However, danger online was understood to go beyond inappropriate sharing and privacy breaches to far more sinister activities, including radicalisation and criminal activity. Social isolation caused by spending time online rather than in face-to-face social interaction was seen as creating vulnerability and a potential pathway into radicalisation for lonely young adults, who could be more willing to connect with unknown others online. Examples given of sophisticated scams and schemes conducted partly or totally online included identity theft, romance fraud, property theft, frauds involving diversion of bank transfers, and creation of fake profiles to meet and 'groom' potential victims online. 'Digital vigilantes' were also seen as creating problems by taking the law into their own hands, for example, by posting pictures of suspected paedophiles online. The stakeholders emphasised how citizens need to be empowered to behave responsibly and in ways that keep them safe online.

Challenges and opportunities of connecting with citizens

Connecting with citizens was a central concern for the stakeholders. They expressed concern that connecting with all citizens to an equal extent is challenging, since citizens have a diversity of skills and interests, as well as a range of attitudes about online communication channels. The stakeholders also speculated about effective delivery of Government and other 'official' communications to citizens. They asked, 'Is an email as good as a letter in the post, or will it be ignored as part of the digital deluge?' without a clear answer.

The stakeholders saw education to raise digital literacy as a central opportunity for government to digitally enable UK citizens. Such education might include providing guidance on how to behave online and how to identify online danger, including a 'primer' on trusting people who were met through online channels. They suggested that easy-to-use online government services could reduce digital literacy challenges faced by citizens. This ease of use could be increased by helping citizens to complete online forms through the use of Artificial Intelligence or 'In-interface' tutorials.

They also described wishing to increase citizen engagement with Government by understanding and engaging with the diverse experiences of digital citizens. Stakeholder suggestions for new ways to engage – particularly around payment of taxes – included moving towards digital communications (and evaluating their effectiveness) such as paperless reminders to all and tailored communications to predicted 'bad payers', and tying Her Majesty's Revenue and Customs (HMRC) services with banks. This discussion leveraged our findings about how citizens express a high level of use and trust of online banking across age groups.

DISCUSSION

Talking to stakeholders from government, industry, and the public sector raised interesting points about their perceptions of UK citizens living as digital citizens today, coping with privacy concerns, vulnerability and danger online, and potential challenges and opportunities of connecting with citizens. We found that citizen and stakeholder concerns were quite different. For example, our project participants placed a strong focus on communicating with friends and family, and maintaining relationships over time and distance. On the other hand, our stakeholders hoped to connect to citizens they (often) had little prior relationship with, and persuade them to use new software for a 'business' purpose such as preparing tax returns or providing other sensitive information to the government.

Our research aligns well with previous work investigating the perceptions of citizens and government stakeholders regarding e-government, although our findings suggest a need to move beyond theorising about stakeholders (e.g., Rowley, 2011) to begin to incorporate the concerns of both parties when seeking digital technology solutions acceptable to the widest possible group.

We found that our stakeholders tended to *underestimate* the willingness and ability of citizens to become involved online with the government and public sector and *overestimate* citizens' vulnerability online. Part of this seemed related to common stereotypes of particular citizen groups – for example, believing that retirees will uniformly find it hard to get online. Our project data did not support the characterisation of older adults as averse to getting online or naive to what they might find there; instead, we found that our participants were extremely cautious about revealing any personal information online because they *were* aware of online risk. This risk awareness of older adults is also supported by several investigations into older adults' privacy behaviours on social media cited above.

Finally, we realised that one of the main benefits of the Workshop for the stakeholders was the opportunity to meet together to talk through a range of concerns about digital technology and its use at various life transitions. Stakeholders were able to hear what others were doing with digital technology at the Workshop, and walk away with ideas of how to improve their own services, and how to harmonise offerings across government departments.

In addition to seeking citizen and stakeholder perspectives, future research should investigate practical strategies for increasing communication between these two groups. The Workshop discussions identified that Government engagement needs to be developed to take into account the diversity of the UK's digital citizens, preferably tailored to individual and (age) group differences, and target information to specific kinds of users by customising the medium and the message. Citizens themselves have a role to play here by engaging in bidirectional communication with Government. If citizens provide feedback to Government on new digital services and are acknowledged, the Government may be seen as more effective and approachable as a result. Suggestions for new ways to engage with citizens – particularly around payment of taxes, that emerged from the Workshop discussions included:

- Evaluating the effectiveness of paper versus digital communications in communicating with UK citizens, including sending paperless reminders out to taxpayers towards financial year-end;
- Tying HMRC (tax) services with banking services;
- Sending automatic, tailored communications to predicted 'bad payers' to encourage payment;
- Deploying new approaches and communication channels specific to young adults – for example, helping them to understand their responsibilities regarding making tax payments.

We also encourage further research into ways to encourage stakeholders to work together to benefit their common clientele – the citizens. Through the Workshop, it emerged that common goals exist within Government departments of protecting UK citizens, and of making citizens more aware of the risks inherent in online life. Although stakeholders identified an existing reticence to share data about citizens, even within the same Government department, there was enthusiasm towards the potential to share knowledge and instigate coherent cross-department initiatives, ensuring future integration and cohesion of data and combined knowledge of UK citizens. Inter-agency cooperation around citizens' engagement with digital technologies was seen as desirable.

By seeking perspectives from citizens and stakeholders, and encouraging the two groups to work together, as well as encouraging stakeholders to work together across government, the public sector, and industry to create a more streamlined e-government experience, we encourage researchers and others interested in this subject matter to address the social implications of digital technology use *along with* statistical data to create better e-governance systems for the future.

CONCLUSION

In this paper, we have provided a summary of the qualitative findings about digital personhood and digital technology use generated from CDL studies of UK citizens after they have experienced one of the three life transitions: leaving school, becoming a parent, and retiring from work. We have further described how these findings were presented to a group of stakeholders in the research for further sense making about the CDL subject matter, reporting on their responses and perspectives.

Findings from the first studies revealed how individual participants were concerned with building an online persona and getting practical things done online. However at the Workshop, stakeholders from Government, the public sector, and industry showed more concern for the privacy implications of citizens' behaviours, and how to reach out to citizens to encourage them to use (and be honest with) government services. Our project is one of the first to incorporate both citizen and stakeholder perspectives on risks and opportunities associated with the digital lifespan in an increasingly digitally mediated age, an area that presents fertile ground for further inquiry.

NOTES

1. http://ec.europa.eu/justice/data-protection/files/factsheets/factsheetdataprotectionen.pdf.

DISCLOSURE STATEMENT

No potential conflict of interest was reported by the authors.

FUNDING

This work was supported by RCUK [grant number EP/L00383X/1]; the third author is additionally supported by the Leverhulme Trust (ECF-2012-642).

NOTES ON CONTRIBUTORS

Kathryn M. Orzech is a research fellow interested in how online life and connectivity change across the physical lifespan. Her research focuses on the intersections of anthropology, digital technology, and health. [email: kathrynorzech@gmail.com].

Wendy Moncur is a Reader in Socio-Digital Interaction at the University of Dundee, UK, and a Visiting Fellow at the University of Technology Sydney, Australia. Her interdisciplinary research focuses on the design of digital technologies to support lived human experience across the lifespan, and is grounded in Human–Computer Interaction. [email: w.moncur@dundee.ac.uk].

Abigail Durrant is a Leverhulme Fellow in Human–Computer Interaction. Her fellowship investigates the role of interaction design in supporting collaboration and communication within inter-disciplinary teams working on complex technology research that engages academia, industry, and the public sector. [email: abigail.durrant@newcastle.ac.uk].

Diego Trujillo-Pisanty is a design researcher and artist focusing on emerging technologies and the impact they have on individuals, societies, and cultures. His work often consists of working electronic objects that prompt reflection. [email: diego@trujillodiego.com].

REFERENCES

1. Ahern, S., Eckles, D., Good, N. S., King, S., Naaman, M., & Nair, R. (2007). Over-exposed? Privacy patterns and considerations in online and mobile photo sharing. In *Proceedings of CHI'07* (pp. 357–366). New York, NY: ACM.

2. Ammari, T., Kumar, P., Lampe, C., & Schoenebeck, S. (2015). Managing children's online identities: How parents decide what to disclose about their children online. In *Proceedings of CHI'15* (pp. 1895–1904). New York, NY: ACM.

3. Anderson, B., & Tracey, K. (2001). Digital living the impact (or otherwise) of the internet on everyday life. *American Behavioral Scientist, 45*(3), 456–475. doi:10.1177/00027640121957295

4. Antti, S., Asko, L., Leif, H., Paula, V., & Jaana, N. (2010). Information technologies and transitions in the lives of 55–65-year-olds: The case of colliding life interests. *International Journal of Human-Computer Studies, 68*(11), 803–821. doi: 10.1016/j.ijhcs.2010.06.007

5. Arnett, J. J. (2000). Emerging Adulthood: A theory of development from the later teens through the twenties. *American Psychologist, 55*(5), 469–480. doi: 10.1037/0003-066X.55.5.469

6. Bartholomew, M., Schoppe-Sullivan, S., Glassman, M., Kamp Dush, C., & Sullivan, J. (2012). New Parents' facebook use at the transition to parenthood. *Family Relations, 61*(3), 455–469. doi: 10.1111/j.1741-3729.2012.00708.x

7. Bartlett, J. (2015, June 16). How we all became obsessed with online privacy. Retrieved from http://www.telegraph.co.uk/news/general-election-2015/politics-blog/11678157/How-we-all-became-obsessed-with-online-privacy.html

8. Brandtzæg, P. B., Lüders, M., & Skjetne, J. H. (2010). Too many facebook "friends"? Content sharing and sociability versus the need for privacy in social network sites. *International Journal of Human-Computer Interaction, 26*(11–12), 1006–1030.

9. Burkell, J., Fortier, A., Wong, L. Y. C., & Simpson, J. L. (2014). Facebook: Public space, or private space? *Information, Communication & Society, 17*(8), 974–985.

10. Cabinet Office, & GO-Science. (2014, July 8). Futures toolkit for policy-makers and analysts – Publications – GOV.UK. Retrieved April 30, 2015, from https://www.gov.uk/government /publications/futures-toolkit-for-policy-makers-and-analysts

11. Charmaz, K. (2011). *Constructing grounded theory: A practical guide through qualitative analysis*. Los Angeles, CA: Sage.

12. Clarke, A., & Margetts, H. (2014). Governments and citizens getting to know each other? Open, closed, and big data in public management reform. *Policy & Internet, 6*(4), 393–417.

13. Connolly, R., Bannister, F., & Kearney, A. (2010). Government website service quality: A study of the Irish revenue online service. *European Journal of Information Systems, 19*(6), 649–667.

14. Daou, A., Karuranga, E., Thiam, F., Mellouli, S., & Poulin, D. (2012). Appropriation of online services by citizens: Innovating through a "bottom-up" approach. In *Proceedings of DG.O'12* (pp. 68–73). New York, NY: ACM.

15. Debatin, B., Lovejoy, J. P., Horn, A.-K., & Hughes, B. N. (2009). Facebook and online privacy: Attitudes, behaviors, and unintended consequences. *Journal of Computer-Mediated Communication, 15*(1), 83–108.

16. Durrant, A., Moncur, W., Kirk, D. S., Trujillo-Pisanty, D., & Orzech, K. M. (2016). *On presenting a rich picture for stakeholder dialogue*. Paper presented at the 2016 design research society, Brighton, UK.

17. Dutton, W. H., & Blank, G. (2013). OxIS 2013 Report: Cultures of the Internet.

18. Engineering and Physical Sciences Research Council. (2015). Skills: Priority areas: Digital identity. Retrieved from https://www.epsrc.ac.uk/skills/students/centres/2013cdtexercise/priorityareas/diid/

19. Fox, J., & Tokunaga, R. S. (2015). Romantic partner monitoring after breakups: Attachment, dependence, distress, and post-dissolution online surveillance via social networking sites. *Cyberpsychology, Behavior, and Social Networking, 18*(9), 491–498.

20. Furlong, A., & Cartmel, F. (1997). *Young people and social change.* Maidenhead, UK: Open University Press.

21. Gibson, L., & Hanson, V. L. (2013). Digital motherhood: How does technology help new mothers? In *Proceedings of CHI'13* (pp. 313–322). New York, NY: ACM.

22. Grimsley, M., & Meehan, A. (2007). E-government information systems: Evaluation-led design for public value and client trust. *European Journal of Information Systems, 16*(2), 134–148.

23. Hoofnagle, C. J., King, J., Li, S., & Turow, J. (2010). *How different are young adults from older adults when it comes to information privacy attitudes and policies?* (SSRN Scholarly Paper). Rochester, NY: Social Science Research Network.

24. Hulme, A. (2014). *Next steps: Life transitions and retirement in the 21st century.* Calouste Gulbenkian Foundation. Retrieved from http://gulbenkian.org.uk/files/01-07-12-Next%20steps%20-%20Life%20transitions%20and%20retirement%20in%20the%2021st%20century.pdf

25. Kawachi, I., & Berkman, L. F. (2001). Social ties and mental health. *Journal of Urban Health : Bulletin of the New York Academy of Medicine, 78*(3), 458–467. doi:10.1093/jurban/78.3.458

26. Kumar, S., Ureel, L. C.II, King, H., & Wallace, C. (2013). Lessons from our elders: Identifying obstacles to digital literacy through direct engagement. In *Proceedings of PETRA'13* (pp. 22:1–22:8). New York, NY: ACM.

27. Lee, D. B., Goede, J., & Shryock, R. (2010). Clicking for friendship: Social network sites and the medium of personhood. *MedieKultur. Journal of Media and Communication Research, 26*(49), 137–150.

28. Lenz, B. (2001). The transition from adolescence to young Adulthood: A theoretical perspective. *The Journal of School Nursing, 17*(6), 300–306.

29. Lindley, S. E., Harper, R., & Sellen, A. (2009). Desiring to be in touch in a changing Communications landscape: Attitudes of older adults. In *Proceedings of CHI'09* (pp. 1693–1702). New York, NY: ACM.

30. McCarthy, J., & Wright, P. (2004). *Technology as Experience.* Boston, MA: MIT Press.

31. Mergel, I. (2014). Social media adoption: Toward a representative, responsive or interactive government? In *Proceedings of DG.O'14* (pp. 163–170). New York, NY: ACM.

32. Ofcom. (2015). *Adults' media use and attitudes Report 2015.* Retrieved from http://stakeholders.ofcom.org.uk/market-data-research/other/research-publications/adults/media-lit-10years/

33. Rodríguez, M. D., Gonzalez, V. M., Favela, J., & Santana, P. C. (2009). Home-based communication system for older adults and their remote family. *Computers in Human Behavior, 25*(3), 609–618.

34. Rosen, J. (2012). The right to be forgotten. *Stanford Law Review Online, 64*, 88.

35. Rowley, J. (2011). E-government stakeholders – who are they and what do they want?*International Journal of Information Management, 31*(1), 53–62.

36. Schensul, S. L., Schensul, J. J., & LeCompte, M. D. (1999). *Essential ethnographic methods: Observations, interviews, and questionnaires.* Walnut Creek, CA: AltaMira Press.

37. Shackleton, P., Fisher, J., & Dawson, L. (2004). Internal and external factors impacting on e-government maturity: A local government case study. *Journal of Information Technology Case and Application Research, 6*(4), 36–50.

38. Smith, J. A., Flowers, P., & Larkin, M. (2009). *Interpretative phenomenological analysis: Theory, method and research.* Los Angeles, CA: Sage.

39. Smith, M. L. (2011). Limitations to building institutional trustworthiness through e-government: A comparative study of two e-services in Chile. *Journal of Information Technology, 26*(1), 78–93.

40. Strauss, A. L., & Corbin, J. M. (1990). *Basics of qualitative research: Grounded theory procedures and techniques.* Thosand Oaks, CA: Sage Publications.

41. Tan, C.-W., Pan, S. L., & Lim, E. T. K. (2005). Managing stakeholder interests in e-government Implementation: Lessons learned from a Singapore e-government Project. *Journal of Global Information Management, 13*(1), 31–53.

42. Trujillo-Pisanty, D., Durrant, A., Martindale, S., James, S., & Collomosse, J. (2014). Admixed portrait: Reflections on being

online as a new parent. In *Proceedings of DIS'14* (pp. 503–512). New York, NY: ACM.

43. Wright, P., & McCarthy, J. (2010). *Experience-centered design: Designers, users, and communities in dialogue.* San Rafael, CA: Morgan & Claypool Publishers.

4

CHAPTER

An In-Depth Study of The ICT Ecosystem in a South African Rural Community: Unveiling Expenditure and Communication Patterns

Carlos Rey-Moreno[a]*, Renette Blignaut[b], William D. Tucker[c] and Julian May[d]

[a] *Department of Computer Science, University of the Western Cape, Private Bag X17, Bellville, 7535 South Africa;*

[b] *Statistics and Population Studies, University of the Western Cape, Bellville, South Africa;*

[c] *Computer Science Department, University of the Western Cape, Bellville, South Africa;*

[d] *Institute for Social Development, University of the Western Cape, Bellville, South Africa*

ABSTRACT

There is no doubt of the contributions made by mobile phones and mobile network operators in increasing access to communications in rural areas of developing countries. Yet how affordable is this ubiquitous access in such an ICT ecosystem? Using data from two stratified random surveys conducted in a South African rural community, this paper provides a unique in-depth picture of the expenditure and communication patterns of its dwellers. Results show a high access ratio of people using mobile phone services weekly and a high proportion of disposable income dedicated to a very constrained set of mobile phone services. Factors such as mobile phone charging and the extra charges added by airtime resellers contribute to increase the communication costs. This data and its analysis can be used by the following: regulators and government agencies to better design their policy implementations to provide universal service and access; competing industry players to understand the dynamics within rural communities to better target their products; civil society organizations to use it as a case in their efforts to make affordable communications a constitutional right.

Keywords: ICT ecosystem, expenditure patterns, mobile phone usage, mobile phone ownership, rural areas, South Africa

1. INTRODUCTION

Providing telecommunication access to rural areas is neither easy nor profitable. As a result of these disincentives, traditional telecommunication companies have had little interest to provide telecom infrastructure in the past (Navas-Sabater, Dymond, & Juntunen, 2002). *Telecommunications and information services for the poor.* (World Bank Discussion Paper No. 432), 1–118.). This lack of private sector action has pushed national governments to devise policy strategies to ensure that parts of their population are not left behind in the digital revolution. Initially, universal access and service plans took different forms in different countries with very uneven success rates (Gillwald, 2005). In most cases, these plans proved to be inadequate to meet the substantial scale of the access problem (Ó Siochrú & Girard, 2005). Due to the inefficiency of their former plans, some governments included universal service and access obligations to mobile network operators when liberalizing the telecommunications market (Westerveld, 2012). In combination with the overwhelming and unexpected uptake of mobile phones by poor people (Aker & Mbiti, 2010), this initiative has mitigated the lack of access in most places, and now mobile phone use and the opportunity to exercise their right to communicate extend to those living below the international poverty lines (Elder, Samarajiva, Gillwald, & Galperin, 2013).

Despite the contribution made by mobile network operators in increasing access to communications in rural areas of developing countries, the development outcomes that increased connectivity promises to bring to underprivileged areas remain unmet (Westerveld, 2012). Digital communication remains unaffordable to most, and both voice (Bidwell et al., 2011) and data communications (Johnson & Roux, 2008; Mathee,

Mweemba, Pais, Van Stam, & Rijken, 2007) are more expensive to rural dwellers when compared to their urban counterparts.

In South Africa, very little is known about the communication expenditure and patterns of use by rural dwellers. Considerable work has been dedicated to understanding mobile phone usage among the youth, specifically students, in urban areas (Donner & Walton, 2013; Kreutzer, 2009). Using the methodology from Kreutzer (2009), Gunzo and Dalvit (2012) presented a study on the ICT usage between school students (grade 2–12) from rural and urban areas in the Eastern Cape. The research reports usage among students is mainly for text messaging, playing games and listening to radio, and a higher uptake of multimedia services by township students. In Bidwell et al., (2011), a more comprehensive age group (ages from 14 to 54) is considered, but the mobile phone usage was primarily focused on investigating the Callback service.

In terms of affordability, whereas in South Africa the national average monthly individual income is R2587 (US$230.3[1]), the average monthly income in rural traditional areas is around R800 (US$71.21)[2], i.e. less than a third of the national average. Thus, for a rural dweller to afford the Organisation for Economic Co-operation and Development's (OECD) lower user mobile basket from the incumbent operator in the area (R90 (US$8.01) for 40 calls/month[3]), they would need to use 11.25% of their monthly income on mobile phone services. Very little is known about the telecommunication expenditure of traditional rural dwellers. In other words, it is uncertain if rural residents spend R90 (US$8.01) on telecommunications a month. Only Bidwell et al. (2011) provide some data and report a weekly expenditure of R13 (US$1.16) in 2010. The authors also report that rural dwellers in these areas incur additional expenses to communicate, such as paying for charging the mobile phone battery and additional mark-up added by the local airtime resellers. Similarly, there is little data on Internet expenditure for rural dwellers. One study argues that 50% of respondents are not using the Internet because they consider it too expensive (Gillwald, Moyo, & Stork, 2012).

This paper bridges this research gap by providing a unique in-depth study of the communication expenditure and patterns of a rural community in South Africa. This research can be used by regulators and government agencies to better design policy implementations to provide universal service and access, by competing industry players to understand the dynamics within rural communities to better target their products and by civil society organizations to use it as a case study in efforts to make affordable communications a constitutional right (Right2know, 2013).

In order to provide a more systemic look at understanding communication expenditure and patterns of usage, this paper uses the concept of the ICT ecosystem (Fransman, 2007). The paper is organized as follows: in Section 2, the concept of ICT ecosystem is introduced along with the different actors who play a role in rural areas. Section 3 describes the methods followed to answer questions about expenditure and communication patterns in a specific rural community. The results are presented in Section 4. In Section 5, the results are discussed, and Section 6 concludes the paper.

2. THE ICT ECOSYSTEM

The use of ecological metaphors to describe complex systems has grown over the past two decades, and this terminology has been linked to Moore's suggestion that businesses operate in symbiotic relationships with one another, with their customers and other economic actors (Moore, 1993, p. 76).

If this proposition is appropriate for ICT, then as with any ecosystem we might expect that the ICT ecosystem would comprise many mutually interacting parts, heterogeneous in their arrangement and characterized by interdependence (Bertalanffy, 1968). These components are arranged in sub-systems, each with their own networks and dynamics (Saaty & Kearns, 1985).

In this paper, the model presented by Fransman (2007), reproduced in Figure 1, is used. Apart from the four groups of players included in Figure 1 (network element providers; network operators; content and application providers; and final consumers), Fransman also includes institutions and organizations in the environment where those players interact. Institutions described are those who "shape the rules of the game according to which the players interact and influence their behaviour" and include standardization and regulation bodies and financial markets. Organizations are described as those who "have the power to change institutions although they themselves are influenced by the institutions they are changing" (Fransman, 2007, p. 2).

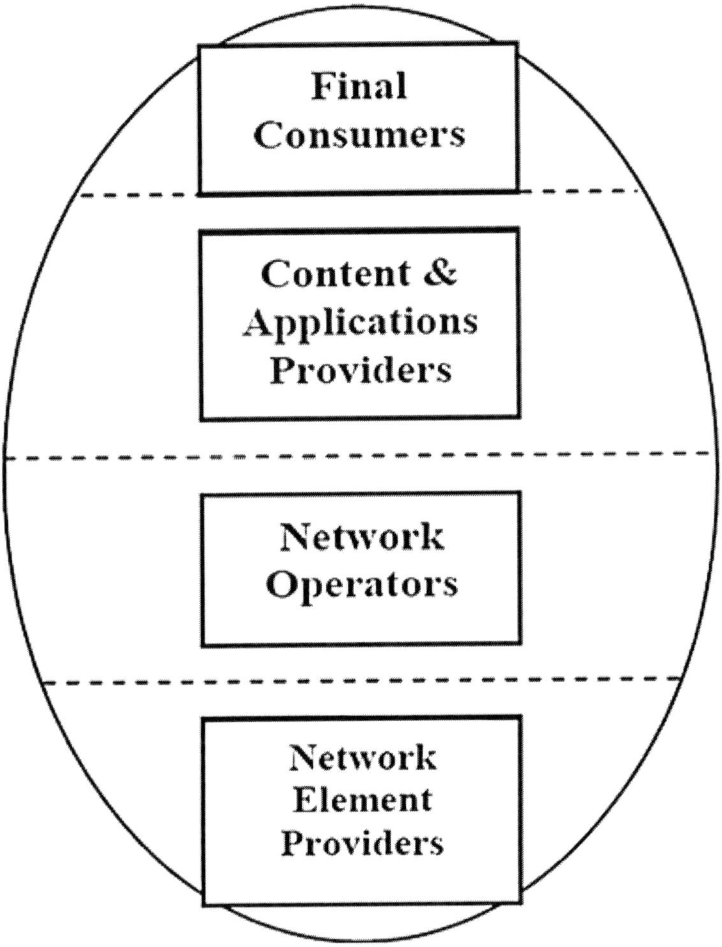

Figure 1. The new ICT ecosystem (Fransman, 2007).

Although other groups of players may also play a role in the ICT ecosystem of rural areas in South Africa, this paper focuses on the two groups of players we believe to be most important for the discussion: the network operators – in particular the mobile network operators (MNOs), which are the predominant access providers used in rural areas – and the final consumers. Similarly, with the institutions and organizations, we only discuss the role played by the regulatory bodies. For the final consumers, we focus on the community where the study has taken place to provide the background for the results presented later in the paper.

2.1. The mobile network operators

In South Africa, there are four mobile network operators (MTN, Vodacom, Cell C and Telkom Mobile) and one virtual operator (Virgin Mobile). However, two operators, MTN and Vodacom, jointly command a market share of more than 80%.

The reduction of the mobile termination rates (MTR) (see below) has allowed smaller operators to reduce their prices in an attempt to gain some market share. Cell C has managed to use it to win a considerable market share previously belonging to MTN (McLeod, 2014), forcing it to bring the prices down in response. Vodacom has managed to maintain its dominant position by using "innovative new ideas" (McLeod, 2015) without having to bring the price of its products down.

Using an international benchmark for price, namely the OECD prepaid low-usage mobile basket (40 calls per month), Research ICT Africa (RIA) has been compiling data about the cheapest low-usage basket offered by each MNO.[4] In its data, the reduction by MTN (from R122.00 (US$10.86) to R90.00(US$8.01)) can be observed clearly. Still, there are no data about the uptake of the different products offered by an operator, and other products such as those implementing "dynamic pricing discounts (where prices are determined on the basis of real-time traffic assessment within a particular cell)" (Gillwald et al. 2012) may have been more popular for the users. In such products, it is difficult to calculate the price per minute. For instance, with MTN Zone the price per minute, both off-net and on-net, is R2.50 (US$0.22), but users get a random discount up to 100%.[5] Making some assumptions, RIA established that when using this product users pay R167.00 (US$14.86) for their low-usage basket.[6]

2.2. The regulatory environment

In a study from 2009, the South African Department of Communications (DoC) compared the cost, usage, access and quality of voice and data services in South Africa with those of five countries with similar economic indicators (Brazil, Chile, Korea, Malaysia and India). Their comparative study concluded that "South Africans talk less on their mobiles than any of the other peer countries" despite having the highest mobile penetration in the group (DoC, 2009). The same study pointed at the high MTR that mobile phone companies were charging others to terminate calls in their networks. These conclusions forced the Independent Communications Agency of South Africa (ICASA) – the national regulatory body – to take action and regulate the MTR, agreeing on a glide path that would reduce the MTR to R0.40 (US$0.036) per call by March 2013.

Although the reduction of the MTR has enabled smaller operators to provide off-net prices more cheaply than the on-net prices of the incumbents, forcing at least one of the latter to bring their prices down to retain customers, prices have remained high when compared to other African countries (Gillwald & Calandro, 2014). It is argued that the effects expected did not happen due to the MTR in South Africa being high compared to their real costs. Additional reductions in the MTR are considered to be required to create the competitive environment that will eventually bring the prices down (Gillwald & Calandro, 2014). Following this recommendation, ICASA designed a new glide path to cut down the MTR to R0.10 (US$0.01) by March 2016. Since March 2014, the MTR for calls from the smaller players (Cell C and Telkom Mobile) to the incumbent providers (Vodacom and MTN) has decreased to R0.20 (US$0.02). Smaller players benefit from an asymmetric rate in order to foster competition, and terminating the calls in their networks cost R0.44 (US$0.04) for incumbents, with a five-year glide path towards symmetry (Gillwald & Calandro, 2014).

2.3. The final consumers

Mankosi is a traditional rural community in the Eastern Cape province of South Africa. It comprises 564 households in 12 villages spread across 30 km^2 of very hilly land. Families are composed of up to five adults and seven children, with a reported household average size of 6.02 people. Thus, Mankosi has an estimated local population of 3395 people, out of which 2070 (60.9%) are 15 or older. These data correlate with the data from StatsSA for the Nyandeni local municipality, where 59.6% of the population are 15 or older. The distribution of the age in the respondents greatly resembles the one observed by the local municipality.[7]

Most of the population (70.6%) in Mankosi is female. This contrasts with the 56.8%[8] female distribution in the local municipality statistics (StatsSA, 2012a, p. 68). The higher concentration of men who stay and work in the urban areas of the local municipality explains this discrepancy.

Families live in homesteads: clusters of thatched, mud-brick rondavels, an occasional tin-roofed 2-room dwelling, an animal corral and a garden for subsistence crops. At the household level, it is only since 2013 that the government has built pit latrines in the yards of all the households. Most affluent people use their flat-roofed housing to collect rain water, although access to water for many is mainly via water points in communal land, and it is very common to see women carrying buckets in their heads (95.4% of the households use this method). With cooking, most families use open fires outside, with wood collected from the nearby indigenous forests, as they can only afford gas or paraffin from time to time. Eighty-four percent in Mankosi do not have electrical lighting infrastructure. Only 2.1% are connected to the grid, and only 13.5% have a generator, solar or gas infrastructure of some sort. This is similar to what has been observed by the authors in other rural communities in the Eastern Cape.

The reported household income per month is R1831 (US$163), coming mainly from government grants and payments from family members who temporarily migrate for work. Thus, the reported individual's monthly income in Mankosi is R388; that is, people in Mankosi live on roughly US$2 a day.[9] This income is mainly used to cover the credit accumulated during the month in the local shop, which receives the outstanding debt the moment the government grant is received. Thus, very few households have savings and see no reason to open a bank account (58.6% of the households do not have one). This income contrasts with the R2587 (US$230.3)[10] of monthly individual income, which is the average of the country (StatsSA, 2012b). Due to the high levels of inequality in South Africa[11] and the average disposable income at local municipality levels (StatsSA, 2012a), many other rural communities are expected to show disposable income levels similar to those of Mankosi.

Regarding formal education, only 13% of the population has completed matriculation (grade 12) or a higher level of education. The reported capacity to read and write isiXhosa is considerably higher (above 70% of respondents report doing so easily) when compared to the ability to read and write English (above 32% of respondents report not ability at all in both), with older people reporting the lowest levels.[12]

In Mankosi, as in 7218 communities in the country (of which 1192 are in the Eastern Cape) inhabitants are governed by a Tribal Authority (ICRPOB, 2011), which in Mankosi consists of the Headman, 12 Subheadmen (one per village) and messengers. The Headman and Subheadmen's homesteads are sites for local administration (Bidwell et al., 2013).

Recently, a non-for profit locally owned co-operative, Zenzeleni Networks Ltd., has been established in the community. Its main goal is to reduce the cost of communications of the community members. Since September 2014, the cooperative is a legal Internet Service Provider (ISP) (Rey-Moreno, Tucker, Cull, & Blom, 2015b) and is offering access to telephony services at a fraction of the price offered by MNOs from 12 public phones installed in the community. In August 2012, adapting the idea introduced in the community by Bidwell et al. (2013), 8 of these public phones started operating as charging stations where community members can charge their phone batteries cheaper than in the options available before then. The records starting in July 2013 show that more than 3300 phones have been charged there since then (Rey-Moreno, Sabiescu, Siya, & Tucker, 2015a).

3. METHODOLOGY

The data presented in this paper (including the context information presented above) come from using a stratified sampling survey of the households in the 12 villages that comprise Mankosi. The total number of households per village was obtained by asking the subheadmen about "pieces of land"[13] that they had allocated in their villages, as they keep a register.

The sample size was determined by setting the confidence level to 95% with a margin of error of 5%. Given a population size of 564 households, a sample of 224 was calculated. For a possible 10% non-response rate, the sample was increased to 250. The sample size within each strata (village) was calculated to be proportionally similar to ensure representativeness within the community surveyed.

The data presented in this paper come from two different surveys conducted in the community:

- The first survey was conducted between July 2012 and January 2013. It was collected in paper format adapting the Household Survey from RIA (Stork & Stork, 2008). It consisted of the Household Roster being answered by the head of the household and an individual part being answered by a randomly selected individual. The data were collected by six different Local Researchers in three different periods. A total of 247 questionnaires were administered.
- The second survey took place between December 2013 and January 2014. It was collected by a team of five Local Researchers using Open Data Kit (ODK). The questionnaire consisted of a reduced version of the first questionnaire, adapted based on communication patterns learned via participant observation and ethnography. A total of 255 questionnaires were administered.

Within each village, each household was selected randomly, choosing the first household found along a footpath and skipping the next. Within the household, the individual interviewed was also randomly selected. Two different techniques were used in the two surveys:

- In the first one, the selection of the participants within the household followed the same procedure used by RIA, asking the date of birth of all the household members older than 15 years old and selecting the individual with the next birthday from the day the questionnaire was being administered.

- In the second one, a random number generated by ODK was used. Every person 15 years old or older present at that moment in the household was given a number, and the person given the number shown by ODK was asked to give consent. If accepted, the questionnaire was administered. If not, the process was repeated again, excluding that person.

Due to inconsistencies discovered during the data-cleaning process of the first questionnaire, in-depth interviews were conducted with the surveyors. Although the randomization of the households was done properly, cultural factors were overlooked. This prevented the proper randomization of individuals[14] inside each household and so the individual data are not representative of the population in Mankosi. Thus, only information from the Household Roster of the first questionnaire is used in this paper.

Additionally, the paper builds on ethnographic data collected by the principal author, who has lived in Mankosi for more than 15 months. During this period, the author has additionally visited nearby communities, and engaged in countless formal and informal conversations with their members about mobile phone ownership, usage and expenditure. This includes structured interviews to understand the sense of ownership of the local cooperative described in Section 2 (Rey-Moreno et al., 2015a) and the co-design of a billing system for the voice services provided by the cooperative (Rey-Moreno, Ufitamahoro, Venter, & Tucker, 2014).

4. RESULTS

4.1. ICT ownership

In terms of ICT ownership per household, 97.7% of the households owned at least one mobile phone, with an average of 2.93 mobile phones per household. This contrasts with the ownership of a computer and a landline, where none of the households owned one.

At the individual level, 81.2% of the people surveyed owned a phone. Of them, 73.4% bought it and 26.6% received it from someone else. On average, people have had a phone for 4.09 years, but with a very high uptake in recent years, as observed in Figure 2.[15]

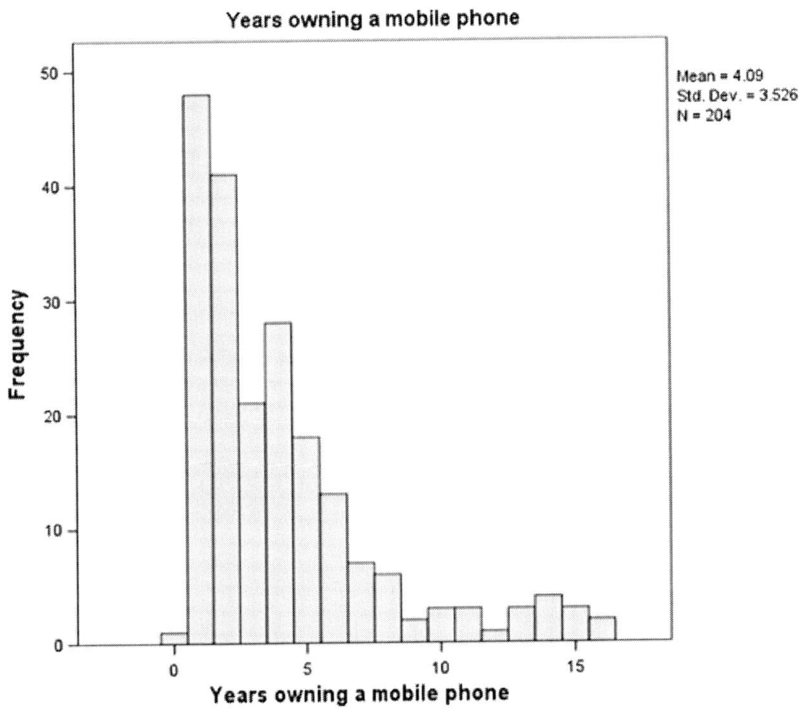

Figure 2. Years owning a mobile phone.

Years owning a phone correlates positively with age, with older people owning phones for longer ($\rho = 0.152$ ($n = 202$), $p < .05$). Additionally, there is a difference in the age ($U = 2997, p = .003$) of the people who have bought their phones (Md = 27) and those who were given phones (Md = 21). Due to access to more employment opportunities, older people are more likely to have the economic resources to buy mobile phones than younger ones.

In relation to preferences about features in their mobile phones, many reported a preference for longer battery life (preferred by 58.0% of individuals). This is followed by the phone providing Internet capabilities (18.3%) and the possibility to play music (15%). As shown in Table 1, older people prefer longer battery life, whereas younger people are more interested in bigger storage capacity, music reproduction and Internet capabilities.

Table 1. Median comparison [Kruskal–Wallis] among categories in variable "Main feature you wish your phone had" and "Age."

Variable	Grouping	Md	*n*	χ^2	*p*
Main feature you wish your phone had	Big(er) storage capacity	21.00	10	33.868	0.00
	Long(er) battery life	32.00	144		
	Music reproduction	22.00	39		
	Internet capabilities	20.00	46		
	Others	24.00	11		

With regard to SIM card ownership, 83.9% of the respondents owned one, in contrast to the 81.2% of people who owned a phone. From those, 93.9% of SIM cards belonged to MTN, 5.6% to Vodacom and 0.5% (1 respondent) to Cell C.

4.2. ICT expenditure

Those owning a mobile phone were asked about the amount of money spent on charging it in the last month. People reported an average expenditure of R10.36 (US$0.92) a month. However, there is a large dispersion in the responses, as seen in Figure 3. Forty-nine percent of the respondents reported R0 of expenditure. This is explained by the different levels of access to a place where phones can be charged for free. Some households are connected to the main power grid or have solar systems specifically installed for charging mobile phones, where not only household members, but relatives and friends also charge phones. Additionally, some locals work in the cottages owned by affluent fishermen from the cities, the main backpackers or other places (including the school and the high school) where one can charge at no cost.

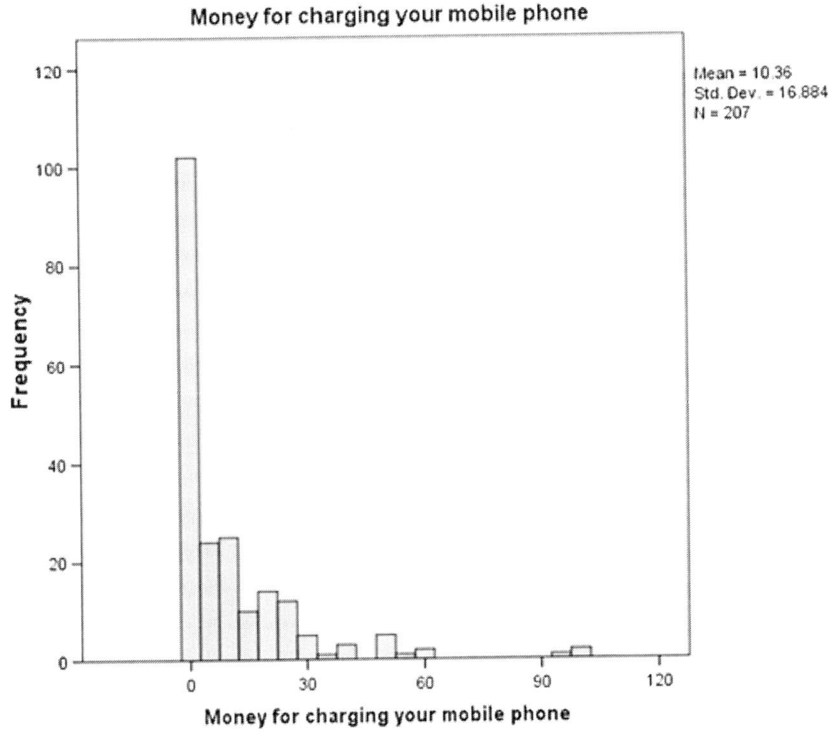

Figure 3. Money spent on charging the mobile phone per month.

Those who reported owning a SIM card were asked how much airtime they purchased in the last week, and how much they paid for it. On average they spent R17.49 (US$1.55) for a total of R15.28 (US$1.36) airtime, reflecting the mark-up reported previously by Bidwell et al. (2011). This mark-up increased on average 17.19% the cost of communication for Mankosi dwellers. This mark-up is common in rural areas and has been observed in many other communities visited, and it is related to the increase on the prices that other products (like milk) suffer in rural areas to compensate for the cost of transport.

It is worth noting that 93.2% of the individuals reported topping-up R40 (US$3.56) or less, including 18% of those who reported not having topped-up their phone in the past week. Thus, on average mobile phone users in Mankosi received only R0.58 (US$0.05) in promotions[16] during the previous month. The money received in promotions correlates positively with the airtime topped-up ($\rho = 0.196$ (206), $p < .01$). However, due to the low amount topped-up, very few people (all of them with MTN SIM cards) received money in promotions (10.7% of those with a SIM card). It is worth noting that the MTN product observed to be used by the majority of people is MTN Zone. With this product, discounts come as a percentage of discount over the normal rates. In the main author's experience – observing the discounts received by others – they are regularly 20%. Sometimes they receive 100% discount of the total cost of the call; when this happens, they tell their friends as it seems to be related with the base station covering the area, and it lasts only a few minutes.

Table 2. Spearman correlations between "Age," "Years owning a mobile phone," "Money spent on charging last month" and "Airtime topped-up last week" (*n* shown in brackets).

Variable	**Age**	**Years owning a mobile phone**	**Money spend on charging last month**	**Airtime top up last week**
Age		.152* (202)	.262** (205)	.215** (204)
Years owning a mobile phone			.161* (204)	.178* (196)
Money spend on charging last month				.208** (199)

*$p < .05$.

**$p < .01$.

There is a correlation between the number of years owning a mobile phone and the money spent on mobile phone services, both having it charged and topped-up, as shown in Table 2. Furthermore, both expenditures correlate with age. Given the big uptake of mobile phones in the community in recent years and the young age of the population, this data suggest that expenditure in mobile phone services in Mankosi will continue increasing in the coming years.

The data in Table 2 also suggests that people spending on one mobile phone service (for example, charging) are willing to spend more on another (for example, airtime). This is further corroborated by the difference in expenditure between those who bought their phones, and those who were given phones as shown in Table 3. People buying their phones clearly spend more on mobile phone services.

Table 3. Median comparisons [Mann–Whitney] between those who bought their phones and those whose phone was given.

Variable	**Grouping**	**Md**	*n*	*U*	*r*	*p*
Money spent for charging the phone last month	Bought	5.0	152	3163.5	.20	.004
	Given	0.0	55			
Airtime topped up last week	Bought	10.0	152	2764.5	.22	.002
	Given	5.0	55			
Days without airtime las week	Bought	3.0	149	2781	.23	.001
	Given	2.0	53			

Additionally, people not only topped-up their own mobile phone, but they also used someone else's mobile phone during the week previous to the questionnaire. Sixteen percent of all respondents reported to have used someone else's phone in the prior week. This group included people not owning either a phone or a SIM card. Thus, 86.66% of the population used mobile phone services during the week prior. Out of those using someone else's phone, 56% (9% of the total) reported topping-up that phone with airtime for their own use, spending on average R6.65 (US$0.59) that week. It is unclear whether this amount was considered in the total amount reported as having been spent on airtime in the previous week. In this study it is considered as if it were.

People not only bought their own airtime, but they were also transferred airtime from wealthier relatives working outside the community. From those owning a SIM card, 54.7% reported having received airtime in the last week, for an average value of R4.87 (US$0.43).

Table 4. Median comparison [Kruskal–Wallis] among categories in variable "Reasons for buying airtime" and "Age".

Variable	**Grouping**	**Md**	***n***	χ^2	***p***
Reasons for buying airtime	Social calling and buzzing (relatives, friends, and other people)	27.00	169	16.556	0.01
	Social Internet (Web surfing and Social Sites: Whatsapp, 2Go, Mixit and Facebook) and others (SMS, social calling)	20.50	24		
	Social Calling and SMS	20.50	12		
	Social calling and special calling (work and other arrangements)	27.50	8		

Respondents were asked about the reasons for buying airtime. Most people (79.5%) used the airtime to make social calls only. As this was an open question, it is worth noting that some of the people reported these social calls to be made in order to be in contact with the people they care about, who are living outside the communities (including wives/husbands, children and other relatives). As observed in Table 4, older people used it for social calling and official calling, whereas younger people used it additionally for SMS, social Internet and music downloads.

The respondents who owned a phone, a SIM card or who topped-up someone else's phone were also asked whether they sacrificed on other things to buy airtime and 41.2% reported affirmatively. From those, 34.5% reported sacrificing on essential food items, 16.7% for essential items for cooking and lighting and an additional 2.4% on both. As observed in Table 5, older people are more likely to sacrifice essential food and cooking and lighting, whereas younger people are more likely to sacrifice more luxurious items.

Table 5. Median comparison [Kruskal–Wallis] among categories in variable “Items sacrificed for airtime” and “Age.”

Variable	**Grouping**	**Md**	*n*	χ^2	*p*
Items sacrificed for airtime	Essential food (milk, tea, bread, groceries and salt)	29.00	31	16.734	0.002
	Luxury (cigarettes and cosmetics)	19.00	10		
	Household, cooking and lighting (matches, candles and paraffin)	28.00	15		
	Luxury food (chips, sweets, drinks, fried chicken and meat)	21.00	15		
	Others	32.00	13		

Considering the money spent for buying airtime in a week and on charging the mobile phone in a month, and extrapolating the money spent on a week to a monthly expenditure,[17] the average monthly expenditure in mobile phone services was R85.21 (US$7.59). With the people older than 15 years using a mobile on a weekly basis showed above, a rural community like Mankosi spends around R153,000 (US$13,620) monthly using their mobile phones. Only 76.76% of this estimated individual monthly amount is for actual airtime used for MNO services.[18]

When comparing this expenditure with the average monthly income reported above, people in Mankosi spend 21.97% of their income on mobile phone use. If the disposable income does not increase, this percentage is likely to be higher with time due to the correlations with age and years owning a phone.

4.3. Mobile phone use

With regard to mobile phone use, the first variable studied was how many times people charged their phones during the last week. People in Mankosi charged their phones at least once a week, with 35.8% charging their mobile phones three or more times a week. The second variable took into consideration the number of days without a charged battery a month; 41.1% of the respondents reported not having had their phone without battery a single day in the last month. Still, the average was 1.95 of days without battery a month.

Then, respondents with phones were asked about the number of days they did not have airtime on their SIM cards last week. It is worth noting that on average people reported being without airtime 2.78 of days a week, i.e. people in Mankosi reported to be on average 39.71% of their time without airtime. So, even though people have their phone charged most of the time, they were unable to make calls or send SMS for a large portion of their time.

The days without battery in the last month correlates negatively with the airtime top-up last week ($-.148, p < .01, n = 202$) and the days without airtime in the last week ($-.199, p < .01, n = 199$). This suggests that the more the phone was charged, the more they spent, and the quicker they did so.

For obtaining the number of calls made during the last week and the length of these calls, interviewers obtained the readings directly from the person's mobile phone. Since not all participants carried their phone with them or had battery in the moment of the interview, only 195 responses for the variable "number of calls made last week" were collected. For the length of the calls, the number was even smaller as some of the local researchers added up the total number rendering values that did not make sense (e.g.

39 hours ...); these numbers were modified to compute as missing. Thus, a total of 141 answers were used to calculate this variable, as well as the average length of each call. According to the data, people in Mankosi made an average of 4.35 calls a week. The average total length of the calls reported was 17.95 minutes, with an average length per call of 3 minutes 20 seconds.[19] A large proportion of people made shorter calls as shown in Figure 4, even though they were made for social calling as reported earlier. The longest calls are associated with users who reported having received big discounts (90–100%) when using MTN Zone, and as it can be observed they represent the minority of the calls.

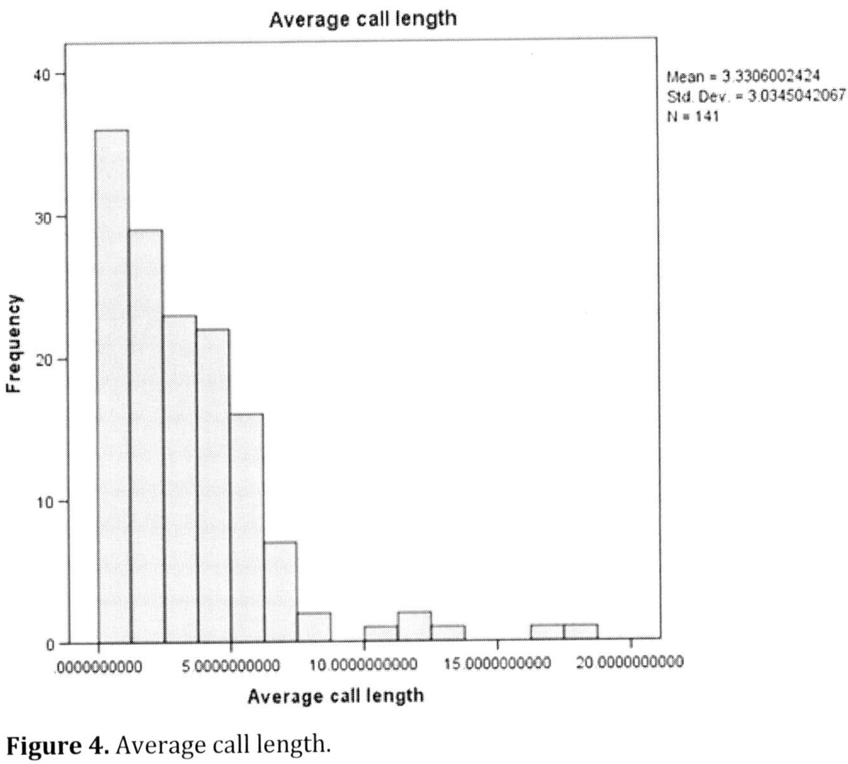

Figure 4. Average call length.

Table 6 presents the strong positive correlations between the amount topped-up in the last week and the number of calls made, total minutes called, average call length, SMS sent last week and money spent on bundles in the last month. The same trend (and correlation) is shown by the other variables related to increased airtime (airtime transferred by others, airtime topped-up in others' phones and airtime received in promotions). These results, although apparently obvious (the more people are willing to expend in communications the more they can communicate) show the important role played by the cost of the airtime (including the mark-up), especially for those who cannot afford it.

Table 6. Spearman correlations for "Days without airtime last week" and "Airtime topped-up last week" with communications-related variables (n shown in brackets).

Variable	**Number of calls made last week**	**Number of minutes called last week**	**Average call length last week**	**PleaseCallMe sent last week**	**SMS sent last week**	**Money spent on bundles last month**
Days without airtime last week	−.260** (185)	−.291** (134)		.147* (203)	−.220** (207)	−.636** (19)
Airtime top up last week	.458** (185)	.563** (132)	.420** (132)		.318** (206)	.786** (19)

*$p < .05$.

**$p < .01$.

This trend is further corroborated by the days without airtime as shown in Table 6. The more the number of days without airtime, the less people communicated. It is worth highlighting that in the absence of airtime, people still used PleaseCallMe.[20]

Additionally, as shown in Table 7, there is a strong positive correlation between communication indicators, suggesting that communication methods reinforce each other. This is relevant to the notion of an ICT ecosystem in which the components (i.e. the communication methods shown in Table 7) form symbiotic relationships (i.e. reinforce each other).

Table 7. Spearman correlations among communications-related variables (n is shown in brackets).

Variable	Number of calls made last week	Number of Minutes called last week	Average call length last week	SMS sent last week
Number of calls made last week		.751** (141)	.257** (141)	.235** (195)
Number of minutes called last week			.768** (141)	.252** (141)
Average call length last week				.160° (141)

*$p < .05$.

**$p < .01$.

°$p = .058$.

Then, people were asked about the number of calls they made to people in Mankosi, the number of SMS and PleaseCallMe sent last week, whether they had had problems with the operators signal in the last week, and whether they had used Internet on their phones in the last month. These questions were only asked to people having a phone, a SIM card or someone who reported having used someone else's phone last week. Thus, a total of 221 answers were considered. The number of valid answers for the number of

PleaseCallMe was later reduced to 217, as there were 4 respondents who reported values over the maximum limit allowed by their mobile phone company.[21]

People reported to call mainly to someone outside Mankosi, with only 26.7% reporting that "Most" or "All" their calls were made to people locally, thus corroborating that they mostly communicate with their relatives outside the community. With regard to the number of PleaseCallMe, 67% of the people reported using the service, with an average of 9.98 sent in the last week. This contrasts with the number of SMS sent last week, where 41.7% reported to have not sent any in the last week, and the average was 1.75. In both cases, older people use these services less (see Table 8 for more details), which is associated to the lower capacity to read and write IsiXhosa and English reported earlier.[22]

Table 8. Median comparison [Kruskal–Wallis] among categories in variable "Age" and services used.

Variable	**Grouping**	**Md**	***n***	χ^2	***p***
PleaseCallMe sent last week	15–29	6.0	131	20.631	0.00
	30–49	5.0	58		
	50+	0.00	24		
SMS sent last week	15–29	1.00	135	28.261	0.00
	30–49	0.00	58		
	50+	0.00	26		

Thirty-nine percent of the respondents reported having problems with the signal of the operator during the last week when making or receiving calls (something the researcher has experienced personally on several occasions).

With regard to Internet use, 22.2% of the respondents reported having used it in the last month. From those, 42.9% reported to have bought data bundles in the last month. Among them, the average expenditure on data bundles in the previous month was R12.15 (US$1.08). There is a strong correlation among the money spent on bundles in the last month with the average call length (.609, $p < .05$, $n = 19$) and the number of SMS sent in the last week (.534, $p < .05$, $n = 20$). This suggests again, as presented in Table 7, that the more people communicate, the more they are inclined to use all means available to do so. Those reporting to have used Internet on the phone in the month previous to the questionnaire were significantly younger (Md = 21.0), than those who have not (Md = 28.0).

As in the case of the PleaseCallMe and SMS, a Chi-square test for independence (with Yates Continuity Correction) indicated significant association between the use of Internet and the ability to read and write English and IsiXhosa, as shown in Table 9. All cross-tabulation between those who used the Internet last month, who answered "Easily" in any of the four variables regarding ability (read IsiXhosa, write IsiXhosa, read English and write English), showed values over 71%.

Table 9. Chi-square test for independence between the "use of Internet on the phone last month" and the ability to read and write English and isiXhosa.

Grouping: use Internet on the phone last month	**Level of ability**	%	*n*	χ^2	*p*
Read IsiXhosa	Easily	85.7	42	8.886	.012
	With difficulty	14.3	7		
	Not at all	0	0		
Write IsiXhosa	Easily	89.8	44	8.448	.015

Grouping: use Internet on the phone last month	**Level of ability**	%	*n*	χ^2	*p*
	With difficulty	8.2	4		
	Not at all	2.0	1		
Read English	Easily	81.6	40	29.638	.000
	With difficulty	12.2	6		
	Not at all	6.1	3		
Write English	Easily	71.4	35	36.552	.000
	With difficulty	22.4	11		
	Not at all	6.1	3		

Finally, people were asked about whether they thought there was something preventing them from communicating more, and if so, to mention the factors. A total of 44.3% of the respondents answered positively. The combination of lack of battery, bad signal and lack of airtime represented 92.3% of the answers, with lack of airtime alone being the factor preventing communications to 46.2% of the respondents.

5. DISCUSSION

This section discusses the nuances of networks and dynamics existing within the consumers' component of the ICT ecosystem. Then, it provides a detailed discussion of the interaction between the three players considered within the ICT ecosystem in rural areas of South Africa: MNO, regulator and consumers.

5.1. Networks and dynamics within the consumer sub-system

Access to mobile services in rural South Africa is very high, with almost all households (97.7%) of Mankosi owning a mobile phone. Individual ownership among people is 81.2%, which is a considerable increase from the 66.6% reported in Bidwell et al. (2011) three years earlier. An additional 5.66% access mobile phones either from using a SIM card on someone else's phone or by directly topping up someone else's SIM card, for a total of 86.6% of people older than 15 accessing mobile phone services on a weekly basis. Shared used of devices is consistent with other findings in rural areas of South Africa (Bidwell et al., 2011; Gunzo & Dalvit, 2012). Access to mobile phones contrasts with the low, or non-existent penetration of other devices, such as landlines and computers.

People in Mankosi report spending disproportionate amounts of their disposable income (21.97%) in communications compared with the national average (2%) (StatsSA, 2012b). As acknowledged by the respondents, this forces them, especially older people, to sacrifice on basic food and household items to use mobile phone services. Other studies concerning poor people in South Africa (Duncan, 2013) and elsewhere in Africa (Diga, 2008; Gillwald & Stork, 2008) have shown similar results.

A significant percentage of the total expenditure is used to keep the mobile phone charged. Considering the data reported in Bidwell et al. (2011), people's expenditure on charging has decreased over the past 3 years. In December 2010, they were expending R5.50 (US$0.49) a week on average (R23.57 (US$2.1) a month), while in December 2013 they reported to be expending R10.36 (US$0.92) a month. This 56% reduction can be explained by a combination of two factors: (1) the installation by Zenzeleni Networks Ltd and previous initiatives (Bidwell et al., 2013) of cheaper and closer charging stations (R3 (US$0.27) per charge instead of R5 (US$0.45) (Bidwell et al., 2011)) and (2) more community members installing their own solar energy solutions to charge phones. Although there is no data to prove this as yet, qualitative findings suggest that the charging stations from Zenzeleni Networks Ltd have also contributed to reduce the number of days without battery per month. Before the installation of the charging stations, a mobile phone could go a week without being charged (Bidwell et al., 2013). The importance for people to have their mobile phones charged should not be disregarded, as shown by the number of days a month people reported not having battery. Mobile phones are not only used to make calls but they are also used as a torch at night, for listening to the radio and music, and, provided they had long periods of time without airtime, receiving urgent information from their relatives (news, job offers, etc). Then, it is no surprise that the main feature people wished to have on their mobile phones was a longer battery life. This again supports the idea of an ICT ecosystem, in which access to battery charging may determine usage and the capacity to communicate and to build knowledge, and how local people are creating ways for this to happen.

Expenditure on airtime has increased by 33.05% when compared to the data collected by Bidwell et al. (2011) three years prior to the results presented in this paper. This corroborates the results obtained with regard to the correlation between years owning a phone and expenditure in mobile phone services, a trend which is expected to continue in the future due to the even higher uptake in recent years. Provided there is a negative correlation between the number of days without battery a month and the airtime expenditure in a week, the installation of the charging stations (which has reduced the number of days without battery) may have contributed to the increased airtime expenditure.

Airtime, as other products resold in rural areas (Dicks & Simons, 2013), includes a mark-up. This mark-up, which was previously identified by Bidwell et al. (2011), has been quantified in this research and increases communication costs by 17.19%. The mark-up and the payment required to charge a mobile phone are not incurred in urban areas and account for 23.24% of the total expenditure in mobile phone services in Mankosi.

The high proportion of disposable income dedicated to communications (which does not even include the airtime transferred to individuals by others) is not correlated with high usage. On the contrary, with this expenditure people were on average able to send 7.49 SMS and make 18.6 calls a month. Similarly, the monthly expenditure in Internet services (R12.50 (US$1.11)) provides very constrained use: 30MB with Vodacom,[23] and 25MB with MTN.[24] In this regard, cost appears as the main factor limiting people to communicate, similar to the results reported in other studies (Elder et al., 2013), followed by lack of battery and bad signal. This constrained use of communications may serve to explain the high number of PleaseCallMe messages sent, which do not require having airtime, and can be personalized (Bidwell et al., 2011). However, its use is further limited by levels of literacy, which results in older people not benefiting from these services. A similar occurrence happens with the Internet services, where mostly young people use them due to the increased level of literacy required.

5.2. Inefficient interactions of players within the ICT ecosystem

To understand the high percentage of income (21.97%) dedicated to mobile phone use, the huge income inequality existing in the country and the high prices of mobile services need to be considered (Gillwald & Calandro, 2014). This value is considerably higher than the 5% that the government is targeting (RSA, 2010). Additionally, usage numbers are considerably below both the low-usage basket created by the OECD, which considers 40 calls a month (Gillwald & Calandro, 2014), and the total length of these calls that the government targets people to speak with the 5% of their disposable income, which considers 90 minutes a month (RSA, 2010). Our results revealed an average duration of only 76.82 minutes per month.

Efforts to reduce these costs are not reaching the people in Mankosi. For instance, all people reported using pay-as-you-go services due to the informal economy, and so they are not able to benefit from the discounts available when using contracts (Elder et al., 2013). Additionally, they top-up small quantities that do not qualify for the free airtime that MNOs randomly give when topping-up bigger amounts. Finally, the small number of big discounts might also be explained by the fact that most of the calls are made during peak-time[25] when it is not profitable for the operators to provide discounts. Products using dynamic pricing, like MTN Zone, with high uptake in rural areas, could provide some relief, but still the percentages of discount observed are not in line with the existent inequality.

Due to the high cost of communications in South Africa, the regulator introduced a reduction of the MTR (this reduction was announced after the data presented in this paper were collected). Although this initiative has already had an effect on cheaper packages available nationwide (Gillwald & Calandro, 2014) with smaller operators offering bigger discounts, the effect of these changes in rural areas is still to be seen; and several factors may hinder this endeavor. The lack of price transparency from incumbent MNOs reported elsewhere (Right2know, 2013) has been corroborated by initial data collected in Mankosi and shows that very few community members are aware of the price they pay per minute (Rey-Moreno et al., 2014). This, together with the low levels of numeracy and literacy reported in other research conducted in Mankosi (Dicks & Simons, 2013), affects consumer's perceptions of price, which may explain not only why 93.9% ownership of SIM cards from the most expensive provider in the country at the moment this study took place (Gillwald & Calandro, 2014), but also why they prefer one of their most expensive products, namely MTN Zone.[26] Additionally, it has been reported that low-income consumers in Mankosi do not base their purchase purely on price. Habit and perception of quality are deeply entrenched with consumer's choices (Dicks & Simons, 2013). Finally, the possibility to access these discounted prices by alternative providers is advertised via channels accessible to very few.[27] In the event of someone being aware of them, it requires him/her to go to the closest city to change to a different provider. Given that people do not go to the city very often, this may prevent them from benefiting from these discounts. Due to the high proportion of users from the same company, airtime resellers mainly sell airtime from MTN, and people with SIM cards from the other companies struggle to find airtime locally.

The lack of airtime from a given provider and the mark-up problems could be solved simultaneously if people used Internet banking, as most people do in urban areas. It has no mark-up associated, and allows topping up a number from any MNO. However this method has several requirements attached: (a) being aware of the service; (b) having a bank account and (c) having access to the Internet. Data about awareness have not been collected, but the low number of households with bank accounts (41.4%, presumably owned by older people) and the low uptake of Internet services (22.2%, mostly young people) limit this possibility.

Similarly, the literature has reported using PleaseCallMe (Bidwell et al., 2011) or using missed calls (Donner, 2007) to avoid costs associated with communications. One would expect that the same might happen with the use of mobile data. In South Africa, airtime can be used to purchase data bundles that, depending on their size, can be up to 720% cheaper.[28] However, only 42.9% of the people using Internet use this mechanism.

6. CONCLUSIONS

This paper has presented an in-depth study of the ownership of mobile phone devices and the expenditure and usage patterns of the members of a rural community in South Africa. Results show a high access ratio (86.6% of people using mobile phone services weekly) and a high proportion of the disposable income (21.97%) dedicated to a very constrained set of mobile phone services (1.75 SMS and 4.35 calls, with a total length of 17.95 minutes, a week). Regarding data, only 22.2% reported having used mobile data in the last month, with a very constrained use of the services (25–30 MB a month).

The internal dynamics within the consumer sub-system, such as the mark-up added to the price of the airtime and the challenges for changing to a cheaper provider, may circumvent the efforts from the regulator to reduce communication costs. In the meantime incumbent MNOs are benefiting extensively from the need of illiterate consumers to communicate. In addition, components of the ICT ecosystem, perhaps taken as given in the urban context, may become binding constraints in the rural context. Access to battery charging is a noteworthy example in this study. Although the time spent without a charge appears minimal, it is likely that consumers manage their usage of ICT in order to minimize the risk of being without access.

Communities of consumers such as Mankosi, where the majority of people live on incomes below national and international poverty levels, are willing to spend more than R150,000 (US$13,6200) monthly on a constrained set of communication services. In it therefore heartwarming that Mankosi is starting to follow other examples in rural areas worldwide (Heimerl, Hasan, Ali, Brewer, & Parikh, 2013) and become producers of their own networks. In this context, community-owned efforts such as Zenzeleni Networks Ltd are worth exploring, and the data presented in this paper provide a solid baseline to measure its impact.

Acknowledgements

We would like to thank Jay, Thulani, Patheka, Phiwe, Nomphelo, Vhiwe, Sibongeseni, Patrick, Mapru and Sintle for assisting with the data collection. We also thank Nic, David, Steve, Christoph and Isabella for the hints during the design of the questionnaire and the discussion around it.

DISCLOSURE STATEMENT

No potential conflict of interest was reported by the authors.

FUNDING

This work is based on research supported in part by the National Research Foundation (NRF) of South Africa (Grant-specific unique reference number (UID) 75191) and the CONFINE Integrated Project FIRE #288535. Any opinion findings and conclusion or recommendations expressed in this material are those of the authors and therefore the NFR does not accept any liability in this regard. We thank Telkom, Cisco, Aria Technologies and THRIP (Technology and Human Resources for Industry Partnership) for their financial support via the Telkom Centre of Excellence (CoE) programme

NOTES

1. The exchange rate used throughout the article is R1.00 = US$0.089; that of the 8 November 2014 http://zar.fxexchangerate.com/usd/388-currency-rates.html.
2. Own calculations based on the data from StatsSA (2012a).
3. Cheapest prepaid product for OECD basket by operator in South Africa in local currency (ZAR). http://www.researchictafrica.net/prices/country.php?c=20. Last accessed 6 August 2015.
4. Cheapest prepaid product for OECD basket by operator in South Africa in local currency ZAR): http://www.researchictafrica.net/prices/country.php?c=20. Last accessed 6 August 2015.
5. Price Plans: https://shop.mtn.co.za/crs/siteInformation /staticPage.jsp?breadcrumb=Pre-Paid%20Plans&siteInformation=PrePaidPlans. Last accessed 6 August 2015.
6. Since MTN does not release ARPU and MoU data, for comparative purposes Research ICT Africa has calculated MTN Zone price based on assumed dynamic discounts and the resulting price is more expensive than other price plans. See OECD low-usage basket price of MTN prepaid products at http://www.researchictafrica.net/prices/operator.php?o=65.
7. Considering the data available per age group in (StatsSA, 2012a, page 68) and in the data collected.
8. Number generated by the author considering the data available on gender per age group in StatsSA (2012a).
9. This does not take account of Purchasing Power Parity which adjusts for the cost of living, and the figure would be closer to US$75 http://zar.fxexchangerate.com/usd/388-currency-rates.html; http://salaryconverter.nigelb.me/).
10. Number generated by the author considering the average annual income household and the average household size (StatsSA, 2012b)
11. GINI Index, World Bank, http://data.worldbank.org/indicator/SI.POV.GINI. Last accessed 8 November 2014.
12. A Kruskall–Wallis analysis shows significant difference between the three capacity levels and age. For "Not at all" medians are above 42 years, and "Easily" below 23 years for the 4 variables studied.

13. Piece of land is the term used among Xhosa people to refer to the space in the community that has been allocated by the Tribal Authority to a given family to build its household.

14. Cultural norms in the community include addressing questions to the head of the household, in this case the eldest woman.

15. For this variable, three values were declared as missing, as the interviewee should have entered the natural year when the phone was obtained, and the value in the cell was 0 instead.

16. Bonus airtime received, many times randomly, after topping-up prepaid airtime.

17. A month is considered to have 30 days, i.e. 4.28 weeks.

18. This amount can be higher if the airtime topped-up on someone else's SIM card was not considered as part of the airtime purchased last week.

19. The average total call-length does not equate the number of calls times the average call-length per call, due to the disparities on the sample size.

20. It allows sending a message through USSD containing the caller's cell-phone number and a 10-character text intended to identify the caller. Users can personalize this text once per day (Bidwell et al., 2011).

21. In the moment, the data were collected the maximum number of PleaseCallMe per day allowed from MTN in a day are 5 and 10 from Vodacom. http://mypleasecallmeback.blogspot.com/2013/02/call-me-back-cell-cvodacommtnvirgin.html Last accessed 8 November 2014.

22. A Kruskall–Wallis analysis shows significant difference in the number of SMS and PleaseCallMe sent among the three ability levels in the four variables regarding reading and writing. People with reported ability "not at all" present in all tests the lowest value.

23. http://www.vodacom.co.za/mobile/main/datapromotion Last accessed 22 August 2014

24. https://www.mtninternet.co.za/getit/Pages/InternetBundles.aspx Last accessed 22 August 2014

25. Most rural activities take place during day light.

26. OECD low-usage basket pricing of different MTN prepaid products http://www.researchictafrica.net/prices/operator.php?o=65 Last Accessed 5 September 2015.
27. Cell-C offered the possibility to call for 50 cents to any network, but it was only advertized via their website. http://www.cellc.co.za/explore/newsroom/cell-c-launches-50c-per-minute-prepaid-call-rate. Last accessed, 8 November 2014.
28. Out of bundle costs are R2(US$0.18) per MB. http://www.vodacom.co.za/vodacom/shopping/plans/1gb-30-day-once-off-prepaid-data-bundle.

NOTES ON CONTRIBUTORS

Carlos Rey-Moreno is a Post-Doctoral fellow in the Computer Science Department at the University of the Western Cape, South Africa. His research interest is focused on multidisplinary approaches to allow disadvantaged communities to reduce their cost for digital communications.

Rénette Blignaut is a professor in Statistics at the University of the Western Cape, South Africa. Her research interests include HIV and AIDS prevention, biostatistics, data mining and science education.

William D. Tucker, also known as Bill Tucker, is an Associate Professor of Computer Science at the University of Western Cape (UWC) in South Africa. His research group is called BANG, Bridging Application and Network Gaps, and conducts applied multi-disciplinary research into information and communication technology for development (ICT4D).

Julian May is the Director of the NRF-DST Centre of Excellence in Food Security at the University of the Western Cape. He has worked on options for poverty reduction including land reform, social grants, information technology and urban agriculture in Africa and in the Indian Ocean Islands. His current research focuses on food security, childhood deprivation and malnutrition.

REFERENCES

1. Aker, J. C., & Mbiti, I. M. (2010). Mobile phones and economic development in Africa.*Journal of Economic Perspectives, 24*(3), 207–232. doi: 10.1257/jep.24.3.207
2. Bertalanffy, L. (1968). *General systems theory.* New York, NY: George Braziller.

3. Bidwell, N. J., Lalmas, M., Marsden, G., Dlutu, B., Ntlangano, S., Tucker, W. D., & Elizabeth,P. (2011). Please call ME.N.U.4EVER: Designing for 'Callback' in rural Africa. In Donald Day(Ed.), *Proceedings of the 10th international workshop on internationalisation of products and systems* (pp. 117–125). Kuching, Malaysia: Product & Systems Internationalisation.

4. Bidwell, N. J., Siya, M., Marsden, G., Tucker, W. D., Tshemese, M., Gaven, N., ... Eglinton, K. A. (2013). Walking and the social life of solar charging in Rural Africa. *ACM Transactions on Computer–Human Interaction, 20*(4), 1–3. doi: 10.1145/2509404.2493524

5. Dicks, E., & Simons, S. (2013). *Value perceptions of fast moving consumer goods: Study of low-income rural consumers in South Africa.* (unpublished Honors dissertation). University of Cape Town, Cape Town, South Africa.

6. Diga, K. (2008). *Workshop on the role of mobile technologies in fostering social development.* Sao Paolo, Brazil: World Wide Web Consortium.

7. DoC. (2009). *International peer benchmarking study on South Africa's ICT sector.* Retrieved September 14, 2015, from https://pmg.org.za/files/docs/091020telecoms.pdf

8. Donner, J. (2007). The rules of beeping: Exchanging messages via intentional "missed calls" on mobile phones. *Journal of Computer-Mediated Communication, 13*, 1–22. doi:10.1111/j.1083-6101.2007.00383.x

9. Donner, J., & Walton, M. (2013). Your phone has internet-why are you at a library PC? Re-imagining public access in the mobile internet era. In Proceedings of the *Human–Computer Interaction–INTERACT* 2013, 347–364. Berlin: Springer.

10. Duncan, J. (2013). Mobile network society? Affordability and mobile phone usage in Grahamstown East. *Communicatio: South African Journal for Communication Theory and Research,, 39*(1), 35–52. doi: 10.1080/02500167.2013.766224

11. Elder, L., Samarajiva, R., Gillwald, A., & Galperin, H. (2013). *Information lives of the poor. Fighting poverty with technology.* Ottawa, Canada: IDRC.

12. Fransman, M. (2007). Innovation in the new ICT ecosystem. *Communication & Strategies,68* (4), 89–110.

13. Gillwald, A. (2005). A closing window of opportunity: Under-serviced area licensing in South Africa. *Information Technologies and International Development, 2*(4), 1–19. doi:10.1162/154475205775249364

14. Gillwald, A., & Calandro, E. (2014). SA prepaid mobile prices plateau. RIA Policy Brief Feb 2014(1) South Africa.

15. Gillwald, A., Moyo, M., & Stork, C. (2012). Understanding what is happening in ICT in South Africa. A supply – and demand-side analysis of the ICT Sector. Evidence for ICT Policy Action Policy Paper 7.

16. Gillwald, A., & Stork, C. (2008). Towards evidence-based policy and regulation: ICT access and usage in Africa. *Research ICT Africa, 1*(2), 45.

17. Gunzo, F., & Dalvit, L. (2012, 28–29 February). *A survey of cell phone and computer access and use in marginalised schools in South Africa.* Proceedings of the M4D 2012, 232–246, New Dheli, India.

18. Heimerl, K., Hasan, S., Ali, K., Brewer, E., & Parikh, T. (2013, 7–10 December). *Local, sustainable, small-scale cellular networks.* Paper presented at the 6th ACM international conference on information and communications technologies and development: Full Papers, 1, 2–12, Cape Town, South Africa.

19. ICRPOB. (2011). Annual Report 2011.

20. Johnson, D., & Roux, K. (2008). *Building rural wireless networks: Lessons learnt and future directions.* Proceedings of the WiNS-DR'08, San Francisco, USA, pp 17–22.

21. Kreutzer, T. (2009). *Generation mobile: Online and digital media usage on cellphones among low-income urban youth in South Africa.* University of Cape Town. Retrieved March 7, 2016, from http://tinokreutzer.org/mobile/MobileOnlineMedia-SurveyResults-2009.pdf

22. Mathee, K., Mweemba, G., Pais, A., Van Stam, G., & Rijken, M. (2007). *Bringing internet connectivity to rural Zambia using a collaborative approach.* Paper presented at the 2nd IEEE/ACM international conference on ICTD, 1–12, Bangalore, India.

23. McLeod, D. (2014). *MTN haemorrhaging market share. 19 February 2014, TechCentral.* Retrieved September 2, 2015, from http://www.techcentral.co.za/mtn-haemorrhaging-marketshare/46537/

24. McLeod, D. (2015). *Cell C enjoying 'massive' growth. 19 August 2015, TechnCentral.* Retrieved September 2, 2015, from http://www.techcentral.co.za/cell-c-enjoying-massive-growth/59102/

25. Moore, J. F. (1993). Predators and prey: A new ecology of competition. *Harvard Business Review, 71*(3), 75–83.

26. Navas-Sabater, J., Dymond, A., & Juntunen, N. (2002). *Telecommunications and information services for the poor.* (World Bank Discussion Paper No. 432), 1–118.

27. Ó Siochrú, S., & Girard, B. (2005). Community-based networks and innovative technologies: New models to serve and empower the poor. In *Making ICT work for the poor.* United Nations Development Program. Retrieved from http://p-ced.com/reference/community-based_nets.pdf

28. Republic of South Africa. (2010). Electronic communications act: Determination: Universal access to and universal provision of electronic communications services and electronic communications network services. *Government Gazette, 85*(32939), 3–6.

29. Rey-Moreno, C., Sabiescu, A. G., Siya, M. J., & Tucker, W. D. (2015a). Local ownership, exercise of ownership and moving from passive to active entitlement: A practice-led inquiry on a rural community network. *The Journal of Community Informatics, 11*(2). Retrieved from http://ci-journal.net/index.php/ciej/issue/view/54

30. Rey-Moreno, C., Tucker, W. D., Cull, D., & Blom, R. (2015b, May 15–18). *Making a community network legal within the South African regulatory framework.* Proceedings of the 7th international conference on information and communication technologies and development, Singapore, 57.

31. Rey-Moreno, C., Ufitamahoro, M. J., Venter, I., & Tucker, W. D. (2014). *Co-designing a billing system for voice services in Rural South Africa: Lessons learned.* Proceedings of the 5th conference of

computing for development (ACM Dev-5). San Jose, CA, USA. 5–6 December.

32. Right2know. (2013). *The right 2 communicate.* Retrieved October 10, 2014, fromhttp://www.r2k.org.za/wp-content/uploads/right2communic8_WEB.pdf

33. Saaty, T. L., & Kearns, K. P. (1985). *Analytical planning: The organisaton of systems, international series in modern applied mathematics and computer science, 7.* Oxford:Pergamon Press.

34. StastsSA. (2012a). *Census 2011 municipal report: Eastern Cape".* Report No. 03-01-50. Statistics South Africa. ISBN 978-0-621-41460-8.

35. StastsSA. (2012b). Income and expenditure of households, 2010/2011. Statistical Release No. P0100. Statistics South Africa.

36. Stork, C., & Stork, M. (2008). ICT household survey methodology & fieldwork. Volume ONE. Policy Paper 1. Research ICT Africa. 2008.

37. Westerveld, R. (2012). Inverse telecommunications: The future for rural areas in developing countries? In T.M. Egyedi, & D.C. Mehos (Eds.), *Inverse infrastructures: Disrupting networks from below* (pp. 187–207). Cheltenham, UK: Edward Elgar Publishing Ltd.

5

CHAPTER

Social Media and Ghana's 2012 Election Petition—A Discussion

Kirchuffs Atengble

Institute for Scientific and Technological Information (CSIR-INSTI), Accra, Ghana

ABSTRACT

For about five decades after independence, the media in Africa had been dominated by traditional mass communication such as radio, television and the print. But with development in internet technologies, coupled with developments in mobile communication devices, such tendencies begun to wane, and the resultant effect was the blend of the old and new media. This paper therefore discusses some of the activities that were threaded on social media platforms, and then identifies particular effects that such activities had on the petition sessions. As social media was deemed to have been critical in the Arab Spring of 2010, and in the general political landscape of the continent (campaigns, political rallies, elections, etc.), it has also been a significant part of the 2012 Election Petition Hearings in Ghana.

Keywords: Social Media, Election Petition, Ghana, Politics, Information in Society

1. INTRODUCTION

Developments in social media adoption and use in Africa have attained tremendous heights, especially within the past decade, as information and communication have been much emphasized (UNDP, 2013; Dutta & Lanvin 2013; BBC World Service Trust, 2010). Not only has such development been keen in agriculture, rural development, economics, education, science and technology, but also in politics (Ross, 2010; International Crisis Group, 2011). From political discussions on traditional media, through electoral campaigns, to popular revolutions in North Africa, social media has consumed political activity on the continent, and there seems not to be an avenue to evade this phenomenon. As a result, stakeholders of political activities on the continent are responding by developing themselves to fit this social media revolution.

For about five decades after independence, the media in Africa had been dominated by traditional mass communication such as radio, television and the print (Karikari, 2007). And therefore political activities tended to concentrate on these media. This has been identified as a catalyst for military overthrows (coup d'états) recorded on the continent; as such media could easily be controlled from a centre (Karikari, 2007). But with development in internet technologies, coupled with developments in mobile communication devices, such tendencies begun to wane, and the resultant effect was the blend of the old and new media (Patnaik, 2011). In this past decade, social (new) media, mostly Facebook, Twitter and YouTube, have complemented the means of communication even by traditional news agencies (Giroux et al., 2013; Chui et al., 2012).

With the developments in social media and African politics, questions arise whether social media did influence hearings of Ghana's 2012 election petition. And if it did, what was its effect on the hearings. Caution is however made that such a discussion is limited to the hearings of the Supreme Court of Ghana, and not in any way connected to the final verdict. This paper therefore discusses some of the activities that were threaded on social media platforms, and then identifies particular effects that such activities had on the petition sessions. The methodology adopted was qualitative and involved data mining (Asur & Huberman, 2010; Ruotsalainen, 2008; Han & Gao, 2009; Baradwaj & Pal, 2011) from popular social media sites; namely Facebook, Twitter, Google Plus, and YouTube.

Review of these social media sites then expanded and redirected the scope to other social media tools such as blogs and wikis. Some websites were also contacted to confirm items mentioned in some discussions on social media. Identifiable social media pages of stakeholders of the election petition hearings, including the petitioning and responding parties, civil society groups, and security agencies were searched for and reviewed. Hash tags and search strings were other techniques used for finding social media content on the proceedings captured on these platforms. Following has tags directed data mining activities to individuals and organizations that created more content on the subject of study. Content analysis was also used to identify and aggregate common trends worth discussing.

2. FINDINGS—SOCIAL MEDIA AND GHANA'S 2012 ELECTION PETITION HEARINGS

Social media is the term commonly given to web and mobile-based channels and tools that allow users to interact with each other and share opinions and content. As the name implies, social media involves the building of communities or networks and encouraging participation and engagement (CIPR, 2011). They serve as tools capable of utility by any individual or institution for the purpose of information provision, generating feedback, promoting discussion (Patnaik, 2011; Zanamwe et al., 2013). Between the period of April 16 and August 29, 2013, many activities threaded on social media sites concerning Ghana's 2012 election petition hearings. These activities were undertaken by various stakeholders of the hearings, individuals and institutions alike. This current study identified common trends among these social media activities as; the provision of public information concerning the proceedings; commentaries and the expression of opinions; advocacy; and humour.

2.1. Provision of Public Information

Social media platforms were used during the landmark election petition hearings for providing public information. To this end, stakeholders including political parties (mostly NDC and NPP), news agencies and civil society groups, employed social media avenues to augment their traditional means of information provision. Any new development during the period was immediately posted on social media sites and links made to official websites. For example there was a tweet from the Ghana police Service that informed of the apprehension of 25 people in Kpassa, a town in the Nkwanta North District of the Volta Region. And the American Embassy in Ghana used social media sites such as Twitter and Facebook to clarify the skeptical reception accorded to its security alert to US citizens in Ghana. Major news agencies such as the Ghana Broadcasting Corporation (GBC) and the MyJoyOnline employed Google Plus and YouTube channels to promote their reach on the petition's coverage.

2.2. Commentaries and Expression of Opinions

Another set of activities that treaded on social media sites concerning the petition hearings was commentaries on the proceedings of the court. Such commentaries were real-time on social media, as the proceedings were simultaneously broadcasted on television and radio. Some commentaries were objective analysis of the proceedings, as others towed usual partisan arguments. Opinions were also expressed with regards to the conduct of some officers of the court, and general proceedings. Through these new media, the average follower of the proceedings could expresses his or her opinion on anything and at anytime and from any part of the globe, provided they had internet access. Examples of such expressions of opinions were recorded on the YouTube channel of MyJoyOnline for Day 4 (22-4-13) and Day 7 (25-4-13).

Many platforms were created using social media to promote discussions on the subject, including a group called Ghana Supreme Court—Election Petition Centre on Facebook, and Twitter discussions that used hash tags #ElectionPetition and #TheVerdict (Credit: Blogging Ghana). Jamila Abdulai, on behalf of Ghana Decides (a Civil Society Organization), also moderated a discussion of the final verdict via Google Plus' Hangout platform. A similar online discussion was moderated by Kobby Blay, and statements from such discussions were tweeted; an aggregate of which is found on their www.storify.com page.

2.3. Advocacy

Probably the most treaded activity on social media for the period; many advocacy campaigns were witnessed on popular social media sites. Various projects and campaigns were launched and promoted, mostly on social media sites, for various causes in relation to the election petition hearing. Whiles some were calling for peace, others were calling for justice. The One Ghana Project for instance was initiated by three students of the School of Communication Studies of the University of Ghana to advocate peace during and after the verdict was announced; and the main platform used was Facebook, posting pictures of public figures who vowed for peace. Star Ghana funded many peace messages that were circulated on the YouTube channels of many individuals and organizations. The Institute for Democratic Governance (IDEG) and Ghana Decides (@GhanaDecides) also channeled their peace advocacy activities across mostly twitter and Facebook accounts. Political parties involved in the legal tussle and their leaders were not left out, but also used their social media presence to calm nerves especially before the verdict.

2.4. Humour

Other than the usual seriousness attached to political activities in Africa, Ghana's 2012 election petition hearings attracted some comic elements that seemed to work magic for the country. Some creative pictures and videos that conveyed different humorous contents were spreading like wildfire on social media platforms. There was a picture on Whatsapp Messenger and other platforms that portrayed skinny ugly-looking soldiers who sought to warn the public of misbehavior; and another that had an old woman sitting with her weapon (dagger) stuck in her cloth and ready for "all die be die". The Black Narrator has a great deal of cartoons work on his Facebook profile. Akosua, from the daily guide newspaper, also has some interesting comical cartoons available on social media.

The direction of this discussion is not to imply an absence of negative and abusive tendencies of some of the activities. Indeed there were elements of personality attacks, hate speeches, and the like[1]. But the focus of this discussion is to identify the major threading issues on social media during the entire spectrum of proceedings on the election petition, and through that identify any effect they had on the hearings. But the focus of this discussion is to identify the major threading issues on social media during the entire spectrum of proceedings on the election petition, and through that identify any effect they had on the hearings.

3. EFFECTS: SOCIAL MEDIA INFLUENCED GHANA'S 2012 ELECTION PETITION HEARINGS

As expected of any social phenomenon, social media threads had some influences on the 2012 election petition hearings, primary of which was the creation of an informed and educated public[2]. The creation of an informed public also had rippling effects on the general petition hearing, of which include a collective creation of Ghanaian knowledge, transparency in court proceedings, and general law and order within the Ghanaian society during the period under discussion.

3.1. A More Informed and Educated Public

Traditionally, journalists would have listened to proceedings and probably interviewed some party-aligned attorneys for comments before reporting events from the Supreme Court. This would have under normal circumstances made public information highly dependent on the interpretations of these intermediaries, and capable of misconstruction. But as video files from the proceedings flooded YouTube channels and discussions threaded on other social media platforms, and with links to daily transcripts of the proceedings, interested parties within Ghana and abroad could review every bit of the courtroom proceedings—even the drama. As a result, no room for misinformation was left open. Social media had virtually filled in the information gap that would otherwise have resulted from enclosed courtroom proceedings. And every section of the listening (following) public had a satisfaction of information need; literate and illiterate alike.

A similar effect of the social media broadcast was an education for many people who were not up-to-terms (conversant) with legal proceedings. Education in this context specifically means public awareness on legal concepts, terminologies, and generally legal proceedings. During the court proceedings, many terms and concepts such as "I put it to you", "I suggest to you", amicus curiae, contempt, councils, order in court, motions, and the like were used by officers of the court, and this trickled into public discourses as members of the listening public were readily digesting such contents, also on social media. In effect, many willing-to-know followers sought understanding of such legal concepts; which also afforded the public an opportunity to discuss legal proceedings. Apart from the non legally-inclined publics, students of law and some attorneys took the opportunity to acquire some experience of landmark proceedings of the sort from asynchronous access to social media videos.

3.2. A Collective Creation of Ghanaian Knowledge

Social media not only provided an avenue to inform and discuss proceedings of the petition hearings, but also provided the people of Ghana an opportunity to collectively record an event with such a national dimension. If not for social media involvement, the responsibility for recording the events would have been primarily left in the hands of clerks of the Supreme Court, and journalists of various media houses. Knowledge on this event would therefore have been monopolized by these few institutions, and any attempt to compile a scholarly work, for example, would therefore have meant contacting such institutions—with its attendant challenges. But social media availability, together with its multimedia knowledge recording capacity, knowledge was collectively created by the people of Ghana. Knowledge created on this landmark event may therefore include the official records of the Supreme Court of Ghana, reports carried out by news agencies, and social media video files, transcripts and commentaries.

3.3. Transparency in Proceedings

As there were more avenues to channel information from the courtroom proceedings, there were fewer rooms for suspicions. It was reckoned prior to the hearings that interested parties to the petition were likely to flood the Supreme Court premises. Truly, this may have been the situation, as suspecting publics would have been interested in first hand information, and therefore pushed for space, at least in the court premises. But as the events were telecasted and streamed on social media, there was little need for such reactions. This was the situation to the extent that, people even neglected television broadcast, as they could follow events later from social media. Public commentators and political activists could therefore not relay inappropriate and inconsistent information to the public, as they were privy to first-hand information.

3.4. General Law and Order

As a result of activities identified in this paper and that which generally took place on social media, law and order was the norm during the fifty (50) days exercise. No member of the public could "pollute" another, and proceedings could continue unabated as members of the public were not in any means attracted to crowd the premises of the Supreme Court. Advocacies made through social media to a great extent also achieved their objectives as they mostly called for a united Ghana after the verdict; peace, and justice. The moments before the verdict and its aftermath showed the general works that had been done by Ghanaians in general, and through social media in particular. For such reasons it can generally be concluded that social media also influenced the maintenance of law and order within the period. This does not in any way indict security provisions made, but rather a compliment to efforts and all other measures put in place to this end.

4. CONCLUSION

It has been discussed in this paper that the relationship between social media and African politics has been close, and may be getting more intricate. Between the fifty (50) days period of the elections petitions hearing at the Su- preme Court of Ghana, many activities threaded on social media sites. These activities resulted in the provision of public information concerning the proceedings, commentaries and opinions expressed, advocacy, and humour. The effect was that these activities had rippling effects, of which included the creation of an informed and educated public on legal proceedings, a collective creation of Ghanaian knowledge, transparency in the proceedings of the court, and general law and order within the Ghanaian society. As social media was deemed to have been critical in the Arab Spring of 2010, and in the general political landscape of the continent (campaigns, political rallies, elections, etc.), it has also been a significant part of the 2012 Election Petition Hearings in Ghana. Once again, the main objective of the study was to provide insight into social media activities during the period, based on findings from data mined. It is hoped that the discussion will contribute to cross-disciplinary social media research in Ghana and the continent at large. It is recommended that further studies be conducted on the issue, employing other methods of data collection and analysis, for the purpose of validation and greater evidence.

ACKNOWLEDGEMENTS

Kirchuffs Atengble (GINKS) is acknowledged, as this discussion was initiated by the organization. Any opinions, findings, and conclusions or recommendations expressed in this paper are those of the author and do not necessarily reflect the views of the GINKS.

REFERENCES

1. Asur, S., & Huberman, B. A. (2010). Predicting the Future with Social Media. http://www.hpl.hp.com/research/scl/papers/socialmedia/socialmedia.pdf
2. Baradwaj, B. K., & Pal, S. (2011). Mining Educational Data to Analyze Students' Performance. International Journal of Advanced Computer Science and Applications (IJACSA), 2, 63-69.
3. BBC World Service Trust (2010). How to Use Communication to Make Aid Effective: Strategies and Principles for Programme-Based Approaches. London: BBC World Service Trust. http://downloads.bbc.co.uk/worldservice/trust/pdf/research_and_policy/communication_to_make_aid_effective.pdf
4. Chartered Institute of Public Relations (CIPR) (2011). Social Media Best Practice Guide (2011 ed.).
5. Chui, M., Manyika, J., Bughin, J., Dobbs, R., Roxburgh, C., Sarrazin, H., Sands, G., & Westergren, M. (2012). The Social Economy: Unlocking Value and Productivity through Social Technologies. New York: McKinsey Global Institute.
6. Dutta, S., & Lanvin, B. (2013). The Global Innovation Index 2013: The Local Dynamics of Innovation. Johnson Cornell University, INSEAD, WIPO.

7. Giroux, J., Roth, F., & Herzog, M. (2013). Using ICT & Social Media in Disasters: Opportunities & Risks for Government. [Background Document]. Third Special Report, Center for Security Studies (CSS), Zurich: ETH.
8. Han, J., & Gao, J. (2009). Research Challenges for Data Mining in Science and Engineering. In Next Generation of Data Mining (pp. 1-18). London: Chapman & Hall.
9. International Crisis Group (2011). Kenya after the Elections. Policy Briefing. Africa Briefing Number 94. Nairobi/Brussels.
10. Karikari, K. (2007). African Media Since Ghana's Independence. In E. Barratt, & G. Berger (Eds.), 50 Years of Journalism: African Media since Ghana's Independence (pp. 10-20). Johannesburg: The African Editors' Forum, Highway Africa and Media Foundation for West Africa.
11. Gyasiwaa, A. (2013). Police Hunts for Rumour Mongers on Social Media. MyJoyOnline.com. http://www.edition.myjoyonline.com/pages/news/201308/112004.php
12. Patnaik, S. (2011). Going Social: Case Studies of Successful Social Media Marketing. In J. Liebowitz (Ed.), Beyond Knowledge Management: What Every Leader Should Know (pp. 58-68). Florida: Auerbach Publications.
13. Ross, A. (2010). Internet Freedom: Historic Roots and the Road Forward. SAIS Review, 30, 3-15.
14. Ruotsalainen, L. (2008). Data Mining Tools for Technology and Competitive Intelligence. Espoo 2008, VTT Research Notes 2451, VTT Tiedotteita.
15. United Nations Development Programme (UNDP) (2013). The Rise of the South: Human Progress in a Diverse World. Human Development Report. New York: UNDP.
16. Zanamwe, N., Rupere, T., & Kufandirimbwa, O. (2013). Use of Social Networking Technologies in Higher Education in Zimbabwe: A Learners' Perspective. International Journal of Computer and Information Technology, 2, 8-18.

6
CHAPTER

Pedagogies for the Open Knowledge Society

Paola Ricaurte

Humanities and Social Sciences, Tecnologico de Monterrey, Mexico City Campus, Calle del Puente 222, Col. Ejidos de HuipulcoCP 14380 Tlalpan, Mexico City

ABSTRACT

Learning challenges within the knowledge society cannot be limited to the technological dimension. Learning and education are embedded in economic, political and cultural contexts. Pedagogies reflect this social condition. Consequently we need to situate the pedagogical practices on a wider debate regarding their place in the global system of knowledge production. Some questions are relevant in assessing these practices: What is the sense of learning in the knowledge society? How this technological paradigm modifies our conception of learning and education? How pedagogies are related to technology and the global system of knowledge production? What principles are associated with these proposals? Through a case study of two learning environments, Wikipedia and Google, we compare the two models and the values, principles and competences associated to each of them.

Keywords: Network society, Information society, Open learning environments, Wikipedia, Google, Emerging pedagogies

To remain human and liveable, knowledge societies will have to be societies of shared knowledge.

(United Nations, 2005, p. 5)

INTRODUCTION

The current state of the development of capital, which some authors have called information society (Masuda, 1981), knowledge society (Drucker, 1993), network society (Castells, 1996), knowledge economy (Powell & Snellman 2004) or, from a critical perspective, cognitive capitalism (Moulier-Boutang, 2012) is characterized by the role knowledge gains as the main value in the accumulation of capital (Vercellone, 2007). While is true that the ways in which we can define this age and its features possess substantial distinctions, authors agree that information and communication technologies constitute an essential dimension in the creation, dissemination and use of the knowledge and, therefore, in society's structural transformation processes (Castells & Cardoso, 2005).

The term knowledge society has been the preferred option by international organizations because it highlights that it is knowledge–the capacity to create new meanings, and not information or technology–, which allows for the improvement of human condition (United Nations, 2005). On the contrary, Castells (1996), Castells and Cardoso (2005) maintains that the interconnected characteristic of society, derived from the possibilities offered by communication technologies, and not just information or knowledge, it's what defines its fast, global and complex nature in every age in history. For Castells the technology-information paradigm is what defines the current type of social organization. This new form of organization reconfigures the mechanisms of power, notions of time and space, institutions, sociality, work, and forms of cultural production and cultural action. To authors like Vercellone (2007) it is important to differentiate that knowledge society is the evolution of a post-Fordism capitalism, which is founded on the exploitation of cognitive work, creating a new space for conflict between work and knowledge. Cognitive capitalism's thesis emerges as a critique of knowledge based economy.

As a result of this new configuration within global social organization new inequalities, new forms of exclusion and new uses of power through knowledge control emerge from those nations who insert themselves in knowledge society. An alternative paradigm is the one proposing to speak about knowledge societies based on a collaborative economy not governed by market laws (Kostakis & Bauwens, 2014) whose characteristics are peer production (Bauwens, 2006; Benkler, 2006; Benkler, Shaw, & Hill, 2015) and a notion of knowledge not as merchandise (Powell & Snellman, 2004) but as a common good which can be defended and shared (Bollier, 2007; Hess & Ostrom, 2007; Ostrom, 1990).

This dynamic techno-social scenario entails deep transformations: on a macro level in the processes of knowledge production and capital accumulation; on a medium level in the transformation of education and educative institutions; and on a micro level in the learning processes, which must be understood as a complex interdependent relationship. Through contrasting two paradigmatic technology platforms of knowledge society, Google and Wikipedia, we propose to show how they respond to two different models of value accumulation that materialize in pedagogical practices. We start from the hypothesis that pedagogic tendencies contribute to shape a given model of knowledge society that reflects the close interconnections between micro, meso and macro social levels.

Education and technology in knowledge economy

According to David and Foray (2003, p. 22 and the following) knowledge economy is characterized by the massive and extremely fast acceleration of knowledge production; intangible capital growth on a macroeconomic level, through the investment in human capital, education and the dissemination of knowledge; innovation; a revolution in knowledge instruments: and information and communication technologies. The link between socio-technical context and education and economy is visible on different levels: a) the educative industry, which hand in hand with the technology industry allows for mass knowledge production; b) development of abilities through training human capital (workers and creators in knowledge economy), establishing and reproducing the rules for knowledge production and dissemination, and defining educative models, curricula and pedagogies that allow to outline teaching and learning (Gurung, 2013), as well as the emphasis on adopting technology and developing digital competences to incorporate and advance in knowledge society; c) investment in innovation and the education's role in the development of creative abilities, the positive valuation of scientific areas and their inclusion in the curricula as priority; the systematization of innovation processes, and scientific and technological progress; d) investing in the development of technology infrastructures that facilitate the creation of networks that allow the flux of information, as well as knowledge creation and transference. Based on these elements it would be possible to establish education as an essential mechanism, part of the generation, transference and appropriation of knowledge and, therefore, as one of the pillars of global economy.

The close relationship between technology and education has created a prolific field of study which, through diverse approaches, looks to characterize the nature, impact and the meaning of the transformations in the learning process. There are numerous studies that approach the problem of infrastructure technologies intervention in the cognitive or pedagogic processes. Others focus on appropriation models and the use of devices or platforms (Internet, mobiles, social networks, applications) in the educative context and its impact on learning, without highlighting the pedagogic processes. It is less frequent to find the argument focusing on the analysis of the relationship between technology systems and knowledge production systems, or ones that propose models to analyze learning in digital contexts in terms of economy, politics and emotion (McCann, 2014). In respect to the questions that we ask ourselves, Postman (1993) considers that what is important about technology is not its efficiency as a learning tool, but the way in which modifies our notion of learning and, on other level, and the way that the ecology of media (McLuhan, 1994) in conjunction is modifying the idea of the educative institution.

In the case of public policy, we see in the governmental and institutional discourse the absence of consideration about linking educative models with technological development and the global ecosystem of knowledge production. In general terms, educative innovation initiatives in educative institutions at any level limit themselves to reproducing the narratives and models designed by the agendas of technology corporations, and in a few occasions they consider original pedagogic proposals that contribute to a critical reflection about learning processes in relation to technology. Exceptionally it is possible to find institutional discourses that on a macro level (as is the case of Uruguay) and micro level (as in some disruptive schools), are interested in framing pedagogic practices and learning in a context that allows to evaluate its influence in the local and global knowledge production systems. Because of this, it is important to highlight that proposals which emphasize the technological dimensions of learning in a network and not its pedagogic dimensions contribute to (or are motivated by) the initiatives of technology corporations that have in the educative institutions and in governments their main clients.

Pedagogic tendencies for the digital Age

The group of socio-technical context transformations and its implications in the field of pedagogy has generated abundant papers in the scholarly literature. There are publications that provide an outlook of the pedagogic tendencies emerging in the digital age (Bates, 2015; Beetham, McGill, & Littlejohn, 2009; Cartelli, 2006; de Oliveira et al., 2015; Limbu, 2014; Yáñez, Okada, & Palau, 2015). Next we'll talk about these tendencies. It is not an exhaustive survey, and its sole purpose is to illustrate certain trends in today's pedagogic debate.

A group of pedagogic approaches that emerged in recent years locates its origin in the social construction of learning (Dewey, 1897; Vigotsky, 2000). They expand classical perspectives on the zone of proximal development (Vigotsky, 2000) in the digital context and include approaches from network society theory (Castells, 1996) and cognition (Downes, 2012, Siemens, 2005). It is also possible to find ramifications in the understanding of learning's social character as a process located (Wenger, 2009) and materialized in the concept of communities of practice (Lave, 1991; Lave & Wenger, 1991; Wenger, 1999) and applied to learning in digital contexts.

Authors such as Cabero (2006), Suárez (2010) and Suárez and Gros (2013) have explored e-learning's advantages and reach, reflecting on internet as an educative environment and the pedagogic use of "social software" (McLoughlin & Lee, 2010). In this tendency, Internet is considered a learning environment that should be approached pedagogically, as proposed by networked pedagogy (Suárez, 2014). Associated to the need to emphasize connected learning we find the research on personal learning environments (PLE) and personal learning networks (PLN), both of which take advantage of technological tools to create environments and networks that potentiate learning (Adell & Castañeda, 2010, 2013; Castañeda & Adell, 2013). Another group approaches this learning methods through specific digital platforms, as MOOC (Massive Online Open Courses), social networks and other applications (Alemán, Sancho & Gómez, 2015; Gallardo, Marqués, & Bullen, 2015; Kigozi, Vesisenaho, Hansson, Danielson, & Tusubira, 2012; Roig Vila, Mengual-Andrés, & Suárez Guerrero, 2014; Santally, 2005).

Another approach comes from a series of meta-theories studying learning processes in a network. This is the case of distributed cognition (Hutchins, 2006; Salomon, 1993) which maintains that cognitive processes may be distributed through the members of a social group. These approaches have used network theory (Barker & Kemp, 1990) and also Latour's (2005) actor-network theory as a pedagogic proposal (Fenwick & Edwards, 2010; Fox, 2002) or as an analytical resource to understand learning processes.

Digital convergence and mobility offered by technology have made way for proposals that exploit these features. Aparici and Silva (2012) in *Pedagogy of Interactivity* explain how interactivity reflects on the value of participation, dialogue and co-authorship, features that are also common to the principles of participative web. McGonigal (2011) proposed to include videogames as pedagogic tools, which allow solving real problems.

In general, on another block, we can identify some trends that reflect on learning in digital environments and that focus in the development of participatory cultures (Jenkins, 2009), collective peer production (Bauwens, 2006; Benkler, 2006; Benkler et al., 2015; Rheingold, 2014), the necessary competences to take advantage of learning possibilities self-suggested by the digital environment (Rheingold, 2014), and common goods (Lafuente, Alonso, & Rodríguez, 2013; Ostrom, 1990). Other authors go back to the foundations of cooperation theories to develop a group of patterns on the learning process in diverse contexts, including digital, as is the case of peer pedagogy (peeragogy) (Rheingold, 2014, Ricaurte, 2013). In the latter, learning is associated to cooperation principles, free software, free culture, and peer production (Corneli, Danoff, Peirce, Ricaurte, & MacDonald, 2015). A different aspect offered is one in which proposals conceive learning and knowledge production as a citizen's process, self-regulated and inscribed in a "makers" culture (Lafuente, 2013; Lafuente et al., 2013; Lafuente & García, 2015), which in a way recovers the spirit of critical pedagogy (Freire, 1970), experiential learning (Kolb, 2014), and even located learning (Lave & Wenger, 1991). This vision emphasizes the need for a form of learning that allows the apprentice to perceive its surrounding needs, to act in consequence, collectively and with self-regulation, hand in hand with the possibilities of digital production through free technologies. With respect to practices and resources, Coughlan and Perryman (2013) argue that Open Educational Practices (OEP) and Open Educational Resources (OER) promote innovative pedagogic models that encourage the creation of communities and collaboration (UNESCO, 2012).

The pedagogical trends reviewed show a vision of the student as an essential agent of the process. Also it is possible to distinguish in them certain common principles as participative culture, network learning, cooperation and openness. However, not all proposals necessarily commit themselves to free culture, through open licenses or the preference of free technology –proposals which could be associated with a vision of an open knowledge society–, over the private. With exception of the pedagogies associated to a concept of knowledge as commons (Bollier, 2007), in general, among the pedagogic approaches, the relationship between their pedagogic goal, the knowledge production system and its implication in the social order is not made explicit.

Wikipedia and Google: Two paradigms in knowledge economy

The emergence of new trends or the reformulation of classical theories makes the universe of possibilities for approaching learning in digital contexts a complex task. However, there are profound implications to making pedagogic decisions outside of knowledge production's global context. We feel it is pertinent to promote an educative culture that doesn't contribute to the concentration of power and capital through knowledge. Because of this, we propose through a case study of two learning environments, Wikipedia and Google, to carry out an analysis that allows us to compare these two models of knowledge society materialization. We chose these two learning environments because they constitute the most representative cases of two opposed paradigms in the Internet universe.

Wikipedia

With almost 375 million visitors each month, Wikipedia is the seventh most visited site in the world (Wikipedia, 2016). It is the largest encyclopedia in the history of humanity and constitutes the largest resource of open knowledge production created through crowdsourcing. Their motto is "Imagine a world in which every single person on the planet is given free access to the sum of all human knowledge. This is what we're doing".

Wikimedia Foundation, to which Wikipedia is subordinate, is a non-profit organization. It supports itself through user donations. Since it isn't a company, one can only estimate its value, which would be along the lines of tens of billions of dollars (Eveleth, 2013). It isn't an organization in the traditional sense; it is a community with several projects: Wikimedia Foundation, Wikipedia Education Program, GLAM (Galleries, Libraries, Archives and Museums), Wiki Initiative, Wikimedia Commons (a repository of images and other resources), and the events of the Wikimedia Foundation (Wikimania, Wikicontests, etc.). Each of these areas has its own projects and a series of associated organizations and initiatives. Their editors are volunteers. The training for the editors (students, teachers or general users) is imparted through manuals and tutorials with Creative Commons licensing, free for downloading. The communities are connected through mailing lists, through the discussion page of the articles, through wiki itself, and others places on the web (social networks, in-person meetings, edit-a-thons, etc.). There's also a volunteer mentor program that supports students or teachers who wish to use Wikipedia as an educative tool.

Google

Google Inc. is the most important Internet company. Their mission "is to organize the world's information and make it universally accessible and useful". It is a multinational technology company specializing in services and products related to the Internet. Their products are adds, searches, cloud computing and software. Since 2015 they are part of Alphabet, a multinational conglomerate that groups Google products and others such as Nest Labs, Jigsaw, Verily, etc. In 2015 the conglomerates earnings reached 75 billion dollars (Wikipedia, 2016). A share on the Stock Exchange costs 742 dollars.

Google's educative program, Google for Education, offers products, training and resources. The products are devices (Chromebook), productivity tools (Drive, Gmail, Calendar, etc.), Classroom (course management platform). Their training is an in depth program in Google tools for teachers at all levels. The courses are guided and designed for the classroom's needs. The content is free, but program certification is not. Certification involves taking multiple choice tests. In case of failing the level's certification test it is necessary to pay again the test fee to receive a new opportunity to take the test again. It consists of four certifications levels: Educator level 1, Educator level 2, Certified Instructor and Certified Innovator. Among the resources, we can find communities of educators that use Google, programs for teachers and students, and instruction on informatics.

These two learning environments, that apparently could be considered open, in reality present profound differences in respect to the place they occupy in the knowledge production system, their pedagogic offer, the values associated to their initiatives and the types of competences they develop. In Table 1 we present the most important differences.

Table 1 Wikipedia and Google: Two Paradigms in Knowledge Economy

Platform	Wikipedia	Google
Type of organization	Foundation	Corporation
	Non Profit	For profit
Education program	-Wikimedia Foundation Education Program -Education Collaborative A group of education leaders around the world who volunteer to mentor education initiatives	-Google for Education Official Google program for educators and educative institutions
Certification	No formal certification	Official certification through payment, with

Platform	Wikipedia	Google
	process	predefined levels and processes by the company
Community	Communities in different spaces (Wikimania, Collaborative Education, Education Program) and other places on the network	Learning communities within their own environment (like Google +)
Privacy	Commitment with personal data privacy	No real commitment with personal data privacy.
Principles	-Open Knowledge -Free software -Free culture (Creative Commons Licensing or public domain) -Remix Culture -Wiki -Collective Knowledge Production	-Service "free of charge", but associated to personal data industry. -Free software, but their algorithm's code, Page Rank, is not open. -It's a business associated to the obtaining of new clients for their platforms. -Hierarchical organization and control of knowledge.
Pedagogic values	Associated with the universal access to open knowledge and the collective production in the different projects of the Wikimedia Foundation (Wikimedia Commons, Wikipedia, Wikibooks, Wikiversity, etc.)	Associated to the use and appropriation of different platforms form the Google ecosystem (Gmail, Youtube, Calendar, Drive, Hangouts, Google+, Blogger).

Platform	Wikipedia	Google
Competences	-Collaborative production -Sharing -Community building -Basic (reading-writing), intercultural and digital competences.	-Collaborative production -Sharing -Building communities -Digital competences associated to their platforms
Prestige	-The more you edit the bigger the recognition one receives from the community -Program of volunteer mentors (education leaders) to promote the use of Wikipedia among students and educators.	-The one who is certified is recognized, the one who isn't is institutionally and/or socially chastised.

DISCUSSION

Through Google's and Wikipedia's platforms it is possible to see the contrast between two models for approaching knowledge production, dissemination and consumption. Both platforms may be considered open learning environments and may be included, for example, as part of the personal learning environments (PLE) of teachers and students (Adell & Castañeda, 2010, 2013; Castañeda & Adell, 2013). However, the choice of one technology over the other implies taking a stand in respect to a specific paradigm in knowledge economy (Powell & Snellman, 2004).

The model of peer knowledge production (Bauwens, 2006; Benkler, 2006; Benkler et al., 2015) which constitutes Wikipedia's essence may be conceived as part of an epistemology and methodology of peer learning (Rheingold, 2014) and peer production (Bauwens, 2006). Using Wikipedia as an open learning environment meets the objective of supporting networks and the necessary global participation structures for the emergence of open knowledge societies. Wikipedia promotes a notion of knowledge as commons. Therefore, ascribes to a culture of free licenses that allow for the use, reuse, and remix of contents that have been generated by everyone and, in consequence, should not be hoarded. Its use as an open learning environment shows possibilities in all three levels previously mentioned. On a micro level: as a pedagogic option for the development of basic, digital and research competences. On an institutional or medium level: since adopting it involves a change in the institutional culture, especially with respect to the evaluation system, the perception of the encyclopedia's validity, the culture of respect towards copyright and the dissemination of knowledge.

For their part, Google has a policy of centralization and knowledge control; while it's true that the company has access, for example, to the largest digital library on the planet, it is them together with the content production industries who decide the limits for their users. Another aspect is that their algorithm for hierarchical organization of content is not open. While its mechanism is known, its code is not open and, therefore, we don't have the possibility to partake in establishing said algorithm. Another relevant aspect is the destiny of personal data and their captive users through their cloud services, like e-mail or Drive, their suite for collective production. Google's influence also appears on the three levels. On a micro level we've seen how users give up their right to their personal data privacy in exchange for their services. On an institutional level, it signs commercial agreements with educative institutions to offer their services and training programs. And on a macro level, it carries out lobbying among politicians and governments for their products and services to be included in digital agendas, or to receive support for those public policies which benefit the company.

CONCLUSION

Through this text we've maintained that the debate around knowledge society and the role of education in it is in permanent tension between two forces: on the one hand, the prevalence of a form of accumulating value, based on intangible and infinite common goods, that is, knowledge as merchandise; and on the other, alternative forms of value production associated to the commons and commons governance (Hess & Ostrom, 2007; Ostrom, 1990). This tension is reflected in the different levels of knowledge society associated with education. On the macro level through the educative and technology industries and public policies; on the medium level, through institutional transformations which define curricula, competences and educative models; and on the micro level, in the pedagogic practices that materialize in the specified decisions about the learning process and its relationship with technology.

The values of openness, cooperation, learning in a network and collective production of knowledge are present in many pedagogic trends that have emerged in recent years. By comparing the cases of Wikipedia and Google, we can find that in spite of some shared features, like digital training for teachers and students, the creation of learning communities or the possibility for developing similar competences, it's also true that they have important differences and that mainly these are because Google initiatives in education are related to the commercialization of their products and to increase their user base. Therefore, the structure and organization of their educative programs are guided by this condition.

We need studies that evaluate the theoretical and practical proposals, the pedagogic discourses and the educative models in the scope of a global system for knowledge production and technology development, characterized by the centralization and restriction of access to contents and the development of personal data industries. The proposals on learning in digital environments cannot be limited to the use of platforms, resources, devices, mobile applications without situating all of these in the material, relational and sociocultural context they come from.

AUTHORS' INFORMATION

Paola Ricaurte holds a Language Science PhD from the National School of Anthropology and History (Mexico). Currently, she is a Professor and Researcher at the Graduate School of Education, Social Sciences and Humanities at Tecnológico de Monterrey, Mexico City Campus. She is the Director of Openlabs, a citizen lab for social and education innovation: http://www.openlabs.mx/.

COMPETING INTERESTS

The author declares that has no competing interests.

REFERENCES

1. Adell, J., & Castañeda, L. (2010). Personal learning environments: a new way to understand learning. In M. Roig & M. Fiorucci (Eds.), *Clues for research on innovation and education quality*. Alcoy: Marfil-Roma TRE.

2. Adell, J., & Castañeda, L. (2013). The pedagogical ecosystem of personal learning environments. In Castañeda, L. & Adell, J.

(Ed.). *Personal learning environments. clues for the educational network ecosystem*. Alcoy: Marfil-Roma TRE.

3. Alemán, L. Y., Sancho, T., & Gómez, M. G. (2015). Indicators of pedagogical quality for the design of a Massive Open Online Course for teacher training. *RUSC. Universities and Knowledge Society Journal, 12*(1), 104–119

4. Aparici, R., & Silva, M. (2012). Pedagogy of interactivity. *Comunicar, 19*, 38. http://dx.doi.org/10.3916/C38-2012-02-05. Accessed 23 Nov 2015

5. Barker, T. T., & Kemp, F. O. (1990). Network theory: A postmodern pedagogy for the writing classroom. In C. Handa (Ed.), *Computers and community: Teaching composition in the twenty-first century* (pp. 1–27). Portsmouth, NH: Boynton/Cook Publishing.

6. Bates, A. W. (2015). *Teaching in a digital age: Guidelines for designing teaching and learning*. British Columbia: BC Open Textbooks. http://bit.ly/1T53L46. Accessed 20 Nov 2015

7. Bauwens, M. (2006). The political economy of peer production. *Post-autistic Economics Review, 37*, 33–44. http://bit.ly/1S368PL. Accessed 20 Nov 2015

8. Beetham, H., McGill, L., & Littlejohn, A. (2009). *Thriving in the 21st century: Learning Literacies for the Digital Age*. The Caledonian Academy, Glasgow Caledonian University. UK: JISC.

9. Benkler, Y. (2006). *The wealth of networks: How social production transforms markets and freedom*. New Heaven: Yale University Press.

10. Benkler, Y., Shaw, A., & Hill, B. M. (2015). Peer Production: A Form of Collective Intelligence. In T. W. Malone & M. S. Bernstein (Eds.), *Handbook of Collective Intelligence*. Cambridge, MA: MIT Press.

11. Bollier, D. (2007). *The growth of the commons paradigm. Understanding knowledge as a commons*. Cambridge, MA: MIT Press.

12. Cabero, J. (2006). Pedagogical foundations of e-learning. *RUSC, Universities and Knowledge Society Journal, 3*, 1. http://bit.ly/1cdgXtd. Accessed 19 Nov 2015

13. Cartelli, A. (2006). *Teaching in the Knowledge Society: New Skills and Instruments for Teachers.* Hershey, PA: Information Science Publishing.

14. Castañeda, L., & Adell, J. (Eds.). (2013). *Personal learning environments. Clues for the educational network ecosystem.* Alcoy: Marfil.

15. Castells, M. (1996). *The information age: economy, society and culture. The rise of the network society* (Vol. 1). Oxford: Blackwell.

16. Castells, M., & Cardoso, G. (Eds.). (2005). *The Network Society: From Knowledge to Policy.* Washington, DC: Johns Hopkins Center for Transatlantic Relations.

17. Corneli, J., Danoff, C., Peirce, C., Ricaurte, P. Y., & MacDonald, L. (2015). *Patterns of Peeragogy.* Pittsburg, PA: International Conference on Pattern Languages of Programs.

18. Coughlan, T., & Perryman, L. A. (2013). Beyond the ivory tower: a model for nurturing informal learning and development communities through open educational practices. *RUSC. Universities and Knowledge Society Journal., 10*(1), 312–326. http://dx.doi.org/10.7238/rusc.v10i1.1586. Accessed 23 Nov 2015

19. David, P. A., & Foray, D. (2003). Economic fundamentals of the knowledge society. *Policy Futures in Education, 1*(1), 20–49.

20. de Oliveira, J. M., Henriksen, D., Castañeda, L., Marimon, M., Barberà, E., Monereo, C., Coll, C., Mahiri, J., & Mishra, P. (2015). The educational landscape of the digital age: communication practices pushing (us) forward. *RUSC. Universities and Knowledge Society Journal, 12*(2), 14–29.

21. Dewey, J. (1897). My pedagogic creed. *School Journal, 54,* 77–80. http://bit.ly/1J4KmcX. Accessed 18 Oct 2015

22. Downes, S. (2012). *Connectivism and Connective Knowledge.* http://www.downes.ca/files/books/Connective_Knowledge-19May2012.pdf. Accessed 15 Nov 2015

23. Drucker, P. F. (1993). The rise of the knowledge society. *The Wilson Quarterly, 17*(2), 52–71.

24. Eveleth, R. (2013). *How Much is Wikipedia Worth? Smithsonian.* http://bit.ly/1ZnS1do. Accessed 15 Nov 2015

25. Fenwick, T., & Edwards, R. (2010). *Actor-network theory in education*. New York: Routledge.

26. Fox, S. (2002). Studying networked learning: some implications from socially situated learning theory and actor network theory. In C. Jones & C. Steeples (Eds.), *Networked learning: perspectives and issues*. London: Springer.

27. Freire, P. (1970). *Pedagogy of the oppressed*. New York: Continuum.

28. Gallardo, E., Marqués, L., & Bullen, M. (2015). Students in higher education: Social and academic uses of digital technology. *RUSC. Universities and Knowledge Society Journal, 12*(1), 25–37.

29. Gurung, B. (2013). Emerging pedagogies in changing contexts: pedagogies in networked knowledge society. *Journal of Global Literacies, Technologies, and Emerging Pedagogies, 1*(2), 105–124.

30. Hess, C., & Ostrom, E. (2007). *Understanding Knowledge as a Commons*. Cambridge, MA: MIT Press.

31. Hutchins, E. (2006). The distributed cognition perspective on human interaction. Roots of human sociality. *Culture, cognition and interaction, 1*, 375.

32. Jenkins, H. (2009). Confronting the Challenges of Participatory Culture. Cambridge, MA: MIT Press.

33. Kigozi, E., Vesisenaho, M., Hansson, H., Danielson, M., & Tusubira, F. F. (2012). Modelling a peer assignment review process for collaborative E-learning. *Journal of Interactive Online Learning, 11*, 2.

34. Kolb, A. (2014). Experiential Learning. 2nd ed. Upper Saddle River, NJ: Pearson.

35. Kostakis, V., & Bauwens, M. (2014). *Network society and future scenarios for a collaborative economy*. London, New York: Palgrave Macmillan.

36. Lafuente, A. (2013). *Citizen science: itineraries of the amateur, activist and hacker. national institute of educational technologies and teacher training. Educalab.* Blog. [online] http://bit.ly/1UJJmQ3. Accessed 15 Oct 2015

37. Lafuente, A., Alonso, A., & Rodríguez, J. (2013). *We are all wise! Citizen science and expanded knowledge*. Madrid: Cátedra.

38. Lafuente, A. Y., & García, M. (2015). *Citizen labs: expanded knowledge, colateral science and experimental politics*. http://bit.ly/1MkpDqD. Accessed 19 Nov 2015

39. Latour, B. (2005). *Reassembling the Social. An Introduction to Actor-Network-Theory*. Oxford: Oxford University Press.

40. Lave, J. (1991). Situating learning in communities of practice. *Perspectives on socially shared cognition, 2*, 63–82.

41. Lave, J., & Wenger, E. (1991). *Situated learning: Legitimate peripheral participation*. Cambridge: Cambridge University Press.

42. Limbu, M. (2014). Emerging Pedagogies in the Networked Knowledge Communities: Interweaving and Intersecting Global Communities in the 21st Century Global Village. In M. Limbu & B. Gurung (Eds.), *Emerging Pedagogies in the Networked Knowledge Society: Practices Integrating Social Media and Globalization*. Hershey, PA: Information Science Reference.

43. Masuda, Y. (1981). *The Information Society as Post-industrial Society*. Bethesda, MD: World Future Society.

44. McCann, A. (2014). *Emotional and Political Possibilities of Pedagogy in Virtual Worlds*. Blogpost, http://bit.ly/1gUkbFP. Accessed 7 Sep 2015

45. McGonigal, J. (2011). *Reality is broken: Why games make us better and how they can change the world*. New York: The Penguin Press.

46. McLoughlin, C., & Lee, M. (2010). Personalised and self regulated learning in the Web 2.0 era: international exemplars of innovative pedagogy using social software. *Australasian Journal of Educational Technology, 26*(1), 28–43.

47. McLuhan, M. (1994). *Understanding media: The extensions of man*. Cambridge, MA: MIT Press.

48. Moulier-Boutang, Y. (2012). *Cognitive capitalism*. Cambridge: Polity Press.

49. Ostrom, E. (1990). *Governing the commons: The evolution of institutions for collective action*. Cambridge: Cambridge University Press.

50. Postman, N. (1993). *Technopoly: The surrender of culture to technology*. New York: Vintage Books.

51. Powell, W. W., & Snellman, K. (2004). The knowledge economy. *Annual Review of Sociology, 30*, 199–220.

52. Rheingold, H. (Ed.). (2014). *The Peeragogy Handbook. A guide for Peer-Learning and Peer Production. V. 2.*. Arlington, MA: Peirce Press.

53. Ricaurte, P. (2013). Peeragogy. In D. Aranda, A. Creus, & J. Sánchez Navarro (Eds.). *Education, digital media and participatory culture*. Barcelona: UOC Press.

54. Roig Vila, R., Mengual-Andrés, S., & Suárez Guerrero, C. (2014). Assessing the pedagogical quality of MOOCs. *Faculty, 18*, 1. http://bit.ly/1gUkbFP. Accessed 18 Oct 2015

55. Salomon, G. (1993). *Distributed cognitions: Psychological and educational considerations*. Cambridge: Cambridge University Press.

56. Santally, M. (2005). From face-to-face classrooms to innovative computer mediated pedagogies: observations from the field. *Journal of Interactive Online Learning, 3*, 4.

57. Siemens, G. (2005). Connectivism: a learning theory for the digital Age. *International Journal of Instructional Technology and Distance Learning, 2*, 1. http://www.itdl.org/Journal/Jan_05/article01.htm. Accessed 15 Nov 2015

58. Suárez, C. (2014). *Network pedagogy. Pedagogy notebooks* (p. 449). http://bit.ly/1vyYibA. Accessed 20 Oct 2015

59. Suárez, C. (2010). *Cooperation as the social condition for learning*. Barcelona: UOC Press.

60. Suárez, C., & Gros, B. (2013). *Networked learning: from interaction to cooperation*. Barcelona: UOC Press.

61. United Nations. (2005). *Towards Knowledge Societies*. Paris: UNESCO Publishing.

62. UNESCO. (2012). *Open Educational Resources*. http://bit.ly/1aPI58f. Accessed 17 Nov 2015

63. Vercellone, C. (2007). From formal subsumption to general intellect: elements for a marxist reading of the thesis of cognitive capitalism. *Historical Materialism, 15*(1), 13–36.

64. Vygotsky, L. (2000). *The development of higher psychological processes*. Barcelona: Crítica.

65. Wenger, E. (2009). A social theory of learning. In K. Illeris (Ed.), *Contemporary theories of learning: learning theorists–in their own words*. New York: Routledge.

66. Wenger, E. (1999). *Communities of practice: Learning, meaning, and identity*. Cambridge: Cambridge University Press.

67. Wikipedia. (2016). About. https://en.wikipedia.org/wiki/Wikipedia:About. Accesed 19 July 2016

68. Yáñez, C., Okada, A., & Palau, R. (2015). New learning scenarios for the 21st century related to education, culture and technology. *RUSC. Universities and Knowledge Society Journal, 12*(2), 87–102.

7
CHAPTER

The Paradox of Mobility in the Kenyan ICT Ecosystem: An Ethnographic Case of How the Youth in Kibera Slum Use and Appropriate the Mobile Phone and the Mobile Internet

Faith N. Kibere*

Department of Media and Communication, University of Leicester, Bankfield House, 132 New Walk, Leicester LE2 7JA, UK

The Kenyan ICT ecosystem has attracted vast global media and policy attention because of notable mobile phone adoption in the country. However, empirical research of how Kenyans use and appropriate new media and ICTs in the diverse contexts within the country remains limited. In order to contribute to the emerging literature on Sub-Saharan Africa ICT ecosystems as well as the Mobility discussions within Mobiles for Development M4D and Information and Communication Technologies for Development ICT4D, this paper discusses an empirical case of how the youth of Kibera use and appropriate the mobile phone and the mobile Internet. The purpose of this critical realist ethnographic research article is to explicate the events in the historical development of the Kenyan ICT ecosystem as well as the components of social and physical structure in Kibera slum along with the relationships between them. This paper argues that the mobile phone eases communication and strengthens existent social ties for the youth of Kibera. However, it cannot bypass the hierarchical nature of Kenya where "class and place of residence are distinctive social markers in the process of social networking" [Wallis, C. (2011). Mobile phones without guarantees: The promises of technology and the contingencies of culture. *New Media & Society, 13*(3), 471–485. Wallis, C. (2013). *Technomobility in China: Young migrant women and mobile phones.* New York, NY: New York University Press]. Therefore, the young Kiberans predominantly use and appropriate the mobile phone to network with those in the same lower income strata. This is because they are widely perceived in Kenyan society as the "other and what does not belong" because they are slum residents [Hall, S. (2013). The spectacle of the other. In S. Hall, J. Evans, & S. Nixon (Eds.), *Representation: Cultural representations and signifying practices* (2nd ed., pp. 223–283). Sage. p. 257].

Keywords: ICT ecosystem, mobile phone, mobility, critical realism, Kenya, ethnography, social class, social networking, slum

INTRODUCTION: THE ICT ECOSYSTEM IN KENYA

The ICT ecosystem is "the policies, strategies, processes, information, technologies, applications and stakeholders that make up a technology environment for a country, government or an enterprise" (Open ePolicy Group, 2005 p. 3). Fransman defines the "ICT sector as an ecosystem comprised of the networked element providers, the network operators, content and application providers and the final consumer" (Fransman, 2010, p. 8). Both definitions aptly capture that ICT usage within a country or environment is interrelated with the institutions that provide, regulate and facilitate the ICT-based services. The Kenyan ICT sector is a good representation of Fransman's conceptualization of the ecosystem. Kenya is one of the most culturally diverse countries in Sub-Saharan Africa with approximately 41 million people drawn from 42 ethnic groups with 70 linguistic variations (Kenya Yearbook, 2012, p. 47). There are also a large number of residents drawn from neighboring countries such as Somalia and Sudan. They account for 15% of the population whilst Indians, Arabs and Europeans account for another one percent of the population (Central Intelligence Agency, 2014).

A historical review of Kenya reveals that legislation passed by the government has been instrumental in promoting the growth of the ICT sector. In 1998, parliament enacted the Kenya Communications Act (KCA) to eradicate the government monopoly in the communications sector. Before the KCA, the government entity Kenya Postal and

Telecommunications Corporation (KPTC) operated as the sole service provider and regulator of telecommunications services. Consequently, when parliament passed the Act, it "facilitated the unbundling of the KPTC into five separate entities" (Munyua & Mureithi, 2007). One of the entities created was the CCK (rebranded as the Communications Authority of Kenya, CA). The body is responsible for licensing all systems and services in the communications sector. It is also responsible for facilitating the development of the information and communications sectors that comprise broadcasting, multimedia, telecommunications, electronic commerce, postal and courier services.

The occurrences in the Kenyan ICT sector illustrate Fransman's proposition that the government institution facilitates change in the broader ecosystem by creating a sound regulatory body. He acknowledges that institutions such as regulators, competition authorities and universities are very significant in the new ICT ecosystem (Fransman, 2010, p. 8). The Kenyan case is a classic example of the government institution facilitating change in the broader ecosystem by creating a regulator. This is because the liberalization of the sector led to the establishment of sound regulatory frameworks that drove the expansion of telecommunications networks by attracting private investment (Kimura, Omole, & Williams, 2010). Consequently, one of the impacts of the increased investment was a higher supply of mobile phone services so that more Kenyans could afford a mobile phone handset and network services.

MOBILE PHONES IN KENYA

The affordability and access to mobile phones were facilitated when leading mobile network operator Safaricom and competitor Airtel (then Kencell) invested greatly in network expansion. Mobile phones were "first introduced in Kenya in 1992 but the cost of handset ownership was too high for many" (Waema & Miroro, 2014, p. 104). The earliest owners of mobile phone handsets were very wealthy businessmen and some government technocrats. Seven years later, in 1999, "less than 1 in 1000 Kenyan adults owned mobile phones. However, by mid-2010, there were 21 million active mobile phone numbers (equivalent to one per adult)" (World Bank, 2010, p. v). According to the CA's (2015, p. 6) last evaluation, the mobile penetration rate was 83.9% (out of approximately 43 million). Kenya's mobile penetration rate is significantly higher than the Sub-Saharan Africa regional average of 38% out of a population of 936.3 million (GSMA, 2014, p. 2). As mobile phone use and adoption continued to rise in Kenya, the innovation of mobile phone money transfer service *M-Pesa* (M is for Mobile and *Pesa* is the Kiswahili word for money) took place.

M-Pesa was launched in Kenya by the mobile network operator Safaricom in March 2007. Within eight months of its launch, over

1.1 million out of 40 million Kenyans had registered to use the service and over US$87million (approximately KSH8.5 billion) had been transferred over the system. By September 2009, the number of registered users rose to slightly over 8.5 million and US$3.7 billion (KSH361.5 billion) had been transferred over the system. (Mbiti & Weil, 2011, p. 1)

The functionalities and economics of *M-Pesa* are vastly explored and well documented in various literature (Aker & Mbiti, 2010; Jack & Suri, 2011; Mbiti & Weil, 2011; Omwansa, 2009). Much of the empirical literature on *M-Pesa*documents the earliest users of the technology service whilst some of the literature cautions against conflating the vast amount of money transferred over the system with actual economic growth. Despite the empirical gaps in the literature, the success of *M-Pesa* is widely touted as one of Kenya's biggest innovation stories. Safaricom has grown tremendously because of the usage of *M-Pesa* and its profit margins have increased steadily. Consequently, it is has amplified the perception that the Kenyan ICT sector is a viable and profitable investment. In addition to provision of an environment for the rapid growth and development of *M-Pesa* and other mobile money competitors (such as Airtel money), the liberalization of the Kenyan telecommunications sector has also facilitated the licensing of numerous Internet Service Providers (ISPs).

The Internet infrastructure in Kenya

The ISPs have invested in various infrastructural developments in the ICT sector to facilitate faster and cheaper connectivity (Waema & Miroro, 2014, p. 105). Of particular significance is that between "mid 2009 and mid-2012, terrestrial fiber networks were laid at the Kenyan coast to support faster and cheaper connectivity" (Miroro & Adera, 2014, p. 71). Kenya has more licensed international gateways than any other country in Sub-Saharan Africa. The localization of connections through the Kenya Internet Exchange Point has facilitated Kenya's emerging Internet leadership as it allows local users to interconnect locally without traffic being pointed back to the US or Europe (CIO, 2014). The 2015 CA annual report indicates that there are approximately 29.6 million Internet users and the number of users is growing steadily. A large number of Kenyans also access the Internet through their mobile phones and there are 19.9 million mobile data subscriptions (CA, 2015, p. 7). Notable scholars forecast that the mobile phone is the "technology of choice for developing countries to reduce their Internet connectivity gap" (Castells, Fernandez-Ardevol, Qiu, & Sey, 2007, p. 245).

Mobile ecosystems in Sub-Saharan Africa

As the "mobile phone diffuses across the globe, a chorus of voices heralds its ability to enable individuals in developing countries to increase their life opportunities and living standards" (Wallis, 2011, p. 472). There are a lot of optimistic projections about the growth of the mobile ecosystems in Sub-Saharan Africa (Ericsson, 2014; GSMA, 2014). One reason for this is the fact that the region has the lowest fixed (land) line penetration rates in the world due to poor infrastructure and past cases of corruption (Versi, 2010 in Porter et al., 2010, p. 1). Therefore, for many Sub-Saharan Africans, their first interactive communications device is the mobile phone. The mobile subscriber base has grown by 18% per annum over the last five years and it is one of the fastest growing regions globally. As of 2013, there were 253 million unique mobile subscribers (and 502 million active SIM connections) in the region (GSMA, 2014). Toward the end of 2013, there were almost 150 million individuals using mobile devices to access the Internet in Sub-Saharan Africa. That figure is equivalent to an overall mobile penetration rate of "17% of the total population compared to a global average figure of just over 30%" (GSMA, 2014, p. 3). Therefore, the widespread usage of the mobile phone has led to the belief that the mobile phone can provide previously unavailable opportunities for communication over long distances or be used as a tool for poverty reduction (Hahn & Kibora, 2008, p. 88).

However, despite the vast optimistic "media reports" about the mobile phone in Sub-Saharan Africa, there is very limited empirical research that discusses the users and beneficiaries of technology (Fox, 2011; *The Economist*, 2009). Ecosystems scholar Toivanen describes the users as the "last mile of the information society" but it is more appropriate to describe them as the first because their role is even more pronounced in mobile telephony, mobile banking and social media access (Toivanen, 2011, p. 25). In the ICT4D and M4D studies reviewed in the next section, it emerges that the motivation to have a mobile phone is not the result of a planned development initiative but it is simply individuals adopting technology that helps with their daily affairs (Ling & Horst, 2011, p. 365). The studies reviewed contribute to the literature on mobility which is the mobile phone's ability to "compress space, distance and time as it eases and facilitates social co-ordination" (McIntosh, 2010, p. 341).

A literature review of mobility studies

In one of the first anthologies of mobile communication in Africa, mobile phones are described as the *new talking drums of everyday Africa* (de Bruijn & Nyamnjoh, 2009). The text contains disparate accounts of the relationship between society and technology in Sub-Saharan Africa. The same team published a second book with more accounts of how the mobile phone and the societies that have embraced it are mutually shaping each other (de Bruijn & Nyamnjoh, 2013, p. 4). One interesting study contained in the anthology is Lamoureaux's (2013) study of Sudanese students. The study reveals that the mobile phone provides a site for new spaces of interaction and some marginalized Nuba student migrants living in the urban capital of Khartoum use the mobile phone to create a place of belonging (Lamoureaux, 2013, p. 178). A qualitative study of young people in Ghana, Malawi and South Africa established that the youth view the mobile phone as an object of desire and a symbol of success. Additionally, the "virtual mobility of the mobile phone also promotes young people's inclusion in existing social networks. However, it is also a likely indicator of sexual liaisons with sugar daddies and the increase of inter-generational tensions" (Porter et al., 2012, pp. 156–159). A study of the youth in Dakar reveals the mobile phone offers young people remarkable opportunities to leapfrog physical mobility constraints as the sending of text messages on mobile phones opens up new corridors of communication between youth, transgressing gender barriers meticulously guarded by parents and other gerontocratic custodians. (Christiansen, Utas, & Vigh, 2006, p. 20 in Porter et al., 2012, p. 146)

The studies reviewed reveal that the mobile phone is a technology with both positive and negative sociocultural outcomes. Therefore, the mobile phone is not just a physical artifact; it is also a sociocultural artifact. Additionally, there are similarities as to how users in the Global South use the mobile phone with the well-researched users in the Global North. For example, the theme of youth using the mobile phone for escape of surveillance and transgressing physical boundaries is well documented in the Global North context (Ito & Okabe, 2005; Kingston, 2004). In addition to the disparate studies, there are some notable studies that reflect the near seamless embedding of ICTs and mobile phones in the context of use. One example is a study of Senegalese boat migrants in Spain. The study probed how the migrants used ICT to stay in touch with their home country. Nyamnjoh (2014) discovers that the phone is the "umbilical cord" that joins mothers and family with their migrant family members (Palen, Salzman, & Youngs, 2001 in Geser, 2004, p. 12). The phone allows the migrants to "embrace their Senegalese identity effectively and reinforce it by protecting the socio-cultural values of their home society" (Nyamnjoh, 2014, p. 160). In Burkina Faso, Hahn and Kibora (2008) discover the domestication and almost seamless embedding of the mobile phone in oral Burkina Faso. They argue that the gadget has been socially integrated into the society in both urban and rural areas (Hahn & Kibora, 2008, p. 96).

Domestication of technologies and appropriation

Hahn and Kibora (2008) employ an approach that is important in the context of this paper: Domestication of technologies (appropriation) approach. In the larger study that this paper is drawn from, the approach is used to make sense of the corpus of data collected. The domestication approach is also referred to as appropriation and it "encompasses the local practices of use which develop around a new object (or medium) anchoring it within particular temporal, spatial and social relations" (Livingstone, 2002, p. 35). It is often used in ICT research within media and cultural studies, to "to analyze the complex process by which ICTs enter and find a place within household routines and practices" (Silverstone, 1994, p. 20). The term domestic is not confined to home and is applicable to other areas of everyday life (Hynes & Rommes, 2006, p. 26). Domestication offers a lens to examine the process of how people purchase and embed artifacts or technologies into their daily lives. The approach is very similar to Bar, Pisani, and Weber's (2007) technology appropriation.

Within the Information Systems (IS) literature, Bar et al.'s (2007) version of technology appropriation is more prominent than the domestication approach. Based on extensive empirical work in Latin America, the scholars define appropriation as the process through which mobile users go beyond mere adoption to make the technology their own and to embed it within social, economic and political practices. The scholars argue that

> the appropriation process is fundamentally political: it is a battle for power over the configuration of a technological system .... Users re-invent the technology while they try out its features, tweak devices and applications so they better answer their needs and develop new social, economic and political practices around the possibilities open by new technological systems. (Bar et al., 2007, p. 2)

Their version was conceptualized to inform an in-depth study of the social, economic and political impact of mobile phones in Latin America. One notable similarity in both versions of appropriation is the acknowledgment that technology use is a process. They also emphasize that users exercise their will and agency to use technologies as they are active consumers and not just passive recipients.

Bar et al. (2007, p. 24) amplify the agency of users as transformative shapers of technology that can actually reconfigure or subvert the producer or manufacturers design. In contrast, the domestication of technologies approach does not explicitly suggest that users are capable of reconfiguring technology. However, the approach does not omit such a possibility. In domestication, the fourth step is that an artifact or technology is "incorporated and ascribed meaning within household rituals and rules" or the daily lives of people (Ward, 2005, p. 151). In the discussion section of this paper, appropriation is revisited in order to examine the incorporation of the mobile phone in the context of Kibera. In all the studies reviewed in the previous section as well as this one, the mobile phone emerges as a sociocultural artifact with both positive and negative implications in diverse contexts. It is evident that technology use and appropriation is a process that will vary from context to context. Therefore, it is important to critically engage with the layers of society that technology permeates and the relationships it has potential to configure. Social markers of difference such as social class, income, area of residence and race emerge as significant in the process of evaluating technology access and use within the ICT ecosystem. However, the markers of difference are not explicitly identified or interrogated in the Sub-Saharan Africa literature.

Social markers of difference in the ICT ecosystem

In the limited African literature, Archambault (2012, pp. 393–394) argues that the use of ICTs such as the mobile phone bridges distances for the low-income youth of Mozambique by opening up a space for global imagined identities. However, the same mobile phone also "betrays the difficulty of the social mobility the youth anticipate in highly unequal Southern Mozambique where ease of communication and networking do not always facilitate social mobility." In the broader Global North literature, social class as a marker of difference is identified as relevant in the process of mobile phone connections and interactions. In a study of China, Wallis discovers that mobile phones enable migrant women to expand their social networks by increasing their strong and weak ties but they do not help them to build connections with those in higher social classes than themselves due to the customary manner in which *guanxi*(social) networks are built and the rigid class and place based distinctions that characterize contemporary Chinese society. (Wallis, 2011, p. 478)

Therefore, social markers of difference such as class, gender, age and rural origin are not overcome by the use and ownership of a mobile phone. In resonance with Wallis (2011), Qiu (2009) argues that usage of ICTs does not always lead to cultural or political empowerment in a highly stratified China because of factors such as gender, education, age, ethnicity and places of origin (Qiu, 2009, pp. 239–243).

In the Indian context, Tacchi, Kitner, and Crawford (2012) explored the relationship between development, gender and technology through a focus on mobile phones and their everyday use by women in rural India. They discovered that mobile phones are potentially transformative but "gender and caste combine to produce formidable constraints" (Tacchi et al., 2012, p. 535). In the US context, boyd (2013) researched teens' meaningful engagement with social media sites MySpace and Facebook. Race emerges as a social marker of difference within the users in the ICT ecosystem. This was manifest when she discovered that white middle-class teens were migrating from social network MySpace to Facebook because they associated the former with being more *ghetto*. The white students from affluent backgrounds preferred to use Facebook rather than MySpace so the teens chose to self-segregate themselves in the two digital sites the same way they do in their physical social environment of school. Therefore, the digital social media sites emerged as an extension of the socially reproduced segregated spaces that they live and school in (boyd, 2013, p. 204). It is evident in the diverse contexts discussed that, ICTs such as the mobile phone and social networking sites facilitate social connections and ease the coordination of activities and relationships amongst users. However, they can also serve to reify one's position and social class in society even further (Wallis, 2013, p. 182). The social spaces created by ICT use can be treated as continuities of the offline (Miller & Slater, 2000Miller, D., & Slater, D. (2000). The Internet: An ethnographic approach. Oxford: Berg., p. 5; Wallis, 2011).

In the Kenyan context, discussions of social or cultural markers of differences such as ethnicity, class or income in the interaction between technology and users within the ecosystem are absent. The optimism generated by the avid usage and success of innovations such as *M-Pesa* has overshadowed discussions on the critical relationship between technology and the very diverse and stratified multicultural society. Additionally, empirical analyses of the social or cultural context of technology use in Kenya are rare. One study that attempts to interrogate the relationship between culture and technology is an ethnographic study of the mobile phone use by the *Giriama* in the coastal town of *Malindi*. McIntosh (2010) explores the use of the local dialect *Kigiriama* in text messaging. She discovers that the phone has been very useful in facilitating social coordination due to the prevalence of phones amongst those 40 and younger. However, she also discovers that mobile phones are viewed with suspicion by the elders due to their foreign origin. They are also perceived as fostering some of the "sexual promiscuity that contributes to social disintegration" (McIntosh, 2010, pp. 341, 348). Therefore, the mobile phone has a dual effect in the society. It fosters social cohesion but it also arouses anxiety because it is a technology that is associated with the former British colonialists and threatening Western values.

The bulk of the empirical literature on Kenyan ICT employs the survey research method as the main form of data collection. Most of the studies are dated as they were conducted over five years ago. One comprehensive study of the Kenyan ICT sector is the Research ICT Africa (2007–2008) quantitative and qualitative study of 400 low-income households in rural and urban Kenya. The study established that the level of education and formal employment of household members positively influenced ICT access and use. The researchers discovered that ICTs contribute to poverty reduction broadly defined as increase in income, participation in governance, enhanced voice, increased access to public goods and services, reduced vulnerability or increase in risk preparedness and increased capacity to cope with, or prepare for and adapt to, natural or economic shocks. They conclude that the use of ICTs as a tool for poverty reduction was limited by high cost of initial purchase and maintenance of ICT equipment (Waema & Miroro, 2014, pp. 113, 126). In another mixed-method study, results from pooled survey data from Nairobi professionals and entrepreneurs in 2002 and 2007 as well as qualitative interviews from 2007 to 2009 show virtual saturation in the diffusion of phones during this period but no direct effect of technology use. They also discovered a network effect of mobile telephony: Increased technological access to existing networks in a context of resource scarcity leads to a strengthening of weak ties and the enhancement of core networks amongst Kenyans (Shrum et al., 2011, p. 614). The other notable studies available are studies about the use of social networking site, Facebook, which is one of most popular sites in Kenya.

Facebook in Kenya

Facebook has approximately 3,800,000 users in Kenya (Digital Rand, 2014). It is very popular amongst the youth who are 18–34 (Njonjo, 2010Njonjo, S. K. (2010). Kenya youth fact book. Institute of Economic Affairs (IEA) and Friedrich-Ebert-Stiftung (FES)., p. 162). Empirical studies of the usage of Facebook in Kenya are focused on cyber café Internet access. A study of select cyber cafés in rural Kenya established that Facebook offers people additional ways to sustain relationships by sharing photos (Wyche, Schoenebeck, & Forte, 2013). In another study of the usage of the networking site amongst 28 young adults from *Viwandani* informal settlement, the researchers discovered that participants connected with old friends, sought employment and remittances from friends and family abroad. Like numerous studies of Facebook use in the Global North, the site is used for connection and maintenance of existing offline relationships (boyd & Ellison, 2007; Ellison, Steinfield, & Lampe, 2007). However, just like the Research ICT survey discussed, the cost of using the Internet as well as limited access to ICT equipment such as computers was identified as a hindrance to widespread usage (Wyche, Forte, & Schoenebeck, 2013, p.

2824). There is a definite gap within the Kenyan and broader Sub-Saharan Africa literatures for more studies that capture the variations of technology use in the diverse contexts. It is necessary to complement the broad quantitative corporate and academic surveys with qualitative interrogation of how social dynamics such as class, income and educational status influence the ICT ecosystem in diverse contexts. Additionally, the voices of technology users within the ecosystem are very limited. Therefore, in order to make a meaningful contribution to the literature, I selected the qualitative research approach of ethnography in order to allow the limited voices of users to contribute to the important discussion on technology appropriation in the Global South.

RESEARCH METHODS

The ethnographic research methods employed to collect data in the villages of Kibera from January to December 2013 were semi-structured interviews, go-alongs and participant observation. In this paper, I focus on data generated from the semi-structured interviews as well as documentary evidence collected in the form of e-mails. Using a non-probability snowball sampling method, I interviewed 12 young men and youth in Kenya is between the 10 young women. The description of ages of 18 and 34.[1] I interviewed youth aged 18–29 and I used an audio voice recorder to record the interviews. During and after the fieldwork phase, I transcribed the interviews into 22 extensive notes on Evernote Application. I broke down the notes into manageable sections and sorted the data in order to "bring similar categories together into broader themes" (Noble & Smith, 2014, p. 2). My objective was to "identify the most complete and logically compelling explanation of the observed events given the specific conditions of the contextual environment" (Wynn & Williams, 2012, p. 799).

I used data from the interviews to illuminate my participant observation field notes and eventually triangulated the different data sources to build a justification for the themes (Creswell, 2009, p. 191). In order to validate the data I collected, I applied methodological triangulation by combining the data from the participant observation (especially the go-alongs) with the data from the semi-structured interviews. Ethnographers commonly triangulate or compare and contrast interview and observation methods to "enhance the quality of their work" (Reeves, Kuper, & Hodges, 2008, p. 513). The contextualization and eventual critique of the world in which the young Kiberans inhabit were not a neat and linear process. In order to fulfill the task, it was necessary for me to undertake an extensive self-audit or reflection. I had to acknowledge the assumptions I had about the research context and "bracket" them from the study (Asselin, 2003, p. 100).

Critical realism

This paper is grounded in the critical realist philosophical perspective. The critical realist perspective is based on the foundational work of Roy Bhaskar on critical realism in the philosophy of science (Bhaskar, 1978, 1979). The philosophy acknowledges partial agreement with positivists that the world contains material conditions upon which certain broad consensus beliefs can be attained, but at the same time, due account must be given to variations in the meanings individuals from different communities may attach to the same social phenomena. (Gunter, 1999, pp. 7–8)

There are no fixed methods to apply critical realism to research. Ethnographic techniques can be used "within the model of critical realism to investigate the nature of generative structures through examination of social phenomena" (Porter, 1993, p. 594). Wynn and Williams (2012) propose five methodological principles to guide a critical realist research. In this particular paper, I focus on two principles. One is the explication of events that entails identifying and abstracting the events being studied, "usually from experiences, as a foundation for understanding what really happened in the underlying phenomena. Secondly, is the explication of structure and context that entails identifying components of social and physical structure along with the relationships between them" (Wynn & Williams, 2012, p. 796).

In alignment with the two principles, I present the physical and social structure of Kibera. This is followed by a discussion of the theme of the paradox of mobility that emerged from the mobile phone use and appropriation of the youthful Kiberans. In the larger research project that this is paper is drawn from, I selected Kibera as a research context because the objective was to assess the optimism surrounding ICT4D and M4D in the Sub-Saharan African context. I purposefully selected a research site that would elicit a great understanding of the research problem. This paper addresses the RQ3 that directed the study.

RQ3. How and why do the youth of Kibera appropriate new media and ICT technologies?

THE EXPLICATION OF THE RESEARCH CONTEXT

Kibera is only five kilometers away from the capital city of Kenya. The Nairobi slum is infamously labeled as Africa's largest slum. It originated over 100 years ago during the colonial era as a "remote military exercise ground for the British army King's African Rifles that comprised an

Islamised mix of Sudanese, Ugandan and Congolese people known as the Nubi" (Parsons, 1997, p. 88). The area is wrongly identified as Kibera instead of the appropriate Kibra due to the diverse diction of the numerous ethnic groups of Kenya. The original and accurate label of Kibra is derived from the Sudanese Arabic word for bushy place or "land of forest" (Constantine, 2014). In 1963, the expansion of Kibera was incentivized by Kenya's independence from the British colonial government when Africans were allowed to migrate to urban areas (Macharia, 1992, p. 225). This led to an influx of people from various ethnic communities drawn from Kenya's 42 tribes (de Smedt, 2009, p. 203). Consequently, the slum's growth is "commonly linked to the failed efforts of the British colonial administration to re-organize it and the colonial government's policy of malicious neglect which entailed withholding municipal services and development as Kibera's population grew and diversified" (Ekdale, 2011). This policy was continued by the independent Kenyan government which also failed to provide low-income housing through private enterprise (Kung'uru & Mwiraria, 1991 Ekdale, 2011, p. 11).

Much of the literature on Kibera is focused on the numerous socioeconomic ills that plague its residents and high-profile visits by celebrities and philanthropists. When the President of the United States Barack Obama was the Senator of Illinois, he visited the slum alongside his wife Michelle Obama in August 2006. The Associated Press described the event as a visit to one of the "world's worst slums where about a third of the capital city's total population, at least 700,000 people, are crammed into a single square mile in the slum of Kibera, with little access to running water and other basic services" (Associated Press, 2006). The high population figures cited in the article and other similar ones have been the source of much debate in Kenya because they contradict the last national census results of 2009. In contrast to the media reports, the government of Kenya census findings indicate that the population of Kibera is 170,070 people set on 2.5 square kilometres (KNBS, 2009). The higher population figures are speculated to be deliberate so as to attract "more aid and profit for some groups" (Rothmyer, 2011). In addition to bloated population figures, much of the literature clusters Kibera as a homogenous area. Kibera is described as a "shanty town of rusting roofs slung across mud, rocks and a rubbish dump" (Ray, 2015).

In reality, Kibera exists as a stratified society distinguished by socioeconomic and infrastructural differences (Desgroppes & Taupin, 2011). Due to the differences evident in the villages, the youth refer to the 13 villages in the informal part of Kibera as ghetto whilst the three in the formal part are described as Kibera as lower Karen suburb. The villages in the ghetto are: *Kianda, Raila, Makina, Kichinjio, Gatwekera, Kisumu*

Ndogo, Kambi Muru, Mashimoni, Laini Saba, Lindi, Silanga, Soweto East and some parts of *Soweto West*. Kibera as Lower Karen suburb denotes Olympic, *Ayany* and *Salama* (or *Karanja* Road) villages that are in the "formal" area. The 22 youth I interviewed were drawn from *Kianda, Makina, Gatwekera, Salama (Karanja road*), Olympic and *Ayany*. It is beyond the scope of this brief paper to illustrate all the villages. In order to provide a comparative description between Kibera as Ghetto and Kibera as Lower Karen, I describe contrasting villages in the next section.

Kibera as Ghetto

I conducted several interviews in *Makina* Village. The amplified space is characterized by a palpable entrepreneurship spirit. It is almost as energetic as Oxford Street, London. Neatly painted shops stocked with diverse wares are packed with customers and smiling salespeople. Restaurants, bars, churches, mosques and schools are scattered in different parts of the village. Even in the midst of crowded rooms and houses, there is a vibrant allure as fashionable young men and women line the streets socializing. Unlike the other ghetto villages, there is very little pollution toward the heart of the village. The shop owners in the village are drawn from diverse communities but the Nubian community owns a lot of the property in the area. Monthly rent can be as high as KSH 20,000 ($200) for a three bedroomed house. *Makina* is documented to have a population density of 25,242 (KNBS, 2009) and a large number of grocery and hardware shoppers frequent the area from surrounding Adams Arcade and Kabarnet Gardens suburbs. The youth from other parts of Kibera such as *Kiand*a were adamant that it was a very unsafe place due to the targeting of profitable businesses by the ethnic gangs. Like other parts of Kenya, most gang membership in Kibera is based on ethnicity: For example, *Mungiki* for the *Kikuyu* and Kamukunji for the *Luo*. The gangs in Kibera control and levy taxes over certain public goods and amenities (water, sanitation, electricity) and they may provide "protection to co-ethnic individuals in cases of inter-ethnic disputes" (Marx, Stoker, & Suri, 2014, p. 8).

Kibera is not a "lawless area" devoid of order and structure (Dickinson, 2002, p. 139). It is administered by formal and informal structures of authority that greatly influence the distribution of land, settlement of disputes and handling of security issues (Bodewes, 2005Bodewes, C. (2005). Parish transformation in urban slums: Voices of Kibera, Kenya. Nairobi: Paulines Publications Africa.; Marx et al., 2014). The formal structure comprises the elected member of parliament, members of the county assembly representatives as well as the government's provincial administration of 4 chiefs and 11 assistant chiefs whilst the informal structure includes a council of elders and the ethnic gangs (Marx et al., 2014, p. 3). The most important sources of influence in the governance of Kibera are the chiefs. Throughout the history of the slum, "chiefs have illegally allocated land to prospective structure owners in exchange for payment. This behaviour exceeds the formal, legal duties of the office of the chief" (Joireman & Vanderpoel, 2010, p. 10). Kibera has a "complex structure of economic stakeholders, who have acquired a degree of social legitimacy to extract profit out of trade of inadequate basic necessities" (Huchzermeyer, 2008, p. 20). Water, electricity, refuse collection and toilet facilities are all available at a fee and residents have the choice to pay "illegal private service providers" or government entities (Odbert & Twigg, 2011). The country's electricity distributor Kenya Power retails legal electricity at a fee and the cost of a legal power connection is about KSH 300 ($3) a month for the average house of 12 by 12 feet. The government's Nairobi Water Company provides water for around KSH 200 ($2) a month or slightly less. Illegal service providers or gangs sometimes tap the genuine electricity and water connections for retail at lower costs to some residents; it is a very dangerous trade that causes accidental fires and tragic deaths in many homes (Mutahi, 2011; Ombati, 2013). In contrast to Kibera as Ghetto, Kibera as lower Karen denotes Olympic, *Ayany* and *Salama (Karanja* road) villages in the formal section of Kibera.

Kibera as lower Karen suburb

It is well documented in the literature that many slum residents usually refer to their neighborhoods by place names or nicknames rather than use pejorative terms like slum or favela (Dürr, 2012, p. 798). The real Karen suburb is one of the oldest and most desirable residential areas in Nairobi named after plantation owner and Danish writer of Out of Africa, Karen Blixen. It is approximately 15 kilometers away from Kibera so it is borrowed as a nickname due to proximity. The suburb offers a blend of colonial bungalows and villas on gorgeous and expansive grounds (Hass Consult, 2014). It is characterized by modern gated housing communities, restaurants, animal parks, art galleries and cultural centers. The first village in lower Karen is Olympic village. Most of the buildings are made of concrete with brick tiled roofs whilst some have corrugated galvanized iron sheet roofing. The area has very well-constructed cybercafés, pubs and restaurants. A variety of vehicles are parked in the area as testament to the well-tarmacked roads. The monthly rent ranges from KSH 7000 ($70) for a one bedroomed house to KSH 30,000 ($300) for a three bedroomed house. Some residents are homeowners and a typical house costs approximately KSH 6 million ($60,000). Unlike the informal ghetto part of Kibera, title deeds for the houses are legally available and it is believed that the Nubian community owns most of the houses in the area.

The price range for housing in the rest of Olympic is almost similar but there are a few smaller mud houses that cost much less. Villages in Lower Karen suburb are interesting because they are typically not slum areas but they are strongly associated with Kibera due to proximity, shared infrastructure and media coverage. Sometimes the youth drawn from the suburban area prefer to describe the area as Kibera slum and most maps will cluster the formal and informal areas together. Kibera is not a simple society to define as it is very diverse. There is hardly any literature of the socioeconomic profiles of the residents of Kibera. There are a few papers that offer sufficient evidence of the diverse nature of the residents but they are not exhaustive (Gulyani & Talkudar, 2010; Research International, 2009). In the course of fieldwork, I documented the education levels, age, interests and job levels of all the participants. That exercise was particularly useful in affirming the diverse nature of Kibera (see the appendix).

Socioeconomic profiles of Kiberans

In 2009, a survey of Kibera established that the average annual income per household was KSH 8500 ($85.00) and an estimate of |KSH 1420 ($14.20) per month as the income per capita. The income was largely derived from wage employment and small business engagement or what are commonly referred to as micro-enterprises (Research International, 2009, p. 26). Another study of 1755 households in Nairobi's slums discovered that 73% of households are poor and 2% are non-poor (Gulyani & Talkudar, 2010, p. 1712). In comparison with the poverty line of KSH 3174 ($31.74) per adult per month (excluding rent), both studies conclude that household (*also micro) enterprises are helping Nairobi's slum residents in their fight against poverty (Gulyani & Talkudar, 2010; Research International, 2009). Kihato asserts that Kiberans have a "reverence for education and they will sacrifice almost anything to send their children to good private schools, pay for extra tuition and buy books" (2013, p. 29). In a case study of private schools in Kibera, Dixon and Tooley (2012, p. 702) discovered that parents prefer to enroll their children in low-cost private schools as opposed to free government schools due to perceptions of higher quality in the former. Research International's report indicates that "three-quarters of slum dwellers in Nairobi have completed primary school with no significant difference between males and females. It also highlights that one of the major pulls to reside in Kibera is the lower rent" (Research International, 2009, p. 20). So how exactly do the youth use and appropriate new media given their context?

DISCUSSION OF RESEARCH FINDINGS

In this section, the research question of how and why Kiberan youth appropriate new media and ICTs is discussed. The most significant form of new media and ICT that the discussion is centered on is the mobile phone and the mobile Internet. This is because the data collection and analysis of the semi-structured interviews revealed that this was the most significant form of new media and ICT to the young Kiberans. The study discovered that the mobile phone and the mobile Internet facilitate contextual mobility. In that form of mobility, "ICTs afford one diversified modalities of interaction and help people to interact easily without contextual constraints" (Kakihara & Sørensen, 2001, pp. 35–36). However, in the process of that effective interaction, the mobile Internet also reveals to the young people of Kibera that they are the "others, the outsiders" (Hall, 2013, p. 257).

Access to mobile phones and mobile Internet

According to the findings from the semi-structured interviews conducted, the most important form of new media and ICT in the lives of the youth is the mobile phone because they can use the Internet on the artifact. All the youth in the sample were mobile phone owners. Eleven young men out of the 12 that I interviewed were smart phone owners. Huawei, Nokia and Samsung were popular choices. Three of the men also had an extra basic phone with no Internet capabilities (a cheap phone popularly known as *kabambe* in Kenya). In contrast, 5 of the 10 women owned smart phones. During times of financial challenges, the youth reported that it was common practice to sell their smart phones or exchange them with *kabambes.* Like the Research ICT Africa study reviewed, "they noted that the high cost of initial purchase, use and maintenance of ICT equipment was a challenge" (Waema & Miroro, 2014, p. 125). However, the flexible data bundles offered by mobile operator *Safaricom* from 2011 were cited to be an incentive for the purchase of a web-enabled phone. The youth frequently mentioned that the purchase of KSH 20 ($0.224) mobile phone top up could enable them to access and surf the Internet very fast. In addition to the data on mobile phone usage and appropriation from the semi-structured interviews, I was also able to acquire complementary documentary data. The purpose of the documentary data is to provide baseline empirical evidence of the mobile phones and Internet-enabled devices used in Kibera.

After one year of negotiation with Safaricom staff members, I received documentary data in the form of two e-mails. In the first e-mail communication I received on 7 March 2014, the Safaricom representative indicated that there are **1,097,315** data-enabled devices in Kibera and another low-income settlement in Nairobi, *Kangemi*. The representative also indicated that the number of mobile phone service subscribers in Kibera and Kangemi is **3,484,330**. The data provided to me clustered the two areas together. Additionally, the data do not distinguish between the use of a mobile phone for Internet access or a mobile dongle device for the same access. However, the representative speculated that most of the Internet-enabled devices reflected in the data were mobile phones. The youth reported that they were avid users of the mobile phone to access the Internet and none of the youth were owners of a Safaricom Internet dongle. They perceived the dongles to be expensive. After I received the first set of data, I requested for another break down of some of the sub-locations or villages of Kibera. The representative was able to furnish me with that data in a separate e-mail on 10 March 2014. The Safaricom representative was only able to provide the data from *Gatwekera*, *Silanga*, *Olympic*, *Lindi* and *Makina* villages. One of the distinctive elements of the data in the e-mail document is that one of the most low-income villages of *Gatwekera* has a higher number of data-enabled devices than the higher income *Makina* village.

Gatwekera has 2331 web-enabled devices whilst *Makina* has 2242 devices. Higher income Olympic has the highest number of web-enabled devices at 4735. The *Safaricom* representative was also able to get some data from the surrounding suburban areas. Some Kiberans are mobile subscribers to other networks such as Airtel and Orange but it was not possible for me to acquire the data from all the mobile service providers. In the context of this study, all the youth were Safaricom subscribers and they cited the popularity of the network. However, they also indicated that other networks were also popular.

Educational status and new media and ICT access

In the course of fieldwork, I also discovered that the most highly educated youth in Kibera had access to a variety of new media and ICT technologies at their work places and in Kibera. The highly educated youth were actively engaged in self-help youth groups such as drug awareness projects in their communities and could access the Internet free from their premises within Kibera. As indicated in the demographic profile (the appendix), most of the youth were enrolled at various reputable colleges and universities within Kenya. Others were in transition from high school to university and college. In the explication of the context, Kiberans are described as people who revere education. In the broader Kenyan society, the pursuit and desire for formal education are very high. This is because it is perceived as a status symbol and key to employment and opportunities. Educational attainment is generally perceived as "one of the means through which youth in the slum can escape the harsh realities of the slum" and acquire a better life through occupational attainment and residential mobility (Kabiru, Mojola, Beguy, & Okigbo, 2013, p. 82). Educational status as a factor in Kiberan ICT use and appropriation resonates with the Research ICT Africa (2007–2008) study reviewed in the literature review. The study established that the level of education and formal employment of household members positively influenced ICT access and use (Waema & Miroro, 2014). In contrast to the highly educated youth I interviewed, the least-educated youth were very preoccupied in seeking employment opportunities and social connections or what they described as *hustling.* Unlike the highly educated, they were not very actively involved in community self-help groups. One of the least-educated young women, Norah was the only one who had never used the Internet.

THE PARADOX OF MOBILITY

The findings discussed in this section relay evidence that the mobile phone eases communication and facilitates the use of mobile Internet access for networking with existent social ties. This resonates with the reviewed study of middle-class Kenyans that established that the mobile phone strengthens existent weak ties (Shrum et al., 2011). Weak ties are an "individual's loosely connected acquaintances who are less likely to know each other than strong ties who exist in a stronger and more connected friendship circle. They are also likely to form a low density network" (Granovetter, 1983, p. 201). Extensive empirical literature proposes that Facebook is used for connection and maintenance of existing offline social ties (boyd & Ellison, 2007, p. 221; Ellison et al., 2007). Similarly, I discovered that youth were actively using and appropriating the mobile phone to maintain their offline weak social ties through Facebook. Additionally, they were predominantly appropriating Facebook through their mobile phones to network with those in their social class and context. Therefore, the social spaces that were created by the mobile phone were simply continuities of the offline (Miller & Slater, 2000, p. 5). The youth have appropriated the mobile phone and objectified it because it facilitates the convergence of social networking sites such as Facebook as well as e-mail facilities and call functions on the mobile phone. It is also a sociocultural artifact that has become what the youth describe as an *important zone of self-expression.* Furthermore, the incorporation of the mobile phone in Kiberan society is embodied in the fourth step of the domestication of technologies approach because it has been "incorporated and ascribed meaning" within the context of the youths' lives (Ward, 2005, p. 151).

The mobile phone as a zone of self-expression

The youth spoke fondly of Facebook and some like Pat [2]who was just about to join a university in USA through a scholarship reported, "I can be myself on Facebook."

John who was a full-time student in a highly reputed local university asserted, "Facebook has made life easier." Liza who was university educated and lived in higher income *Ayany* used Facebook for socioeconomic purposes. She praised the site as it allowed her to "market herself" by posting and promoting her video productions. She discussed her network of friends with passion and reported that rivals used Facebook to spy on her network of friends. "My business competitor likes what I like on the site to try to get my network of friends," she said. After the interviews, we would sit and chat as we enjoyed some fried flat bread *(chapati)* and it was in those moments the youth would revisit some of the interview discussions. It was evident the mobile phone had "collapsed distance and time" whilst affording them ease of communication (McIntosh, 2010). However, the youth also indicated that their *zone of self-expression* and *office* could not bypass the "hierarchical nature" of Kenyan society of who you know and where you come from matter very much (Hofstede, 2015).

The hierarchical nature of Kenyan society

It is not well documented in the literature but Kenya is a very "hierarchical society" where titles such as *mheshimiwa* (your honor), professor, engineer, doctor and graduate are perceived very highly (Hofstede, 2015). Additionally, the place of residence, ethnic affiliations and educational status are used as social markers that potentially lead to inclusion or exclusion in many Kenyan social networks. In this discussion, I focus on the place of origin and residence as a distinctive social marker that leads to exclusion within Kenyan social networks. During the semi-structured interview of Norah, I discovered that the single mother was intimidated by Internet access and the social spaces created by the use and appropriation of ICTs. One reason she was ambivalent to socialize online and offline is that she preferred to converse in local language Kiswahili rather than English. She had attended primary school but was not proficient in English; she perceived that English was more commonly used online than Kiswahili. In contrast, the more educated youth proficient in English as well as Kiswahili lauded the new social spaces that mobile phones facilitate on social networking sites such as Facebook. However, they reported that they used and appropriated the sites whilst in fear of their perceived lower status in society. For example, Liza reported that her close friends did not want to be identified as residents of Kibera on Facebook. She reported that Kiberans would hide their residential areas online and use code language to discuss Kibera. This is a habit she disclosed that took place in offline day-to-day Kiberan life. They youth were ambivalent to be identified as residents of Kibera because they perceived it was a low-income slum with a negative image. Most of the youth had experienced social exclusion in higher middle- to upper-class social networks within Kenyan society. Susan recounted an experience that confirms the practice of inclusion or exclusion based on places of origin.

I have been stigmatized when I was in school. I used to attend Loreto Msongari located in a high class suburb, so form I and 2 I used to lie to people. I told them I lived in Kilimani (an upper middle class suburb) and they considered me to be cool. One day I decided to stop lying and I told them the truth and I became the not so cool friend. Hata sikuwa na mabeshte! (I never had friends!)

Tracy lost her job as a part-time waitress for revealing that she was from Kibera as the owner of the restaurant in the city center in Nairobi was wary that she would be a security threat. Susan and Brian also noted that in some forums they attended, some people would clutch their handbags tighter when they mentioned they were from Kibera.

Diane observed, "women really fear saying they are from Kibera but I don't care! When we go clubbing and they have to be dropped by the guys they ask them to drop them at the mall because it's posh."

In addition to ambivalence about Facebook use and appropriation, some of the youth perceived the social networking site Twitter useful but intimidating. For example, Brian indicated that the site is much more intimidating than Facebook and is more suited to middle- to upper-class-educated Kenyans.

Mi hupenda Facebook sana, kuchat na mabeshte..mi huona that Twitter ni ya watu wasomi ... mi huona hivyo (I like Facebook very much, to chat with my friends but I see like Twitter is for the learned, the rich, that's what I feel).

Moha also noted, "Hiyo Twitter ni ya watu serious si ya ujinga mob kama Facebook (Twitter is for serious and established people, not for stupidity or for common people like Facebook)."

Paul reported, "Twitter si ya ujinga, mi huitumia kufuata watu wa maana (Twitter is not for stupidity, I use it to follow people of substance)."

In light of the empirical discussion, the youths' use and appropriation of social networking sites such as Facebook and Twitter emerge as an "extension of the socially reproduced segregated spaces that they live and school in" (boyd, 2013, p. 204). Their place of residence and origin as the slum of Kibera as well as their perceived lower status leads to their fear of social exclusion in the spaces created by the use and appropriation of the mobile phone and the mobile Internet. In the rare instances that the youth networked offline and online with people drawn from a higher social class or strata, they were often foreigners. They were drawn from countries such as the USA, Finland, Norway or China. The only youth who reported that the mobile phone was helping them to connect to new social ties were part-time tour guide Jackson (27), part-time university student and entrepreneur Francis (28) and volunteer Susan (28).

The mobile phone as an office

Jackson and Francis described their mobile phones as an *office* because various foreigners who wanted to visit Kibera often contacted them. The two youth had previously participated very actively in community projects so they had very many social ties from donor communities and international slum volunteers. Consequently, their phones enhanced their communicative abilities by helping their international networks to communicate with them. In some cases, foreigners familiar with Kibera would distribute their contacts and Facebook details to other social ties who were planning a visit. Similarly, Susan had a vast network of foreign social ties that she acquired when she volunteered at an international first-aid organization. Therefore, she used her social networking sites to chat with the international contacts. However, none of the youth were actively using their social networking sites and mobile phone to maintain social ties with higher class or strata Kenyans. This is because the young people's place of residence as Kibera slum in a hierarchical Kenya restricts their appropriation of the mobile phone for the creation and extension of social ties with higher class and higher income Kenyans from upper-class gated communities or middle-income housing estates. The perceived lower class of the youth and their place of residence and origin act as constraints to the kind of social ties that they can form within Kenyan society. This is because their place of origin is perceived as dangerous and as they are widely perceived in Kenyan society as the "other, the deviant and what does not belong" because they are slum residents (Hall, 2013, pp. 257–258). Consequently, the contextual mobility that the phone facilitates so well by compressing distance, time and space is limited by the existence of the residential and social segregation they experience offline.

CONCLUSION

Academic discussion of social classes and stratification within Kenyan society is absent from the ICT ecosystem literature. The policy and media discussions around ICT and mobile ecosystems in Sub-Saharan Africa are heavily focused on infrastructure and access. This paper concludes that the mobile phone eases communication and strengthens existent social ties for the youth of Kibera. However, the youth predominantly use the mobile phone to network with those in the same class and place of residence. More than "71% of the urban population in Kenya resides in a slum" (Mutisya & Yarime, 2011Mutisya, E., & Yarime, M. (2011). Understanding the grassroots dynamics of slums in Nairobi: The dilemma of Kibera informal settlements. International Transaction Journal of Engineering, Management, & Applied Sciences & Technologies, 2(2), 197–213., p. 197). However, it is also the same population that makes up majority of the "mobile phone inclusive market" (Karugu & Mwendwa, 2008). Therefore, the issue of class and stratification within the Kenyan ICT ecosystem definitely deserves more interrogation so that the diverse user needs in Sub-Saharan Africa countries are met. In a transitional continent filled with diverse cultures and economies, such research can effectively facilitate how countries position themselves within the global ICT ecosystem (Toivanen, 2011, p. 35). A critical realist approach such as the one employed in this paper reveals that the interaction between technologies and low-income contexts is not always a seamless interaction. Future research can interrogate how the social, cultural and political structures within which users of technology in Sub-Saharan Africa are embedded in affect how they use and appropriate technology.

In this particular case, there is evidence that the mobile phone and the Internet are potentially transformative technologies that provide social spaces for enhanced social networking. However, the use and appropriation of the same technologies also reveal that online interactions and networks are replicated from existent offline interactions. Technology is not obligated to seamlessly integrate into contexts and spaces. It can in fact "amplify the structural inequalities" and divisions that already exist within that society (Toyama, 2011, p. 75). Therefore, it is necessary that each and every context within the ecosystem is analyzed with user-centered methodologies so that the interactions between "networked element providers, the network operators, content and application providers and the final consumers' are enhanced" (Fransman, 2010, p. 8).

DISCLOSURE STATEMENT

No potential conflict of interest was reported by the authors.

REFERENCES

1. Aker, J. C., & Mbiti, I. M. (2010). Mobile phones and economic development in Africa. *The Journal of Economic Perspectives, 24*(3), 207–232.

2. Archambault, J. S. (2012). Travelling while sitting down: Mobile phones, mobility and the communication landscape in Inhambane, Mozambique. *Africa, 82*(03), 393–412.

3. Asselin, M. E. (2003). Insider research: Issues to consider when doing qualitative research in your own setting. *Journal for Nurses in Professional Development, 19*(2), 99–103.

4. Associated Press. (2006, August 27). Senator Obama visits Nairobi slum Illinois democrat, wife take HIV tests to highlight Africa's AIDS pandemic. *NBC News.* Retrieved from http://www.nbcnews.com/id/14527295/ns/world_news-africa/t/sen-obama-visits-nairobi-slum/#.VOIAM0LlfzI

5. Bar, F., Pisani, F., & Weber, M. (2007, April). *Mobile technology appropriation in a distantmirror: Baroque infiltration, creolization, and cannibalism.* Paper presented at Seminariosobre Desarrollo Económico, Desarrollo Social y Comunicaciones Móviles en AméricaLatina, Buenos Aires, Argentina. Retrieved September 6, 2007, from http://arnic.info/Papers/Bar_Pisani_Weber _appropriation-April07.pdf

6. Bhaskar, R. (1978). *A realist theory of science.* Hemel Hempstead: Harvester Press.

7. Bhaskar, R. (1979). *The possibility of naturalism.* Brighton: Harvester Press.

8. Bodewes, C. (2005). *Parish transformation in urban slums: Voices of Kibera, Kenya.* Nairobi: Paulines Publications Africa.

9. boyd, D. (2013). White flight in networked publics? How race and class shaped American teen engagement with MySpace and Facebook. In L. Nakamura & P. Chow-White (Eds.), *Race after the Internet* (pp. 203–222). New York: Routledge.

10. boyd, D. M., & Ellison, N. B. (2007). Social network sites: Definition, history and scholarship. *Journal of Computer-Mediated Communication, 13*(1), 210–230.

11. de Bruijn, M., & Nyamnjoh, F. (Eds.). (2009). *Mobile phones: The new talking drums of everyday Africa*. Bamenda, CMR: Langaa.

12. de Bruijn, M., & Nyamnjoh, F. (Eds.). (2013). *Side@ways: Mobile margins and the dynamics of communication in Africa.* Bamenda, CMR: Langaa RPCIG.

13. Castells, M., Fernandez-Ardevol, M., Qiu, J. L., & Sey, A. (2007). *Mobile communication and society: A global perspective (Information revolution and global politics)*. Cambridge: MIT Press.

14. Central Intelligence Agency. (2014). *The world fact book: Kenya.* Retrieved from https://www.cia.gov/library/publications/the-world-factbook/geos/ke.html

15. Christiansen, C., Utas, M., & Vigh, H. (2006). Introduction. In C. Christiansen, M. Utas., & H. Vigh (Eds.), *Navigating youth, generating adulthood: Social becoming in an African context*(pp. 9–30). Uppsala: Nordiska Afrikains Institutet.

16. CIO. (2014, September 9). Kenya leads African countries in bandwidth per person, lowest costs and fastest speeds. *CIO East Africa*. Retrieved from http://cio.co.ke/news/main-stories/kenya-leads-african-countries-in-bandwidth-per-person,-lowest-costs-and-fastest-speeds

17. Communications Authority of Kenya. (2015). *Quarterly sector statistics report: Fourth quarter sector statistics report for the financial year 2014/15 (April-June 2015*). Retrieved from http://www.ca.go.ke/images/downloads/STATISTICS/Sector%20Statistics%20Report%20Q3%202013--2014.pdf

18. Constantine, G. (2014). *Nubians: Then and now.* Retrieved from http://www.nubiansinkenya.com/index.php#mi=1&pt=0&pi=2&s=0&p=0&a=0&at=

19. Creswell, J. W. (2009). *Research design: Qualitative, quantitative, and mixed methods approaches* (2nd ed.). Thousand Oaks: Sage.

20. Desgroppes, A., & Taupin, S. (2011). *Kibera: The biggest slum in Africa?* Retrieved from http://halshs.archives-ouvertes.fr/docs/00/75/18/33/PDF/Amelie_Desgroppes_Sophie_Taupin_-_KIBERA.pdf

21. Dickinson, W. (2002). Population matters – a visit to the Kibera slums, Nairobi, Kenya. *Social Contract, 12*(2), 139–140.

22. Digital Rand. (2014). *State of social media in Kenya*. Report A002, 4 September. Retrieved from http://digitalrand.com/resources/reports/A002.pdf

23. Dixon, P., & Tooley, J. (2012). A case study of private schools in Kibera: An update. *Educational Management Administration & Leadership, 40*(6), 1–17.

24. Dürr, E. (2012). Urban poverty, spatial representation and mobility: Touring a slum in Mexico. *International Journal of Urban and Regional Research, 36*(4), 706–724.

25. Ekdale. (2011). *Kibera history*. Retrieved from http://www.brianekdale.com/uploads/kiberahistory.pdf

26. Ellison, N. B., Steinfield, C., & Lampe, C. (2007). The benefits of Facebook "friends": Social capital and college students' use of online social network sites. *Journal of Computer Mediated Communication, 12*(4), 1143–1168.

27. Ericsson. (2014). *Sub-Saharan Africa: Ericsson mobility report*. Retrieved from http://www.ericsson.com/res/docs/2014/emr-june2014-regional-appendices-ssa.pdf

28. Fox, K. (2011, 15 July). Africa's mobile economic revolution. *The Guardian*. Retrieved from http://www.theguardian.com/technology/2011/jul/24/mobile-phones-africa-microfinance-farming

29. Fransman, M. (2010). *The new ICT ecosystem: Implications for policy and regulation*. Cambridge: Cambridge University Press.

30. Geser, H. (2004). Towards a sociology of the mobile phone. In: *Sociology in Switzerland: Sociology of the mobile phone*. Retrieved from http://socio.ch/mobile/t_geser1.pdf

31. Granovetter, M. (1983). The strength of weak ties: A network theory revisited. *Sociological Theory, 1*(1), 201–233.

32. GSMA. (2014). *Sub-Saharan Africa mobile economy 2013.* Retrieved from http://www.gsmamobileeconomyafrica.com/SubSaharan%20Africa_ME_Report_English_2013.pdf

33. Gulyani, S., & Talkudar, D. (2010). Inside informality: The links between poverty, microenterprises and living conditions in Nairobi's slums. *World Development, 38*(12), 1710–1726.

34. Gunter, B. (1999). *Media research methods: Measuring audiences, reactions and impact.* London: Sage.

35. Hahn, P. H., & Kibora, L. (2008). The domestication of the mobile phone: Oral society and new ICT in Burkina Faso. *The Journal of Modern African Studies, 46*(1), 87–109.

36. Hall, S. (2013). The spectacle of the other. In S. Hall, J. Evans, & S. Nixon (Eds.), *Representation: Cultural representations and signifying practices* (2nd ed., pp. 223–283). London: Sage.

37. Hass Consult. (2014). *Residential areas of Kenya.* Retrieved from: http://www.hassconsult.co.ke/index.php?option=com_janews&view=janews&Itemid=161

38. Hofstede, G. (2015). *What about Kenya? Power-distance.* Retrieved from http://geert-hofstede.com/kenya.html

39. Huchzermeyer, M. (2008). Slum upgrading in Nairobi within the housing and basic services market: A housing rights concern. *Journal of Asian and African Studies, 43*(1), 19–39.

40. Hynes, D., & Rommes, E. (2006). Fitting the Internet into our lives: IT courses for disadvantaged users. In T. Berker, M. Hartmann, Y. Punie, & K. Ward (Eds.), *Domestication of media and technology* (pp. 123–142). Maidenhead: Open University Press.

41. Ito, M., & Okabe, D. (2005). Intimate connections: Contextualizing Japanese youth and mobile messaging. *The Kluwer International Series on Computer Supported Cooperative Work, 4*, 127–145.

42. Jack, W., & Suri, T. (2011). *Mobile money: The economics of M-PESA* (No. w16721). National Bureau of Economic Research.

43. Joireman, S., & Vanderpoel, S. R. (2010). *In search of order: Property rights enforcement in Kibera settlement, Kenya. World Bank Resources.* Retrieved from http://siteresources.worldbank.org/EXTARD /Resources/336681--1236436879081/5893311--1271205116054/Joireman.pdf

44. Kabiru, C. W., Mojola, S. A., Beguy, D., & Okigbo, C. (2013). Growing up at the "margins": Concerns, aspirations, and expectations of young people living in Nairobi's slums. *Journal of Research on Adolescence, 23*(1), 81–94.

45. Kakihara, M., & Sørensen, C. (2001). Expanding the "mobility" concept. *ACM SIGGroup Bulletin, 22*(3), 33–37.

46. Karugu, W. N., & Mwendwa, T. (2008). *Vodafone and Safaricom Kenya: Extending the range and reliability of financial services to the poor in rural Kenya.* Retrieved from http://growinginclusivemarkets.org/media/cases/Kenya_MPESA_2008.pdf

47. Kenya National Bureau of Statistics. (2009). *Kenya population and housing census.* Ministry of Planning, National Development and Vision 2030.

48. Kenya Yearbook. (2012). *Kenya yearbook 2011/12.* Nairobi: Author.

49. Kihato, W. C. (2013). The bookseller of Kibera. In: The individual writers. (Eds.), *Writing invisibility. Conversations on the hidden city* (pp. 22–33). Johannesburg: Mail & Guardian. Retrieved from: http://cdn.mg.co.za/content/documents/2013/08/30/migration_book.pdf

50. Kimura, K., Omole, D. W., & Williams, M. (2010). Engaging the private sector to upgrade infrastructure. In K. Kimura, D. W. Omole, & M. Williams (Eds.), *ICT in sub-Saharan Africa: Success stories* (pp. 339–352). World Bank. http://siteresources.worldbank.org/AFRICAEXT/Resources/258643--1271798012256/ICT-19.pdf

51. Kingston, J. (2004). *Japan's quiet transformation: Social change and civil society in the twenty-first century.* New York: RoutledgeCurzon.

52. KNBS. (2009). *The 2009 Kenya population and housing census.* Nairobi: Kenya National Bureau of Statistics.

53. Kung'uru, J., & Mwiraria, M. (1991). *NGO support to informal settlements: A case study of Kibera, Nairobi.* Nairobi: UNDP – World Bank Water and Sanitation Program, Regional Water and Sanitation Group – Eastern and Southern Africa.

54. Lamoureaux, S. (2013). Identities of place: Mobile naming practices and social landscapes in Sudan. In M. De Bruijn, I. Brinkman, & F. B. Nyamnjoh (Eds.), *Side@ ways: Mobile Margins and the dynamics of communication in Africa* (pp. 179–199). Leiden: Langaa, African Studies Center.

55. Ling, R., & Horst, H. (2011). Mobile communication in the Global South: An introduction. *New Media and Society, 13*(3), 363–374.

56. Livingstone, S. (2002). *Young people and new media: Childhood and the changing media environment.* London: Sage.

57. Macharia, K. (1992). Slum clearance and the informal economy in Nairobi. *The Journal of Modern African Studies, 30*(2), 221–236.

58. Marx, B., Stoker, T., & Suri, T. (2014). *There is no free house: Ethnic patronage and property rights in a Kenyan slum.* Retrieved from http://www.mit.edu/~tavneet/Marx_Stoker_Suri.pdf

59. Mbiti, I., & Weil, D. N. (2011). *Mobile banking: The impact of M-Pesa in Kenya* (No. W17129). National Bureau of Economic Research.

60. McIntosh, J. (2010). Mobile phones and Mipoho's prophecy: The powers and dangers of flying language. *American Ethnologist, 37*(2), 337–353.

61. Miller, D., & Slater, D. (2000). *The Internet: An ethnographic approach.* Oxford: Berg.

62. Miroro, O. O., & Adera, E. O. (2014). Political economy of ICTs and their effect on poverty. In O. E. Adera, T. Waema, J. May, O. Mascarenhas, & K. Diga (Eds.), *ICT pathways to poverty reduction: Empirical evidence from East and Southern Africa* (pp. 53–76). Ottawa: Practical Action, IDRC.

63. Munyua, A., & Mureithi, M. (2007). *Kenya ICT action network.* Global Information Society Watch, 164–166.

64. Mutahi, P. (2011, October 27). Police reforms have to address the place of gangs and informal security group. *Business Daily.* Retrieved from http://www.businessdailyafrica.com/Opinion-and-Analysis/Police-reforms-have-to-address-the-place-of-gangs-/-/539548/1263122/-/item/1/-/otqn68z/-/index.html

65. Mutisya, E., & Yarime, M. (2011). Understanding the grassroots dynamics of slums in Nairobi: The dilemma of Kibera informal settlements. *International Transaction Journal of Engineering, Management, & Applied Sciences & Technologies, 2*(2), 197–213.

66. Njonjo, S. K. (2010). *Kenya youth fact book.* Institute of Economic Affairs (IEA) and Friedrich-Ebert-Stiftung (FES).

67. Noble, H., & Smith, J. (2014). Qualitative data analysis: A practical example. *Evidence Based Nursing, 17*(1), 2–3.

68. Nyamnjoh, H. M. (2014). *Bridging mobilities: ICTs appropriation by Cameroonians in South Africa and the Netherlands*. Bamenda/Leiden: Langaa/African Studies Centre.

69. Odbert, C., & Twigg, B. (2011, April 4). Power to the people: The black market for electricity in Kibera. *Forbes*. Retrieved from: http://www.forbes.com/sites/megacities/2011/04/04/mee t-chris-dancercommunity-organizer-and-stolen-electricity-supplier/

70. Ombati, C. (2013, September 4). Ransom gangs hit Nairobi, says National Crime Research Centre report. *The Standard Digital News*. Retrieved from http://www.standardmedia.co.ke/?articleID =2000092752&story_title=ransom-gangs-hit-nairobi-says-national-crime-research-centre-report&pageNo=3

71. Omwansa, T. K. (2009). *M-PESA: Progress and prospects*. Retrieved from http://erepository.uonbi.ac.ke/bitstream/handle/11295/478 13/FULL%20TEXT.pdf?sequence=1

72. Open ePolicy Group. (2005). *Roadmap for open ICT ecosystems*. Cambridge, MA: Berkman Center for Internet and Society. Retrieved from http://cyber.law.harvard.edu/ publications/2005/The_Roadmap_for_Open_ICT_Ecosystems

73. Palen, L., Salzman, M., & Youngs, E. (2001). Discovery and integration of mobile communications in everyday life. *Personal and Ubiquitous Computing*, *5*(2), 109–122.

74. Parsons, T. (1997). "Kibra is our blood": The Sudanese military legacy in Nairobi's Kibera location, 1902–1968. *The International Journal of African Historical Studies*, *30*(1), 87–122.

75. Porter, S. (1993). Critical realist ethnography: The case of racism and professionalism in a medical setting. *Sociology*, *27*(4), 591–609.

76. Porter, G., Hampshire, K., Abane, A., Munthali, A., Robson, E., Mashir i, M., & Tanle, A. (2012). Youth, mobility and mobile phones in Africa: Findings from a three-country study. *Information Technology for Development*, *18*(2), 145–162.

77. Porter, G., Hampshire, K., Abane, A., Robson, E., Munthali, A., Mashir i, M., & Tanle, A. (2010). Moving young lives: Mobility, immobility and inter-generational tensions in urban Africa. *Geoforum*, *41*(5), 796–804.

78. The power of mobile money: Mobile phones have transformed lives in the poor world. (2009, September 24). *The Economist*. Retrieved from http://www.economist.com/node/14505519

79. Qiu, L. J. (2009). *Working class network society. Communication technology and the information have-less in urban China*. Cambridge: The MIT Press.

80. Ray, J. (2015, January 15). Inside Kibera slum, Africa's biggest shantytown. *ITV News*. Retrieved from http://www.itv.com/news/2015-01-15/inside-kibera-slum-africas-biggest-shanty-town/

81. Reeves, S., Kuper, A., & Hodges, B. D. (2008). Qualitative research methodologies: Ethnography. *BMJ, 337*, 512–514.

82. Research International. (2009). *Urban poverty and vulnerability in Kenya. Background analysis for the preparation of an Oxfam GB urban programme focused on Nairobi*. Retrieved from http://www.irinnews.org/pdf/urban_poverty_and_vulnerability_in_kenya.pdf

83. Rothmyer, K. (2011, March 17). Hiding the real Africa: Why NGOs prefer bad news. *Columbia Journalism Review*. Retrieved from http://www.cjr.org/reports/hiding_the_real_africa.php?page=all

84. Shrum, W., Mbatia, P. N., Palackal, A., Dzorgbo, D., Duque, R., & Ynalvez, A. M. (2011). Mobile phones and core network growth in Kenya: Strengthening weak ties. *Social Science Research, 40*(2), 614–625.

85. Silverstone, R. (1994). *Television and everyday life*. London: Routledge.

86. de Smedt, J. (2009). Kill me quick: A history of Nubian gin in Kibera. *The International Journal of African Historical Studies, 42*(2), 201–220.

87. Tacchi, J., Kitner, K. R., & Crawford, K. (2012). Meaningful mobility: Gender, development and mobile phones. *Feminist Media Studies, 12*(4), 528–537.

88. Toivanen, H. (2011). From ICT towards information society. *Policy strategies and concepts for employing ICT for reducing poverty*. Retrieved from: http://www.vtt.fi/inf/pdf/workingpapers/2011/W158.pdf

89. Toyama, K. (2011, February 8–11). *Technology as amplifier in international development*. Proceedings of the 2011 iConference, Seattle, WA, pp. 75–82, ACM.

90. Versi, A. (2010). Making the connection. *Msafiri*, pp. 88–96.

91. Waema, T., & Miroro, O. O. (2014). Access and use of ICT and its contribution to poverty reduction in Kenya. In O. E. Adera, T. Waema, J. May, O. Mascarenhas, & K. Diga (Eds.), *ICT pathways to poverty reduction: Empirical evidence from East and Southern Africa* (pp. 102–131). Ottawa: Practical Action, IDRC.

92. Wallis, C. (2011). Mobile phones without guarantees: The promises of technology and the contingencies of culture. *New Media & Society*, *13*(3), 471–485.

93. Wallis, C. (2013). *Technomobility in China: Young migrant women and mobile phones*. New York, NY: New York University Press.

94. Ward, K. (2005). The bald guy just ate an orange. In T. Berker, M. Hartmann, Y. Punie, & K. Ward (Eds.) , *Domestication of media and technology* (pp. 145–163). Maidenhead: Open University Press.

95. World Bank. (2010, December). *Kenya economic update: Kenya at the tipping point? With a special focus on the ICT revolution and mobile money*. Retrieved from http://siteresources.worldbank.org/KENYAEXTN/Resources/KEU-Dec_2010_Powerpoint.pdf

96. Wyche, S. P., Forte, A., & Schoenebeck, Y. (2013, April). Hustling online: Understanding consolidated Facebook use in an informal settlement in Nairobi. In *Proceedings of the SIGCHI conference on human factors in computing systems* (pp. 2823–2832). New York: ACM. Retrieved from http://dl.acm.org/citation.cfm?id=2481391

97. Wyche, S. P., Schoenebeck, S. Y., & Forte, A. (2013, February). *Facebook is a luxury: An exploratory study of social media use in rural Kenya*. Proceedings of the 2013 conference on Computer supported cooperative work, pp. 33–44, ACM. Retrieved from http://dl.acm.org/citation.cfm?id=2441783

98. Wynn, D.Jr., & Williams, C. K. (2012). Principles for conducting critical realist case study research in information systems. *MIS Quarterly*, *36*(3), 787–810.

APPENDIX. DEMOGRAPHIC PROFILE OF THE SAMPLE

Name, age and gender	Education level	Profession	Village
1. Francis (28) (M)	University	Part-time university student NGO work Small-scale business	
2. Sam (28) (M)	University	NGO work	Karanja
3. Diane (20) (F)	College	Part-time college student Family business worker at bar and mobile money shop	Kianda
4. Moha (19) (M)	High school	Youth group coordinator Family business work	Kianda
5. Coby (24) (M)	College	Full-time college student Self-help group participant	Kianda
6. Rythmix (28) (M)	College	Barber shop owner DVD shop owner Reggae artiste Volunteer in self-help group	Makina
7. John (21) (M)	University	Full-time university student Local youth politician	Makina
8. Naima (20) (F)	University	Full-time university student Volunteer at self-help group	Makina
9. Susan (28) (F)	College	Volunteer at local groups Part-time jobs when available	Fort Jesus (Olympic)
10. Brian (28) (M)	Primary school	Gym instructor Manual part-time jobs based on availability- such as cleaning and	Gatwekera

		building	
11. Felista (20) (F)	University	Full-time university student	Kianda
12. Pat (20) (F)	University	Full-time university student	Kianda
13. Liza (25) (F)	University	Videographer, designer and part-time university student	Ayany
14. Tracy (23) (F)	University	University student and waitress	Makina
15. Neema (23) (F)	High school	Adult literacy courses Housewife	Kianda
16. Norah (18) (F)	Primary school	Assistant at family-owned grocery shop	Kianda
17. Jackson (27) (M)	College	Slum tour guide Self-help group coordinator Activist	Makina
18. Collo (25) (M)	High school	PlayStation shop owner	Karanja
19. Paul (19) (M)	High school	Self-help group volunteer Actor	Kianda
20. Tom (22) (M)	Primary school	Any job that is available (Hustling)	Kianda
21. Karen (21) (F)	College	Hustling	Kianda
22. Dave (28) (M)	College	Job based on availability Youth group member Media trainee	Olympic

Key:

(F) For female

(M) For male * Pseudonyms are used to protect the identities of the residents due to ethical considerations of privacy

8

CHAPTER

There iIs an App for That? The Impact of Community Knowledge Workers in Uganda

Bjorn Van Campenhout *

Development Strategy and Governance Division, International Food Policy Research Institute, Kampala, Uganda

ABSTRACT

Rapidly expanding mobile network coverage in developing countries offers new ways to reach poor farmers in isolated places. This article explores the potential for Information and Communication Technology to provide agricultural information and extension services to smallholder farmers, and links this to empirical insights of an intervention in Uganda, where community knowledge workers rely on mobile devices to deliver context- and time-sensitive data to farmers. Consistent with theoretical insights related to information inefficiencies, we find that the intervention leads farmers to move away from low-risk low-return crops toward more commercially oriented commodities. Our analysis also suggests that, for the case of maize, the intervention causes farmers to sell less on the market, but at significantly higher prices.

Keywords: Agricultural extension, market information, Uganda, mobile applications

INTRODUCTION

The use of Information and Communication Technology (ICT) for development has attracted considerable attention in development policy and research (Aker & Mbiti, 2010; Jensen, 2010). There are different reasons as to why expectations are high. Mobile phone networks, and with it mobile phone-based services such as mobile money, are expanding extremely fast in developing countries. Large transaction costs due to, for example, poor road infrastructure give mobile-based technology a competitive edge. Portable devices are also well adapted to situations where power supply is erratic. Increasingly, both private and public initiatives start to use this infrastructure to provide services to their clients.

Within this context, the Grameen Foundation, a global nonprofit organization, initiated an innovative project to deliver agricultural extension and marketing information to smallholder farmers in Uganda. They equipped locally recruited villagers (referred to as community knowledge workers (CKWs)) with an Android smartphone, pre-loaded with an in-house developed mobile application. CKWs can use this 'app' to search for up-to-date and location-sensitive information related to farming and commodity marketing. In this way, the Grameen Foundation aims to build a scalable network of resident rural information providers who use smartphones to help close critical information gaps faced by smallholder farmers.

In this article, we explain how ICT-based initiatives such as the CKW project impact upon the livelihoods of small-scale farmers in developing countries. To do so, we first describe the particular challenges smallholder farmers face by drawing on the theory of information inefficiencies, and explain how ICTs can address some of the issues. We then present an empirical assessment of the causal impact of the CKW intervention, which was designed to mitigate some of the information inefficiencies among smallholder farmers in Uganda. Consistent with the theory, our findings suggest that the CKW initiative led to a change in the crops that farmers grow, shifting to more risky but rewarding crops. Our analysis also suggests that equipping farmers with actionable marketing information is associated with an average increase in the price at which farmers sell maize of about 12%. The analysis also suggests important spillover effects. We do not find an effect on maize productivity.

The article is organized as follows. The next section presents the theory in which the CKW model is grounded, and reviews the literature on the impact of other ICT-based interventions that use a similar rationale. We then present the CKW model, explain the identification strategy and describe the project area and data used. Next, we present the results and discuss them in the light of our theoretical framework. A final section concludes.

The potential of ICTs to remove information inefficiencies

On the face of it, much of the observed behavior of smallholder farmers in developing countries is difficult to comprehend using standard economic theory. For instance, textbook economics would predict that farmers specialize in the crops in which they have a comparative advantage, which is at odds with the observation that most smallholders have similar, yet diversified crop portfolios. In addition, it has been noted that farmers often respond to price signals in ways that differ from those predicted using standard micro-economic demand and supply analysis (de Janvry, Fafchamps, & Sadoulet, 1991). Despite low levels of agricultural productivity, many studies find a reluctance in the adoption of profitable productivity-enhancing technologies. However, Schultz (1964) notes that, given the constraints these farmers face and the environment in which these farmers operate, farmers are surprisingly efficient. Subsequent work on agricultural household models that explicitly accounts for the dual role of farm households as both producers and consumers of often the same consumables and production factors has led to a substantial literature that is able to explain smallholder behavior as a rational response to missing, incomplete or inefficient markets (Singh, Squire, & Strauss, 1986).

We focus on four stylized characteristics of smallholder farming behavior that are important constraints to more general economic development: crop portfolios that are skewed to low-risk low-return crops, low prices accepted by farmers for their produce, under-adoption of improved agricultural technologies and low levels of market participation. In this section, we will discuss each of these factors in turn, explaining the theory pertaining to the causes and consequences of this behavior, reflecting on how ICTs may affect these causes and consequences, and reviewing the evidence so far on the effectiveness of ICTs to change this behavior.

Crop choice, risk and ICT

Smallholder farmers engage in activities of an inherently risky nature. Income derived from agriculture is dependent on climatic events. In addition, in many developing countries farmers have to cope with policy shocks (Dercon, 2002). Farmers also tend to live in disease-prone environments, affecting both labor and livestock. Smallholder farmers also face considerable price uncertainty, exacerbated by shallow markets and high seasonality (Van Campenhout, Lecoutere, & D'Exelle, 2015). With formal insurance lacking, farmers have to rely on informal ways to protect themselves against downside risk. Risk sharing is difficult due to the covariance of risk in agricultural societies and so most farmers rely on self-insurance.

One such self-insurance risk management strategy is through the composition of the crop portfolio. Dercon (2002) notes that a common strategy for households is to alter the income risk they face through diversifying their income sources. Since agriculture is the main source of income for smallholder farmers, this behavior is replicated at the crop choice level. In addition, there is generally a positive relationship between returns to a crop and its riskiness, resulting in some degree of income skewing among the poorest, where poor farmers specialize in low-risk low-return activities (Dar, de Janvry, Emerick, Raitzer, & Sadoulet, 2013). For instance, Dercon (1996) finds poor Tanzanian households grow proportionally more sweet potatoes, a typical low-return low-risk crop.

However, crop choice cannot be viewed independently from other risk management or risk-coping activities. For example, if a household has built up sufficient liquid assets to cope with the consequences of risk ex-post, households may be able to take on more risky activities with a higher expected return. The intricate relationship between different risk management strategies means that changes in the crop portfolio will depend on other options available to farmers to better manage ex-ante risk. This suggests a range of solutions to pull poor farmers out of their low-risk low-return cropping patterns.

Offering formal insurance, such as index-based weather insurance, is likely to encourage farmers to shift into more risky activities (Karlan Osei, Osei-Akoto, & Udry, 2014). Credit is also assumed to be an important risk management strategy, increasing households' ability to shift into riskier forms of agriculture (Eswaran & Kotwal, 1989; Udry, 1990). Precautionary savings have also been found to have a direct effect on crop choice as a risk management strategy (Dercon, 1996). Income diversification into activities in which returns are uncorrelated with returns to agricultural activities also influences a household's ability to deal with risk. Investments in risk-reducing technology and infrastructure, both at the individual and at the more aggregate level, are likely to affect optimal crop allocation. Examples are better storage technologies at the farm level (Ricker-Gilbert & Jones, 2015), or better feeder roads at the more aggregate level (Dorosh, Wang, You, & Schmidt, 2012).

One can think of various ways in which ICTs can alter the above ex-ante risk management strategies available to households. ICTs may facilitate access to agricultural knowledge and best practice through call centers or information presented online. Foster and Rosenzweig (2010) model technology adoption through extension, and show how learning can reduce uncertainty about the profitability of a new technology. Access to price information in other markets, as well as contact information of traders and agro-processors, is likely to substantially reduce price risk. Weather forecasts will also affect the (perceived) riskiness of crops. ICTs may also make risk coping through sale of assets more effective. When common shocks occur, many households may decide to sell off assets at the same time, leading to a collapse of the market. This negative correlation between food prices and asset prices greatly reduces the effectiveness of savings as a risk management strategy (Dercon, 2002). Providing information of markets in regions not affected by the shock may prompt arbitrage, attenuating the price movements and correlation (Van Campenhout, 2007).

Prices, bargaining power and ICT

Farmers located in rural areas often receive very low prices for the crops they sell (Stifel & Minten, 2008). This is partly because these farmers are often poorly informed about prevailing market prices of the crops they produce (Fafchamps & Hill, 2008). In fact, most of the trading involves a bilateral bargaining process, which may result in prices that differ significantly from prevailing market prices. Furthermore, rural markets, where most of the production happens, are often only weakly integrated with urban consumption markets, and high transaction costs of moving crops from supply areas to consumer centers mean prices in towns are often substantially higher.

ICTs provide a suitable platform to disseminate location- and time-specific price information. Such information would enable farmers to engage in spatial arbitrage. The reduction in search costs means that farmers can now choose among a larger set of markets and sell their product in the market where demand is highest and supply lowest, resulting in a better price (net of additional transaction costs incurred) than the one he or she would obtain if he or she would just sell in the nearest market. Jensen (2010) notes that such a process of arbitrage has welfare effects beyond those farmers that engage in arbitrage, generating higher prices for producers and lower prices for consumers.

Price information may also affect the distribution of welfare among agents. In most cases, farmers do not sell directly to consumers. Agricultural commodity value chains typically include assembly traders, brokers, wholesalers, retailers and agro-processors. Along this chain, it is likely that barriers to entry exist, leading to localized monopsonies. This may result in lower prices for farmers. The ability of ICTs to reduce search costs effectively increases the amount of competition between actors along the value chain. In addition, empowering the farmer with price information at different points along the value chain increases the bargaining power of the farmer. Often, the mere threat of doing so will be sufficient to make middlemen increase their reservation price.

There are a number of studies that investigate the effect of ICT on overall market efficiency and arbitrage. Jensen (2007) finds that the introduction of cell phones in Indian fishing communities reduces price dispersion between markets and eliminates waste, consistent with increased spatial arbitrage. Aker (2010) finds that the introduction of mobile phones increases grain market integration in Niger, while Aker and Fafchamps (2015) find that producer price dispersion decreases after the introduction of mobile phones, but only for perishable crops. There are also quite a number of studies on the effect of ICT use for price information provision on surplus sharing. Svensson and Yanagizawa (2009) find large effects of price information delivered through radio, resulting in a 15% increase in the price farmers receive. Goyal (2010) looks at the provision of price information using Internet kiosks in the context of Indian soybean growers who sell to traders in wholesale markets. She finds that price information translates to a 1–3% increase in farmer prices. Fafchamps and Minten (2012) investigate the impact of an SMS-based price dissemination service in India and find no effect on the price reportedly received. Courtois and Subervie (2015) find that price information disseminated though the well-known eSoko project resulted in a 10% increase in the price of maize and 8% increase in the price of groundnuts in Northern Ghana. Futch and McIntosh (2009) do not find price effects resulting from the introduction of mobile phones in Rwanda.

Technology adoption, knowledge gaps and ICT

Another characteristic of smallholder farmers is the under-adoption of yield-increasing investments amid low and stagnating agricultural yields, despite the fact that these investments seem to be highly profitable (Duflo, Kremer, & Robinson, 2008). Experts believe a lack of knowledge about (improved) crops and farming techniques is an important constraint. It is argued that if this knowledge gap can be narrowed through agricultural extension, this may lead to significant improvements in rural welfare.

We see different areas where ICTs may outperform traditional extension services. One of the main threats to sustainability of agricultural extension services is high recurrent costs (Quizon, Feder, & Murgai, 2001), and mobile technology provides a cost-effective way to transfer information to remote locations (Nakasone, Torero, & Minten, 2014). Anderson and Feder (2004) mention that timeliness of information is also key to the success of extension systems. Again, ICTs may be particularly suited to disseminate reminders at particular points in time through, for example, SMSs. But also access to consumer markets is important for the sustainability of improved technology adopted through agricultural extension, as in better integrated markets, returns to increased output diminish less rapidly than in locally segmented markets characterized by more price inelastic demand (Gabre-Madhin, Barrett, & Dorosh, 2002). The use of ICTs to obtain price information mentioned above may thus also affect technology adoption.

There appears to be relatively less ICT-based initiatives that provide information on agricultural practices and inputs, compared to those that disseminate information on prices, weather forecasts, or buyer and seller information. Aker (2011) suggests this may be because such information is more nuanced and difficult to convey. To our knowledge, there are no studies as yet that investigate the effect of extension information delivered through ICTs on agricultural knowledge, practices and outcomes.

Market participation, transaction costs and ICT

Finally, high levels of production for subsistence, or low levels of market participation, are also a hallmark of smallholder farmers in the developing world. This feature can be explained by appreciating the particular nature of farm household models and the context in which they operate. Farm households both produce and consume commodities. Production decisions are made on the basis of profit maximization given available technology, and consumption decisions are made on the basis of utility maximization subject to a budget constraint. If a full set of markets exists, all production factors and consumption items can be traded at a marginal cost, which results in a clear separation between consumption and production. Renkow, Hallstrom, and Karanja (1986) show that, when markets fail, separability breaks down, and consumption decisions may constrain decisions on the production side.[1] de Janvry et al. (1991) explain how transaction costs drive a wedge between prices at which farmers sell and buy, effectively making markets selectively fail for certain products and certain households. The above leads to particular market participation patterns, where some households self-select out of markets (Key, Sadoulet, & de Janvry, 2000; Renkow et al., 2004).

ICTs can have a substantial impact on market participation patterns by reducing transaction costs. By reducing the price bands, the likeliness that markets fail for particular goods and households reduces. Price information in spatially distinct markets also dramatically reduces search costs as farmers do not have to physically go to other markets to discover the price.

Transportation costs are likely to be lower when one can choose between many more transport service providers.

To our knowledge, there is currently only one study that directly looks at the effect of ICTs on market participation. Muto and Yamano (2009) use panel data to investigate if the introduction of cell phone coverage in Uganda increases market participation. They find that after the expansion of coverage, sales of banana in remote communities increased, but not for maize. From this, they conclude that mobile phone coverage expansion seems to induce market participation of farmers in remote areas who produce perishable crops.

The Grameen Foundation's CKW project in Uganda

The CKW initiative was designed to improve the lives of smallholder farmers by improving access to agriculture-related information and extension, delivered through resident community members. In facilitated sessions, villagers are asked to choose an individual from their community, who they deem suitable to take up the role of a CKW. This person is then screened by Grameen, provided with an Android smartphone and trained on how to use it. The smartphone is preloaded the CKW Search app, which allows CKWs to look up information requested by farmers about farming and crop marketing. These questions range from local weather information over market prices to crop and livestock management.

The CKW Search application was developed in Uganda at Grameen's Applab and provides information that can broadly be classified into four categories. First, up-to-date prices for different crops in different locations can be requested. Second, three-day weather forecasts, along with seasonal forecasts, can be retrieved. Third, there is an extensive knowledge base on farmer best practices. Finally, the Uganda National Agro-Inputs Dealers' Association provides information of farm input suppliers across the country that includes their locations down to sub-counties and contact information. There is also an extensive directory of traders, brokers, transporters and agro-processors that can be accessed using the app.

Research methods – difference-in-difference and fixed effects

To investigate if the CKW works as intended, simply comparing outcomes before and after the introduction of a CKW does not suffice. Over time, many other factors may have changed as well, making it impossible to disentangle the effect of the intervention from other factors such as, for instance, the weather. However, comparing outcomes from famers who were exposed to CKWs (henceforth also referred to as the 'treated' group) to that of farmers who have not been exposed to CKWs (a 'control' group) at one point in time is also not satisfactory: unless the intervention was randomized, chances are that households with particular characteristics have a higher chance of ending up in one of the two groups. For example, it may be that Grameen Foundation tries to optimize the expected benefits of the CKW project by searching for locations that are particularly isolated from the rest of the world. In this case, the treated group is likely to be disadvantaged to begin with. The fact that a particular project will 'attract' particular groups is known as selection bias.

We will use a difference-in-difference (DD) estimator, which controls for selection bias by comparing the change in the outcome variable over time in the control group to the change in the outcome variable over time in the treatment group (Angrist & Pischke, 2008). In other words, the approach looks at the difference in outcomes between treatment and control, but it accounts for the fact that the outcomes may have been different to begin with. Therefore, the method needs observations on the outcome variable before the intervention (a baseline) and after the intervention (an end line) for both a treated group and a control group. If the treatment group has fared differently over time from the control group, this is interpreted as an impact.

Central to the DD approach is the parallel trend assumption. This assumption states that the group that received the intervention would have followed the same trend over time than the control group if the intervention would not have occurred. While we cannot test this assumption directly, we will increase the credibility of DD in two ways. First, we try to choose a control group that is sufficiently similar to the treatment group by sampling control villages from areas reasonably close to the treatment villages to ensure they are the same in terms of market access, climate, geology, etc. We also statistically test if the treatment and control population differ in a range of characteristics (Table 2). Second, we add more time periods to our DD (Angrist & Pischke, 2008). In particular, instead of just using one pretreatment baseline, we add a second one. If we find that the difference in outcome for the control group between the first pretreatment and the second pretreatment baseline is not significantly different from the difference in outcome for the treatment group between the first pretreatment and the second pretreatment baseline, we may feel more comfortable with the assumption of a parallel trend.

One of the attractive features of a DD study is that it does not necessitate panel data (Meyer, 1995). Since one assumes the potential omitted variable bias works at the treatment level (the treated group versus the non-treated group), one can simply compare the aggregates (e.g. the means or medians of the two groups). This can be done with repeated cross sections, as long as the before and after control groups are comparable. While repeated cross-sections reduce on tracking costs and eliminate the problem of panel attrition, it prevents us from controlling for household-specific time-invariant (potentially unobservable) effects.

Above, we defined the intervention as the mere presence of a CKW at the parish level and selection bias is likely to work mostly at a more aggregate level. However, within the treatment group, we will also look at treatment intensity (i.e. do households that interact more with a CKW have different outcomes from those that interact less), where the selection bias may be situated at the household level. For instance, it may be that certain households are more likely to contact the CKW because of individual-specific characteristics (e.g. a particular farmer likes the stay current on local weather conditions, because his deceased father told him how he lost half of his crop due to an adverse weather shock). To control for this, we will estimate a fixed effects (FE) regression with household-level FEs. It will also be informative to compare the results of a specification that controls for household specific effects to the DD specification, as this will allow us to say something about spillover effects.

Study areas and data sources used

We decided to sample from three districts in Uganda, one in the East and two in the North. In the Eastern region, CKWs were trained and introduced in Kapchorwa district in 2010. Kapchorwa is on the slopes of Mount Elgon, bordering Kenya, but also covers the extensive Kapchorwa plains. In the Northern region, the project started in October 2010 in Gulu district and in December 2010 in neighboring Oyam district. Grameen collected data on a variety of outcomes before the intervention (i.e. data on the 2009/10 agricultural season), and we revisited a random subsample of these households. We revisited 137 households in Kapchorwa and 133 in Gulu and Oyam.

For our DD estimator, we also need information on people who have not been exposed to the intervention. These control districts were selected such that they resemble the population that received a CKW intervention as closely as possible. As one control area, we selected Sironko, a neighboring district that shares the same geology as Kapchorwa. We sampled from Lira district to assemble a control group that resembles as closely as possible that of Gulu and Oyam. The geographical location of the districts is illustrated in the map in Figure 1. We interviewed 142 households in Lira and 158 households in Sironko.

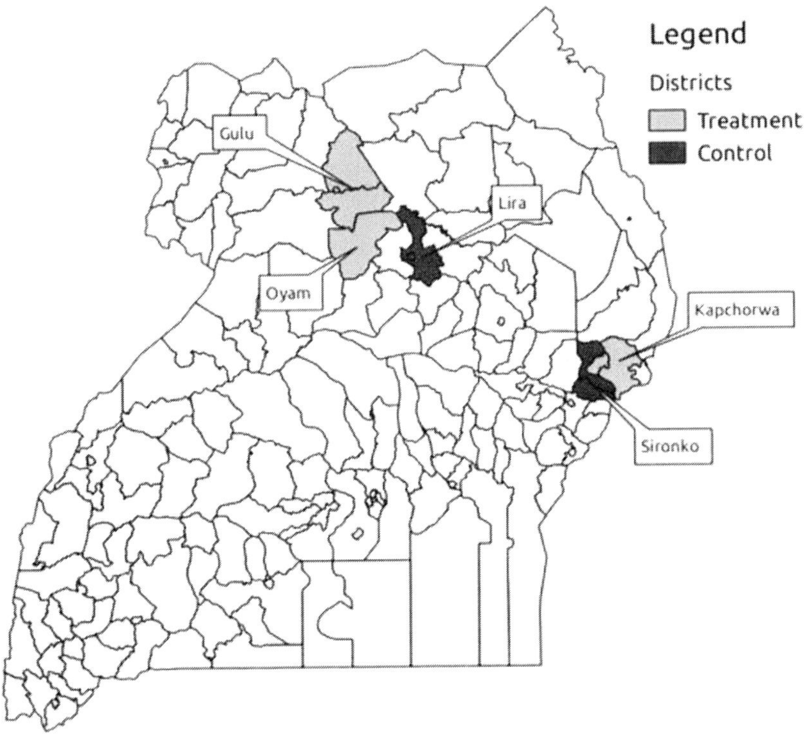

Figure 1. Sampling areas used.

Unfortunately, the baseline survey Grameen did was only done in the treated population and so we have no baseline data in the control areas. Bamberger (2009) describes different techniques for the reconstruction of baseline data and argues that the use of secondary data drawn from representative household surveys provides the most credible pretreatment counterfactual. The Uganda National Household Survey (UNHS) of 2005/2006 covers more than 6000 households nationwide. It is collected by the Uganda Bureau of Statistics with support from the World Bank. The UNHS interviewed 136 households in Lira and 52 households in Sironko, which we update to 2009/2010 to match the baseline data from the Grameen.[2]

To assess the robustness of our findings, we also construct an alternative baseline for which we rely entirely on the UNHS 2005/2006. That is, we also use outcome variables from the UNHS 2005/2006 as the baseline for the districts that were treated (Gulu, Oyam and Kapchorwa), instead of using the data collected by Grameen on the 2009/2010 season. As such, we have a baseline treatment-control sample taken from the UNHS 2005/2006 and an end-line treatment-control sample which we collected ourselves in 2012. We thus end up with three waves of cross-sectional data in both treatment and control areas. Two of these are from before the intervention (one done in 2005/2006 and one done in 2009/2010) and one is from after the intervention (the one documenting the 2011/2012 agricultural season). Table 1 shows the sample sizes by region, treatment regime and time of the survey.

Table 1. Sample design.

North		East		Total			
Gulu/Oyam (T)	Gulu (C)	Kapchorwa (T)	Sironko (C)	T	C	Total	
2005/2006	294	136	36	52	330	188	518
2009/2010	1147	136	842	52	1989	188	2177
2011/2012	133	142	173	158	306	300	606
Total	1574	414	1051	262	2625	676	3301

Table 2 lists some key household characteristics of the farm households that were included in our UNHS baseline survey. For most of the characteristics, there is no significant difference between farm households that will be treated and those that will serve as control groups. Some of these characteristics will also be used in the DD regressions with additional controls below.

Table 2. Means of key variables.

Control	Treatment	Test statistic	*p*-value	
household size	6.13	6.49	−1.36	.174
rural (%)	76.50	83.63	3.63	.057
male headed (%)	77.04	71.64	1.60	.205
age head	42.12	43.17	−0.80	.421
head no formal education (%)	41.81	41.33	−0.09	.985
no of rooms in house	4.24	3.99	−0.13	.198
land size	2.92	4.08	−3.04	.002

RESULTS

We will now estimate the impact of the CKW intervention on four outcome variables informed by the theoretical framework (crop mix, productivity, marketing and price). We will start by estimating DD models. These models will then be augmented with other control variables to check the robustness of our findings. Finally, we will look at treatment intensity and spillover effects.

Difference in differences

Impact on crop portfolio

We first investigate if the introduction of a CKW in the communities leads to changes in the crops that farmers grow. In order to answer this question for a range of important agricultural commodities, we estimate the change in the percentage of households that report growing a particular crop over time for the households that did receive a CKW. This difference is then contrasted to the change in percentage over time for the sample of farmers that did not receive information and extension from CKWs. These DDs for a range of crops are displayed as bar charts in Figure 2, with the whiskers indicating the standard errors of the estimates.

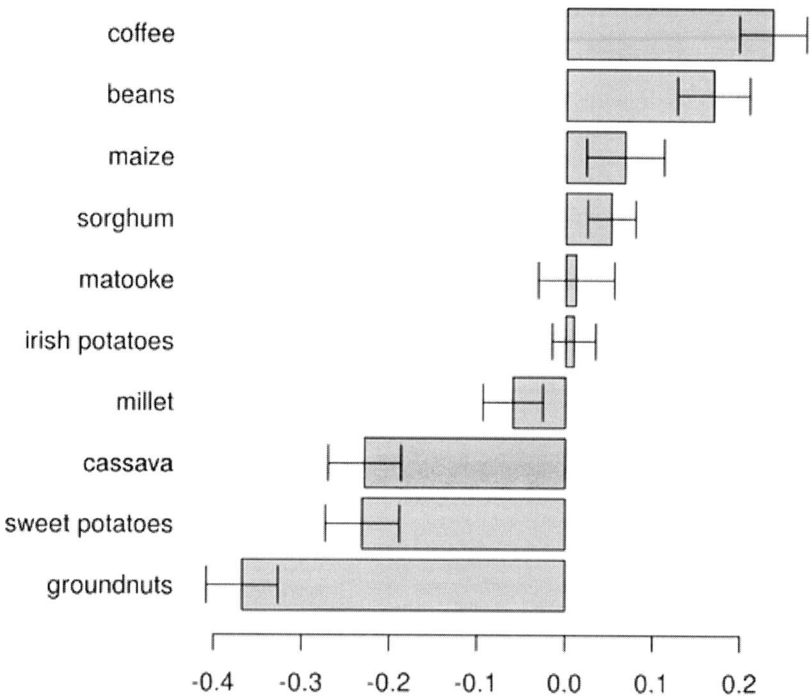

Figure 2. Change in proportion of farmers reporting growing crop attributed to community knowledge workers.

The results suggest that access to information and extension through CKWs leads to a significant increase in the percentage of farmers growing coffee, beans, maize and sorghum. The importance of millet, cassava, sweet potatoes and groundnuts seemed to have declined as a result of the CKW intervention. No statistically significant change in the share of farmers growing matooke and Irish potatoes was observed. The observed pattern suggests a tendency to move away from low-risk low-return crops to more risky but more commercially oriented crops.

It may seem that the increase in the importance of sorghum in the farmer's crop portfolio runs against the general trend of farmers moving away from low-risk low-return crops to more commercially oriented crops. However, more and more, sorghum is becoming valued as a key ingredient in the commercial production of beer. For instance, in 2003, Uganda's Nile Breweries, which is owned by SABMiller, the world's second-largest brewer, introduced Eagle, a beer that is brewed from sorghum (Economist, 2003). The weak Ugandan shilling and high excise duties, as well as a new brewing method that avoids the long and costly malting process, give locally sourced sorghum a clear advantage over imported barley. The resulting lager is two-thirds cheaper and closer to the traditional beer Ugandans are accustomed to, reportedly making it a huge success. This development may have increased the commercial value of a traditional food security crop in Uganda in recent years (Ebiyau, Arach, & Serunjogi, 2005).

Changes in maize productivity

Secondly, we examine if CKWs have improved maize productivity. To do so, we compare amounts harvested (in kg) per area of cultivation (acres) before and after the intervention, both for a sample of treated farmers and a sample of farmers who make up the control group. The results for our analysis are presented in Table 3.

Table 3. The effect of CKW on maize productivity.

Dependent variable: productivity				
(1)	(2)	(3)	(4)	
treat	116.039**	86.002*	144.867***	174.793***
	(45.447)	(50.500)	(51.832)	(56.866)

Dependent variable: productivity				
(1)	(2)	(3)	(4)	
mid				169.505**
				(66.447)
post		5.111	219.726***	180.884***
		(59.172)	(52.252)	(56.930)
east	385.710***	497.201***	355.947***	477.575***
	(50.741)	(22.622)	(35.466)	(20.635)
treat:mid				−87.820
				−74.888
treat:post		29.878	−28.786	−58.886
		−68.415	−65.532	−72.173
Constant	381.020***	295.451***	182.773***	133.842***
	(48.746)	(49.640)	(44.721)	(47.713)
Observations	389	1711	668	1,990

Dependent variable: productivity				
(1)	(2)	(3)	(4)	
R-squared	0.143	0.232	0.265	0.248
Adj. *R*-squared	0.138	0.230	0.261	0.246
Residual std. error	448.172	455.162	395.329	438.248
F statistic	32.183***	128.694***	59.889***	109.270***

Note: Standard errors are in parentheses.

**p* < .10, ***p* < .05, ****p* < .01.

The first model in the table only uses end-line data and is presented as a reference. It not only estimates the difference between treated and control farmers in maize productivity, but also controls for location (East or North). We find that farmers in control villages in the North have an average yield of 381 kg per acre. In the East, average productivity is significantly higher, as was expected by the better agro-ecological conditions near Mount Elgon. Here, farmers harvest on average 766.73 kg of maize from one acre. More important, however, is that we also find a significant and positive additional treatment effect: living in a treated area increases maize productivity by 116 kg per acre.

The second model (2) in Table 3 presents a basic DD model. It regresses the outcome variable (here maize productivity) on a constant, a treatment dummy, an end-line dummy and an interaction term of the treatment and the end-line dummies.[3]3. We also control for region in all specifications.View all notes The second model differs from the third model only in the choice of the baseline data source. While the second model uses the 2009/2010 agricultural season as the baseline, the third model uses the

2005/2006 agricultural season as the baseline. The results, regardless of what baseline data are used, suggest that the difference in productivity found in model (1) is not caused by the CKW intervention, reflected in the fact that the coefficient on the treatment and end-line interaction dummy is not significantly different from zero. The treated population has, on average, a higher productivity than the control population, irrespective of the timing of the measurement. Depending on what baseline has been used, productivity in the treated group is 86 or 145 kg per acre higher to begin with than productivity in the control group. We do, however, find a significant and positive trend in maize yields if we look at differences over the 2005/2006 to 2011/2012 period. But this increase is general and hence cannot be attributed to the CKW intervention.

The fourth model combines the two different baseline data sets into a DD model with multiple pre-intervention time periods. The dummy variable *mid* takes the value of one if the observation comes from the 2009/2010 baseline, and zero otherwise. The dummy variable *post* takes the value of one if the observation comes from the 2011/2012 end line, and zero otherwise. A credible treatment effect would manifest itself through a posttreatment interaction that is significantly different from zero together with a mid-treatment interaction that is not significantly different from zero. However, we find no significant treatment effect for maize productivity. Irrespective of the time period, treated farmers always have a higher average maize productivity than farmers in the control group. Also, while average productivity was significantly higher in both 2009/2010 and in the end-line data compared to the 2005/2006 baseline, these increases are not significantly different between the control and treatment samples.

There may be different reasons why we fail to find a significant effect on maize productivity. First of all, measuring productivity is notoriously difficult. It requires estimates made by the farmer on the area that is planted with maize and the amounts harvested. Both these quantities may be subject to substantial measurement error, reducing precision and making it more difficult to detect a statistical significant effect (Aker, 2011). In addition, it will take some time before the impact of extension information is reflected in productivity measures. For example, it may take several years of fertilizer application before soil regains sufficient nutrients and micro-organisms to lead to significantly higher yields. Our end-line survey was done only two years after the intervention.

Changes in share of maize sold

We now ask ourselves if the increased extension and marketing information brought by the CKWs have changed farmers' participation in the market as suppliers of maize. Table 4 shows the results for the share of the total harvest that is sold. As in the previous section, the first model compares the post-intervention situation of farmers who have access to a CKW to those who have not, controlling for region. We see that, on average, farmers sell about 38% of the maize they harvest. There seems to be no significant difference between farmers in the North and farmers in the East. In addition, there also seems to be no impact of the CKWs, as the difference between treated and control populations is not significantly different from zero.

Table 4. The effect of CKW on maize on the share of maize harvest sold.

Dependent variable: share sold				
(1)	(2)	(3)	(4)	
treat	-4.322	4.684	10.509**	12.054***
	(3.032)	(3.636)	(5.159)	(4.215)
mid				0.000
				(5.103)
post		20.834***	13.874***	20.523***
		(4.008)	(5.121)	(4.052)
east	1.619	-12.075***	0.561	-11.510***
	(3.237)	(1.091)	(2.707)	(1.070)

Dependent variable: share sold				
(1)	(2)	(3)	(4)	
treat:mid				−7.552
				(5.585)
treat:post		−9.517**	−14.871**	−16.865***
		(4.345)	(5.941)	(4.852)
Constant	38.029***	26.742***	24.893***	26.660***
	(3.155)	(3.571)	(4.427)	(3.612)
Observations	372	1,827	526	1,981
R-squared	0.006	0.097	0.017	0.092
Adj. *R*-squared	0.001	0.095	0.009	0.089
Residual std. error	29.165	22.841	28.234	23.107
F statistic	1.171	48.969***	2.199*	33.265***

Note: Standard errors are in parentheses.

$^{*}p < .10$, $^{**}p < .05$, $^{***}p < .01$.

Models (2) and (3) provide estimates for standard DD regressions. As before, model (2) uses 2009/2010 data as the baseline and model (3) the UNHS 2005/2006. We now find a positive common trend. Between the surveys before the intervention and the survey after the intervention, the share of maize sold has increased by 13–20 percentage points, depending on which baseline is chosen. More importantly, we find a significant and negative treatment effect. The presence of a CKW reduces the share of the harvest of maize that is sold in a significant way. There is some indication that the treated population sold a larger share than the control population before the intervention, and that this difference if offset by the intervention.

The final model again looks at all available data in one nested model. We find that there is a significant increase in the share of maize sold over time, but this effect is entirely attributed to the period between 2009/2010 and 2011/2012. In addition, there seems to be evidence of a selection effect, as treated farmers have a significantly higher share of sales. Most importantly, we find again that the CKW intervention reduced the share of the maize harvest sold on the market. As anticipated, the interaction between the indicator for the baseline of 2009/2010 (mid) and treatment status is not significant, making the parallel trend assumption plausible. However, the interaction between the dummy variable indicating the end-line and the treatment status is significantly negative.

The fact that we find a negative treatment effect on maize market participation runs counter to what we expected given the theory. As explained above, the CKW intervention is expected to decrease transaction costs, making market participation (both selling and purchasing) more likely. There are different possible explanations for why we find a negative impact on the share of maize sold. The decision of a particular household to participate in the market for a commodity depends not only on the price band created by the transaction costs, but also on the location of the household-specific shadow price of the commodity (Mather, Boughton, & Jayne, 2013). A decrease in the transaction cost may be offset by an increase in the shadow price, due to an outward shift of the household's demand schedule or an inward shift of the household's supply. Within the context of household agricultural models, it may be that the CKW intervention relaxes the budget constraint (for instance, through increased marketing of another commodity such as coffee), which may shift the demand curve for maize outward, leading to less market participation. In addition, Renkow (1990) shows how an inventory creates effects that are similar to wealth effects, again likely to shift consumption outward. Such an explanation is consistent with the findings of Muto and Yamano (2009), who find that the introduction of mobiles in Uganda increases market participation for bananas, which is perishable, but not for maize.

Price received for maize sold

Farmers were asked about the highest price they received for maize sold over the last agricultural year. The results of four regression models similar to the ones for maize productivity and shares sold are reported in Table 5. The first model again only looks at end-line data. We find that, at the time of the end line in 2011/2012, the average price was about UGX100 higher in the Mount Elgon area than in the North, irrespective of the treatment status. Thus, farmers in the control population received on average UGX513 per kg of maize sold in the North, and UGX613 in the East. We also find a significantly positive effect of the treatment status. Treated farmers received on average about UGX91 per kg of maize more.

Table 5. The effect of CKW on the price received for maize.

Dependent variable: price received				
(1)	(2)	(3)	(4)	
treat	91.831***	-9.505	18.435	1.344
	(24.982)	(11.649)	(16.652)	(17.398)
mid				207.898***
				(17.106)
post		188.757***	359.635***	395.171***
		(18.290)	(19.316)	(19.667)
east	100.538***	-37.272***	50.511***	-33.802***
	(26.879)	(8.308)	(14.044)	(7.267)

Dependent variable: price received				
(1)	(2)	(3)	(4)	
treat:mid				−11.203
				(20.536)
treat:post		85.510***	67.651***	75.060***
		(23.383)	(24.209)	(25.554)
Constant	513.112***	426.892***	190.699***	217.896***
	(26.761)	(10.893)	(14.013)	(14.206)
Observations	346	2,181	793	2,628
R-squared	0.067	0.173	0.618	0.306
Adj. *R*-squared	0.061	0.171	0.616	0.305
Residual std. error	230.471	187.905	165.096	174.436
F statistic	12.240***	113.788***	319.280***	192.893***

Note: Standard errors are in parentheses.

$^{*}p < .10$, $^{**}p < .05$, $^{***}p < .01$.

Again, models (2) and (3) present DD models to account for possible selection bias in model (1). Model (2), which uses the more recent 2009/2010 baseline data, estimates baseline price levels at about UGX427 per kg of maize. Unsurprisingly, we find that, for all farmers irrespective of the treatment status and location, prices rose significantly between the baseline of 2009/2010 and the end line of 2011/2012. This general maize price inflation of about UGX189 represents a 44% increase. However, on top of this 44% increase, farmers who had access to a CKW received an additional premium of about UGX85, such that the price at which they sold was on average about UGX700. When not interacted, the treatment status is not significantly different from zero, meaning that the allocation of the treatment appears to have been independent of the initial prices farmers received for their maize. Model (3) uses the relatively older baseline constructed from the 2005/2006 UNHS. This fact is reflected in a much lower baseline price level and a much higher absolute general price increase between the baseline and the end line. The conclusions of a significant treatment effect and no selection bias remain the same as in model (2). The CKW premium is slightly lower at about UGX68 per kg.

Model (4) presents the results for the pooled sample, with all three surveys included. The average price of maize in 2005/2006 is estimated to be about UGX218 in the North and about UGX184 in the East. These prices have increased by about UGX208 by 2009/2010 due to general maize price inflation. By 2011/2012, they have increased by about UGX395. There is no significant difference between the treatment and the control group before the intervention. The fact that the interaction between the treatment indicator and the time indicator for the 2009/2010 survey is insignificant lends support to the hypothesis of a parallel trend. We find a significant positive treatment effect as the interaction between the treatment indicator and the dummy indicating the end line shows.

While there is some evidence that market information dissemination does play a role in Uganda (Svensson & Yanagizawa, 2009), studies that focus on direct price information dissemination initiatives through mobile phones such as Fafchamps and Minten (2012) often do not find significant effects on prices. The fact that the CKW intervention leads to a more optimistic conclusion than the other studies may be because the CKW model attempts to make information actionable by complementing it with additional information such as a trader directory. For instance, price information is useless if there are no alternatives to the trader the farmer is bargaining with. Additional data suggest that people rely more on price information conveyed by friends, which suggests that the introduction of a CKW raises the credibility of price information that is circulating within communities and social networks. This highlights the public nature of (price) information and may help explain why the CKW model may outperform initiatives that rely on private ownership of mobile phones.

Regression adjusted models

The models presented in the previous sections found no impact from the CKW project on maize productivity, a negative impact on share sold and a positive impact on the price at which farmers sold this maize. However, apart from a dummy indicating survey site (North or East), these models did not allow for other sources of variation in the dependent variables. Therefore, in this section, we control for other variables which may reduce the residual variance. Adding additional controls also investigates the robustness of our treatment effect identified in the previous section to the inclusion of additional household characteristics, such as household size, education of the household head, etc. We add these controls to our preferred specification, which is the DD model that uses the UNHS 2005/2006 as a baseline (i.e. model (3) in the previous sections).

The results of the regression adjusted DD regression are presented in Table 6. For instance, model (1) in this table expands model (3) in Table 3. Maize productivity, measured as kg harvested per acre, seems to be negatively correlated to the age of the household head. According to expectations, we find that farmers who have less education have significantly lower productivity. The questions on clothes and shoes can be regarded as indicators of well-being; hence, it is expected that they are positively related to productivity. We find indeed that coefficients on both dummies are positive, but for shoes it is not significant. What is most important is that after the inclusion of additional controls, our earlier conclusions remain valid: Maize productivity, for which conditions are more conducive in the east, has increased over time. Productivity is higher in the treatment group, but since this was also the case before the intervention, this higher productivity cannot be attributed to the intervention. There is no effect of the CKW intervention on maize productivity.

Table 6. DD regressions with additional controls.

Dependent variable						
prodacre	Sellshar e	price	sellshare	Price		
(1)	(2)	(3)	(4)	(5)	(6)	
treat	141.284 ***	11.492**	16.582	11.926**	15.718	−4.071
	(51.881)	(5.171)	(16.717)	(5.319)	(17.456)	(33.775)
post	183.807 ***	12.178**	344.579 ***	12.291**	344.858 ***	289.161 ***
	(53.738)	(5.242)	(20.450)	(5.375)	(21.244)	(36.720)
east	346.243 ***	0.626	39.560** *	0.916	39.432** *	33.921*
	(36.770)	(2.830)	(14.446)	(2.877)	(14.973)	(20.419)
hhsize	6.697	0.1	5.404**	0.058	5.356**	8.352***
	(5.720)	(0.454)	(2.124)	(0.459)	(2.187)	(3.225)
agehead	−2.147*	−0.150	0.166	−0.147	0.182	0.426
	(1.130)	(0.094)	(0.433)	(0.095)	(0.447)	(0.672)
femhead	3.780	−12.274 ***	−6.475	−11.670 ***	−8.201	−10.443

Dependent variable						
prodacre	Sellshare	price	sellshare	Price		
(1)	(2)	(3)	(4)	(5)	(6)	
	(42.873)	(3.564)	(15.566)	(3.614)	(16.280)	(26.759)
prim	−64.176*	−1.395	−33.056**	−1.496	−32.982**	−15.845
	(35.525)	(2.822)	(14.273)	(2.835)	(14.622)	(19.830)
hasshoes	59.68	3.124	25.884*	3.015	25.325*	31.634
	(38.119)	(3.118)	(14.717)	(3.142)	(15.169)	(22.400)
hasclothes	159.457***	5.383	−9.783	5.176	−8.855	5.514
	(52.657)	(4.156)	(18.215)	(4.236)	(18.924)	(31.334)
land				0.068	−0.242	−0.329
				(0.093)	(0.551)	(0.620)
sellshare						0.759**
						(0.338)
treat:post	−47.65	−19.079***	68.272***	−19.520***	69.407***	93.667**

Dependent variable						
prodacre	Sellshare	price	sellshare	Price		
(1)	(2)	(3)	(4)	(5)	(6)	
	(66.242)	(5.972)	(24.537)	(6.112)	(25.257)	(40.776)
Constant	136.365	29.398***	181.917***	29.332***	181.631***	134.512**
	(86.980)	(7.173)	(30.544)	(7.238)	(31.418)	(52.639)
Observations	658	515	790	510	765	470
R-squared	0.295	0.065	0.63	0.062	0.626	0.526
Adj. *R*-squared	0.284	0.047	0.625	0.041	0.620	0.513
F statistic	27.082***	3.523***	132.483***	2.969***	114.464***	42.239***

Note: Standard errors are in parentheses.

*$p < .10$, **$p < .05$, ***$p < .01$.

Model (2) in Table 6 adds controls to model (3) in Table 4. Only one of the added controls seems to be significant. Female-headed households sell a significantly lower share of their maize on the market, a common finding in the literature on smallholder market participation (Mather et al., 2013). All other estimates remain similar to the ones found in Table 4. In fact, the negative treatment effect seems to become even more important after controlling for household characteristics. Model (3) corresponds to model (3) in Table 5. While it is unclear why household size is positively related to the price at which households are able to sell their maize, the effect of the education level of the household head is as expected: heads who have no or only primary education sell their maize at lower prices. Households that report having at least one pair of shoes for each household member report higher prices. Most importantly, controlling for these characteristics does not change our conclusion that the CKW intervention increased the price at which farmers are able to sell their maize.

The remaining three models add two variables to the regressions for share sold and price at which maize is sold. We find no effect on share sold or price from the area of land that is used for growing maize, and the treatment effects remain virtually the same. In the last column, we add the share of the maize harvest sold as an explanatory variable for the price received. We find that farmers who are more engaged in business also receive higher prices. Again, the treatment effect is robust to the inclusion of this variable. The results in Table 6 show that the impact (or the lack thereof) of the CKW intervention we identify is unlikely to be due to specification issues.

Treatment intensity and spillover effects

Table 7 investigates if there are additional effects associated to repeated direct, personal interactions with a CKW, over and above the ones found in the preceding sections. In particular, we added the number of interactions each farmer had with a CKW and estimated a FE model. The estimates are not significantly different from zero for any of the outcome variables, signifying that what matters is that one lives in an area served by a CKW, not how often one interacts. This suggests substantial spillover effects. There are different ways in which these effects can come about. Information may be perceived as non-rival and public, and information looked up on the smartphone during a CKW farmer interaction may be shared within the communities very quickly. It may also be that farmers who interact with a CKW are seen as model farmers and their practices are copied by other farmers.

Table 7. Treatment intensity.

Dependent variable			
Prodacre	sellshare	Price	
(1)	(2)	(3)	
freq	−6.463	−0.505	1.025
	(7.227)	(0.369)	(3.408)
post	−56.621	−17.167***	285.078***
	(52.414)	(2.861)	(30.298)
Observations	338	431	369
R-squared	0.027	0.233	0.463
Adj. *R*-squared	0.009	0.096	0.163
F statistic	1.642	27.024***	55.977***

Note: Standard errors are in parentheses.

*$p < .10$, **$p < .05$, ***$p < .01$.

The fact that we find effects if we control for selection bias at the treatment level but not at the household level may also provide an explanation why we find an effect of the intervention, and other studies such as Fafchamps and Minten (2012) that evaluate similar interventions do not. Fafchamps and Minten (2012) treat the information as private and randomize at the household level. If, in reality, this price information is shared with the neighbors, there will be spillover of the treatment to the control group. This will result in a smaller difference in the outcome variables between the two groups.

CONCLUSION

In many developing countries, the majority of households seem to be stuck in subsistence agriculture. Most households produce a mix of predominantly food security crops mainly for own consumption, using rudimentary technology on small plots. At the same time, at least some farmers seem to have engaged in successful crop intensification, using improved seeds and modern inputs such as inorganic fertilizers obtained through the market. These farmers also provide the bulk of agricultural produce to the market. The fact that some farmers seem to be able to transform into successful commercial farmers while other are left behind suggests the existence of market failures.

The literature on agricultural economic development identifies a range of market inefficiencies that may result in some farm households self-selecting into subsistence, such as land and labor market inefficiencies or incomplete credit markets. We argue that ICTs are especially promising in overcoming information inefficiencies, which are special in that they are also likely to affect many of the other market inefficiencies (e.g. output market inefficiencies or risk market inefficacies). In particular, we explain how ICTs can induce farmers to grow more cash crops by reducing risk. We also point out the potential of ICTs in increasing the bargaining power of the smallholder vis-à-vis traders and middlemen. ICTs may also be a powerful vehicle for agricultural extension information delivery to spur technology adoption and increase yields. Finally, we also point out that ICTs are important in reducing transaction costs, affecting smallholder market participation.

The potential of ICTs to lift smallholders out of the subsistence is then illustrated through an empirical investigation of the CKW project of the Grameen Foundation. This innovative project, rolled out in 2010, set out to deliver extension and marketing information to smallholder farmers in rural villages in some districts in Eastern and Northern Uganda. They equipped locally recruited individuals with an Android smartphone. These devices were running a mobile application called 'CKW search' that can be used to search up-to-date and location-specific information related to farming and product marketing.

We find that, consistent with the theory, the availability of a CKW in the village of a farmer increases the percentage of farmers who grow high-value crops such as coffee and maize. At the other end of the spectrum, we find that CKW presence reduces the percentage of farmers who report growing food security crops, such as groundnuts, cassava and sweet potatoes. It seems that the access to information on agriculture and commodity marketing through a CKW induces farmers to adjust their crop portfolio, moving from low-risk low-return to higher return crops, but with a larger variance. The CKW project may thus be useful to mitigate a particular type of poverty trap, whereby poor farmers choose low-risk low-return crops because they cannot afford risk, which in turn keeps them poor.

While we expected that ICTs would be effective in reducing knowledge gaps and accelerating technology adoption, we do not find evidence that the presence of a CKW alters maize productivity. To test if ICTs are effective in reducing transaction costs, we looked at the share of the maize harvest that is marketed by the farmers. Here we find that the intervention reduces the share of maize marketed by about 15–20%. Finally, we looked at the effect on the price farmers get for the maize they sell, as theory predicts that ICTs may be effective in improving the bargaining power of the smallholder farmer. We find that farmers who have access to a CKW receive prices that are on average about 12–16 higher. FE models exploring effects from the number of interactions between a CKW and a farmer suggest no additional effects, which may indicate substantial sharing of the extension and marketing information obtained from the CKW.

NOTES

1. An extreme example would be where the market for the primary staple food is missing, such that the farm household has no other option than produce for auto-consumption. However, in other setups, where for example there are labor market and/or land market constraints, the optimal production decisions may be conditional on consumption decisions.
2. An online technical appendix explains how we deal with the fact that the UNHS data refer to the 2005/2006 agricultural season, while the baseline data from the CKW areas refer to the 2009/2010 agricultural season.
3. We also control for region in all specifications.

DISCLOSURE STATEMENT

No potential conflict of interest was reported by the author.

NOTES ON CONTRIBUTOR

Bjorn Van Campenhout is a Research Fellow at the International Food Policy Research Institute based in the Kampala office in Uganda. His research focuses on smallholders farmer interaction with both input and output markets and how this affects production, revenues and ultimately well-being. He obtained his PhD from the KULeuven in Belgium [email: b.vancampenhout@cgiar.org].

REFERENCES

1. Aker, J. C. (2010). Information from markets near and far: Mobile phones and agricultural markets in Niger. *American Economic Journal: Applied Economics, 2*(3), 46–59.
2. Aker, J. C. (2011). Dial 'A' for agriculture: A review of information and communication technologies for agricultural extension in developing countries. *Agricultural Economics, 42*(6), 631–647.
3. Aker, J. C., & Fafchamps, M. (2015). Mobile phone coverage and producer markets: Evidence from West Africa. *World Bank Economic Review, 29*(2), 262–292.
4. Aker, J. C., & Mbiti, I. M. (2010). Mobile phones and economic development in Africa. *Journal of Economic Perspectives, 24*(3), 207–232.
5. Anderson, J. R., & Feder, G. (2004). Agricultural extension: Good intentions and hard realities. *World Bank Research Observer, 19*(1), 41–60.
6. Angrist, J. D., & Pischke, J. S. (2008). *Mostly harmless econometrics: An empiricist's companion.* Princeton, NJ: Princeton University Press.
7. Bamberger, M. (2009). Strengthening the evaluation of programme effectiveness through reconstructing baseline data. *The Journal of Development Effectiveness, 1*(1), 37–59.

8. Courtois, P., & Subervie, J. (2015). Farmer bargaining power and market information services. *American Journal of Agricultural Economics, 97*(3), 953–977.

9. Dar, M. H., de Janvry, A., Emerick, K., Raitzer, D., & Sadoulet, E. (2013). Flood-tolerant rice reduces yield variability and raises expected yield, differentially benefitting socially disadvantaged groups, Science Reports 3, Article number 3315.

10. Dercon, S. (1996). Risk, crop choice, and savings: Evidence from Tanzania. *Economic Development and Cultural Change, 44*(3), 485–513.

11. Dercon, S. (2002). Income risk, coping strategies and safety nets. *World Bank Research Observer, 17*(2), 141–166.

12. Dorosh, P., Wang, H. G., You, L., & Schmidt, E. (2012). Road connectivity, population, and crop production in Sub-Saharan Africa. *Agricultural Economics, 43*(1), 89–103.

13. Duflo, E., Kremer, M., & Robinson, J. (2008). How high are rates of return to fertilizer? Evidence from field experiments in Kenya. *American Economic Review, 98*(2), 482–488.

14. Ebiyau, J., Arach, T., & Serunjogi, K. L. (2005). Commercialisation of sorghum in Uganda. *African Crop Science Conference Proceedings, 7*, 695–696.

15. *Economist*. (2003). *The Eagle has landed*. Retrieved from March 29, 2016, from http://www.economist.com/node/1915350

16. Eswaran, M., & Kotwal, A. (1989). Credit as insurance in agrarian economies. *Journal of Development Economics, 31*(1), 37–53.

17. Fafchamps, M., & Hill, R. V. (2008). Price transmission and trader entry in domestic commodity markets. *Economic Development and Cultural Change, 56*(4), 729–766.

18. Fafchamps, M., & Minten, B. (2012). Impact of SMS-based agricultural information on Indian farmers. *World Bank Economic Review, 26*(3), 383–414.

19. Foster, A. D., & Rosenzweig, M. R. (2010). Microeconomics of technology adoption. *Annual Review of Economics, 2*, 395–424.

20. Futch, M. D., & McIntosh, C. T. (2009). Tracking the introduction of the village phone product in Rwanda. *Information Technology and International Development, 5*(3), 54–81.

21. Gabre-Madhin, E., Barrett, C. B., & Dorosh, P. (2002). Technological change and price effects in agriculture: conceptual and comparative perspectives. In T. Bonger, E. Gabre-Madhin, & S. Babu (Eds.), *Agriculture technology diffusion and price policy* (pp. 3–33). Washington, DC: Ethiopian Development Research Institute and the International Food Policy Research Institute.

22. Goyal, A. (2010). Information, direct access to farmers, and rural market performance in Central India. *American Economic Journal: Applied Economics, 2*, 22–45.

23. de Janvry, A., Fafchamps, M., & Sadoulet, E. (1991). Peasant household behaviour with missing markets: Some paradoxes explained. *Economic Journal, 101*(409), 1400–1417.

24. Jensen, R. (2007). The digital provide: Information (technology), market performance, and welfare in the South Indian fisheries sector. *The Quarterly Journal of Economics, 122*(3), 879–924.

25. Jensen, R. (2010). Information, efficiency and welfare in agricultural markets. *Agricultural Economics, 41*(S1), 203–216.

26. Karlan, D., Osei, R., Osei-Akoto, I., & Udry, C. (2014). Agricultural decisions after relaxing credit and risk constraints. *Quarterly Journal of Economics, 129*(2), 597–652.

27. Key, N., Sadoulet, E., & de Janvry, A. (2000). Transactions costs and agricultural household supply response. *American Journal of Agricultural Economics, 82*(2), 245–259.

28. Mather, D., Boughton, D., & Jayne, T. S. (2013). Explaining smallholder maize marketing in southern and eastern Africa: The roles of market access, technology and household resource endowments. *Food Policy, 43*, 248–266.

29. Meyer, B. D. (1995). Natural and Quasi-experiments in Economics. *Journal of Business & Economic Statistics, 13*(2), 151–161.

30. Muto, M., & Yamano, T. (2009). The impact of mobile phone coverage expansion on market participation: Panel data evidence from Uganda. *World Development, 37*(12), 1887–1896.

31. Nakasone, E., Torero, M., & Minten, B. (2014). The power of information: The ICT revolution in agricultural development. *Annual Review of Resource Economics, 6*, 533–550.

32. Quizon, J., Feder, G., & Murgai, R. (2001). Fiscal sustainability of agricultural extension: The case of the farmer field school approach. *Journal of International Agricultural and Extension Information*, *8*(1), 13–24.

33. Renkow, M. (1990). Household inventories and marketed surplus in semi subsistence agriculture. *American Journal of Agricultural Economics*, *72*, 664–675.

34. Renkow, M., Hallstrom, D. G., & Karanja, D. D. (2004). Rural infrastructure, transactions costs and market participation in Kenya. *Journal of Development Economics*, *73*(3), 349–367.

35. Ricker-Gilbert, J., & Jones, M. (2015). Does storage technology affect adoption of improved maize varieties in Africa? Insights from Malawi's input subsidy program. *Food Policy*, *50*, 92–105.

36. Schultz, T. W. (1964). *Transforming traditional agriculture* (p. 212). New Haven, CT: Yale University Press.

37. Singh, I. J., Squire, L., & Strauss, J. (Eds.). (1986). *Agricultural household models: Extension, application and policy* (p. 335). Baltimore, MD: Johns Hopkins University Press.

38. Stifel, D., & Minten, B. (2008). Isolation and agricultural productivity. *Agricultural Economics*, *39*(1), 1–15.

39. Svensson, J., & Yanagizawa, D. (2009). Getting prices right: The impact of the market information service in Uganda. *Journal of the European Economic Association*, *7*(2–3), 435–445.

40. Udry, C. R. (1990). Credit markets in Northern Nigeria: Credit as insurance in a rural economy. *World Bank Economic Review*, *4*(3), 251–269.

41. Van Campenhout, B. (2007). Modelling trends in food market integration: Method and an application to Tanzanian maize markets. *Food Policy*, *32*(1), 112–127.

42. Van Campenhout, B., Lecoutere, E., & D'Exelle, B. (2015). Inter-temporal and spatial price dispersion patterns and the well-being of maize producers in Southern Tanzania. *Journal of African Economies*, *24*(2), 230–253.

9
CHAPTER

What are you doing on Tinder? Impression Management on a Matchmaking Mobile App

Janelle Ward

Department of Media and Communication, Erasmus School of History, Culture and Communication, Erasmus University Rotterdam, Rotterdam, The Netherlands

ABSTRACT

Mobile dating applications such as Tinder have exploded in popularity in recent years. On Tinder, impression management begins with a motivation to download the app, the choice of one's profile photos and an assessment of the expectations of potential Tinder matches. These processes occur in a technologically mediated environment of reduced cues and increased control, local proximity and a reduced filtering process. My focus in this paper is this first stage of impression management, which consists of both impression motivation and impression construction. Specifically, what are the pre-match impression management practices of Tinder users? I present the results of interviews with Tinder users in the Netherlands. Participants were recruited via a Tinder profile that advertised the study using the University emblem and a brief description. Interview questions focused on user understandings of self-presentation practices and profile construction. The interviews also examined how users evaluated their potential matches. Results show users' motivations for using Tinder range from entertainment to ego-boost to relationship seeking, and these motivations sometimes change over time.

Profile photos are selected in an attempt to present an ideal yet authentic self, and chosen as an illustration of not only one's desirability but also of other indicators such as education level. Tinder users 'swipe' not only in search of people they like, but also for clues as to how to present themselves in order to attract others like them. This research offers insight into user experiences and perceptions within the still under-researched area of inquiry.

Keywords: Impression management, self-presentation, mobile technology, dating apps, Tinder

INTRODUCTION

Whether seeking a passionate love affair, looking for a spouse, or wanting a casual encounter, people have long devised methods of presenting themselves in the most attractive way to potential partners. Mediated possibilities to connect with others have evolved from newspaper advertisements to teletext to online dating websites, and the latest offerings are matchmaking mobile applications like Tinder.

Impression management or self-presentation exists both face-to-face and in mediated forms. Before a first date, a man shaves, applies cologne, and dresses in his finest. On Tinder, a woman selects an attractive picture of herself as her main photo. In a virtual environment, as in real life, impression management starts prior to a face-to-face meeting or even prior to interpersonal communication. On Tinder, this impression management begins with choosing one's profile photos. It continues with what Leary describes as impression monitoring: 'For people to engage in self-presentation, they must monitor, at one level or another, how they are being perceived and evaluated by others' (1995, p. 47). Tinder users do this by assessing the expectations of potential matches.

There is tremendous research regarding self-presentation and romance in the last decade (Ellison, Heino, & Gibbs, 2006; Hall, Park, Song, & Cody, 2010; Manning, 2014), examining dating websites such as Yahoo! Personals (Ellison, Hancock, & Toma, 2012), Match.com (Gibbs, Ellison, & Heino, 2006), and OKCupid (Zytko, Jones, & Grandhi, 2014). Recent work examines Grindr, the matchmaking app geared toward men seeking men (Birnholtz, Fitzpatrick, Handel, & Brubaker, 2014; Blackwell, Birnholtz, & Abbott, 2015; Brubaker, Ananny, & Crawford, 2016; Gudelunas, 2012). Research is emerging on Tinder user awareness of privacy issues (Farnden, Martini, Raymond, & Choo, 2015; Stenson, Balcells, & Chen, 2015). Dating apps present a novel technological environment for impression management, mainly due to issues of reduced cues and increased control, local proximity, and a reduced filtering process. These issues will be explained in more detail in the coming section. First, I provide some detail on Tinder.

Tinder

Though dating websites still account for the largest market share,[1] dating apps have increased in popularity in recent years. In comparison to dating websites, dating apps ask users to provide limited information for potential matches, namely, a number of photos and an optional small amount of text (Blackwell et al., 2015; Gudelunas, 2012). I distinguish Tinder from dating websites because it is a location-based dating platform available only as a mobile app. Further, Tinder does not ask users to answer compatibility questions and does not allow detailed filtering techniques, features common to dating websites. On Tinder, the first impression users have of a potential match is her/his main profile photo. If a user is interested in seeing more, s/he can tap the profile, which will reveal additional photos, optional text, and shared Facebook friends and Facebook likes.[2] Users swipe left to reject and right to accept a potential match. If the right swipe is mutual, it is a match, and Tinder allows users to chat within the app.

Tinder was launched in October 2012 and has achieved global popularity. It has more than 50 million global users in 196 countries, with 9 billion matches since its inception.[3] Globally, Tinder users login an average of 11 times a day and spend between 7 and 9 minutes swiping during a single session. Women browse profiles for 8.5 minutes at a time versus 7.2 for men (Bilton, 2014). In the Netherlands, there were an estimated 1.5 million users in 2014 (Eigenraam & Zandstra, 2014). Though the app itself is relatively new, the concept of meeting a romantic partner online is not. According to Statistics Netherlands, between 2008 and 2013, 13% of Dutch people met their partners online, and half of these met on dating sites.[4]

My motivation for conducting this study, presented as a research question, is to explore: *What are the pre-match impression management practices of Tinder users?* For this paper, I draw on the impression management literature, keeping in mind the relevance of a technologically mediated dating environment. I first present theoretical considerations, followed by a description of the interviews I conducted with Tinder users. Interview analysis is followed by a conclusion and discussion.

IMPRESSION MANAGEMENT ON DATING APPS

Goffman (1959) classically argues that individuals attempt to control or guide others' impressions by manipulating setting, appearance, and behavior. According to Leary and Kowalski (1990), there are two key processes in impression management. First, there is impression motivation, 'when people become motivated to engage in particular self-presentation behaviors' (Leary, 1995, p. 53). Past research has established that in the context of mediated dating environments, users are highly motivated to control the impression they create (Ellison et al., 2012; Koestner & Wheeler, 1988; Kramer & Winter, 2008; Toma, Hancock, & Ellison, 2008; Zytko et al., 2014). This high motivation can be illustrated in how users are sometimes tempted to present themselves in idealized ways. For example, researchers have found that women have the tendency to decrease their reported weight, while men increase their reported height, and men are more likely to exaggerate their income levels (Feingold, 1990; Gonzales & Meyers, 1993; Hall et al., 2010; Harrison & Saeed, 1977; Toma & Hancock, 2010). The second process of impression management is impression construction: when people explicitly choose the impression they want to make and decide the method they will use to create it. Researchers have elaborated on a number of these construction methods. For example, Leary (1995) discusses self-descriptions, attitude statements, social associations, and deception. Tinder users engage in impression construction when deciding which pictures and text to include, and which strategies to use during this process.

The environment, however, is key: Goffman's initial work on self-presentation focused on face-to-face communication. Numerous scholars have adapted Goffman's ideas to electronic environments, though still in an interpersonal context (e.g., Miller, 1995; Papacharissi, 2002; Tufekci, 2008). According to boyd and Ellison (2007), a social network site is a

> web-based services that allow individuals to (1) construct a public or semi-public profile within a bounded system, (2) articulate a list of other users with whom they share a connection, and (3) view and traverse their list of connections and those made by others within the system. (p. 211)

Dating apps like Tinder differ in the sense that connections are not public. However, the profile itself is semi-public and centered around the user, offering the opportunity to connect with other users. In Goffman's terms, Tinder provides the user a stage where s/he can perform.

In this paper, I focus on self-presentation in earlier stages of impression management: In a setting like a dating app, where the potential of romantic encounters loom, users must first be motivated to create a profile before interacting with others. At the same time, others' impressions are key. On Tinder, users are only provided with positive reinforcement in the form of a mutual match. Users are not explicitly aware of who has rejected them.

Within this environment, users construct a profile, observe how others have constructed their profiles, and choose potential matches based on these profiles. On Tinder, this process takes place in an environment that is defined by (1) *reduced cues and increased control* in profile construction; (2) *local proximity* of matches; and (3) a *minimal filtering process*, where Tinder users are exposed to all other users in a geographical, age, and sex-defined area, and must navigate through these potential matches by swiping.

Reduced cues and increased control

Dating app users operate in a reduced cue environment, where cues are static and not dynamic. Walther's hyperpersonal model emphasizes that in such an online environment, individuals have increased control over self-presentation. Communication is asynchronous and it cannot rely on nonverbal communication cues, which are harder for individuals to control. Thus, users can more easily adapt their self-presentation in an online environment like Tinder as compared with face-to-face communication (Walther, 1996). This is the case on more general social networking sites such as Facebook (Lampe, Ellison, & Steinfield, 2007) and particularly true in relation to online dating (Ellison et al., 2006, 2012; Hall et al., 2010; Manning, 2014; Toma & Hancock, 2010). Ellison et al. (2006) discuss how online daters are able to optimize their self-presentation and establish credibility in this environment by balancing 'accuracy with self-promotions and desirability' (p. 430). Hardey (2002) also notes 'users feel obliged [to] anchor their on-line identity in their off-line embodied self' (p. 579).

Though Tinder mimics dating websites in some ways, it reduces these self-presentation cues further. Users can only provide a limited amount of information to potential partners, namely, a number of photos and a small amount of text (Birnholtz et al., 2014; Gudelunas, 2012; Handel & Shklovski, 2012). Tinder users are, therefore, left with a few carefully chosen photos and an option for brief text.

Local proximity

With location-aware mobile devices, we can now connect with people in our close vicinity for particular purposes. Dating apps are also referred to as 'location-based real-time dating' applications (Handel & Shklovski, 2012) or 'People-Nearby Applications' (Van de Wiele & Tom Tong, 2014) as they draw on the location of the user in order to provide matches in one's geographic proximity. Location-based dating apps may facilitate users meeting face to face and potentially forming a relationship (Blackwell et al., 2015; Ellison et al., 2012), which could increase impression motivation.

Further, due to the issue of proximity, especially in the case of location-based dating apps, there may be less of a tendency to deceive potential matches, as there is a real chance that they will meet face to face and form a relationship (Ellison et al., 2012). Researchers describe this as identifiability or the ease with which an online identity can be connected to a known person (Blackwell et al., 2015; Woo, 2006). Due to this possibility, Blackwell et al. (2015) say users have 'an incentive to present in an attractive, but plausible, light' (p. 6).

Minimal filtering process

When it comes to choosing romantic partners, filtering works to screen potential contacts. Focusing on how people choose sexual partners online, Couch and Liamputtong (2008) describe filtering as 'simple assessments of attractiveness and geography and physical proximity ... identity, including appearance, personality, sexual tastes and preferences, and risk management' (p. 273). Best and Delmege find that in an online dating environment that offers a 'plethora of choice ... filtering strategies are adopted spontaneously and refined conscientiously by participants' (2012, p. 253). This process is often more complex on dating websites, in which users are allowed to additionally screen potential matches on height and weight (Hancock, Toma, & Ellison, 2007), race (Lin & Lundquist, 2013), and education level (Skopek, Schulz, & Blossfeld, 2011).

In relation to dating websites, Best and Delmege (2012) describe the filtering process as starting with an initial screening, where users choose potential romantic partners based on search criteria. Then, users interact with the preselected potential romantic partners via messaging. On Tinder, filtering operates by allowing users to determine with whom they would like the possibility to chat, but users are provided only with geographical proximity, age, and sex as criteria, in contrast to more detailed filtering options on dating websites (Hamilton, 2016). This adaptation provides further incentive for research into the impression management practices of dating app users.

METHOD

With these theoretical considerations, my research tries to answer the following question: *What are the pre-match impression management practices of Tinder users?* In line with Leary and Kowalski's (1990) concept of impression management, my goal is to, first, understand Tinder users' motivations for downloading and using the app and, second, explore how Tinder users construct her/his profile and swipe potential partners. I examine these queries through interviews with Tinder users in the Netherlands.

Similar to Blackwell et al. (2015), participants were recruited via Tinder profiles that advertised the study using the University emblem and a brief description. Hamilton and Bowers (2006) suggest that researchers should 'select the most appropriate Internet site to place an announcement of the study' (p. 825). Two profiles with the username 'TinderStudy' were created: One male and one female, both with a reported age of 25 years. This strategy allowed me to access both male and female Tinder users, contributing to the aim of interviewing an equal number of each. Participants could email or contact me through Tinder with questions or to participate (see Figure 1).

Figure 1. Invitation to participate.

Within 2 weeks, 11 male participants contacted me and I conducted the interviews in that time frame. Female users proved more difficult to recruit: In the initial two-week period, I received only three responses via the TinderStudy profile, and resorted to snowball sampling my previous interviewees, both female and male. This issue is addressed again in the discussion section.

I conducted 21 semi-structured interviews with 11 men and 10 women, aged 19–52 years, which lasted between 45 and 107 minutes. As all interviewees were based in urban surroundings, interviews took place in a city convenient for the interviewee. The interviews were conducted between October and December 2014, at a location chosen by the interviewee (a bar or a café). Face-to-face interviews can be successfully conducted on Internet-based research topics (Kazmer & Xie, 2008). Interviewees had used Tinder between two months and one year, and most were active users at the time of the interview. The interviews were conducted in English. Before scheduling the interview, I confirmed with each interviewee that s/he was able to communicate fully in English. All interviewees were asked the same questions from a semi-structured interview guide prepared for the research. Like in Ellison et al. (2012), the open-ended interview questions focused on user understandings of profile construction. The interviews also examined how users evaluate and choose potential matches. Participants were assured of their anonymity and were informed that any quotes taken from their interviews in written or spoken reports would use pseudonyms and exclude potentially identifying details. Interviews were audio recorded and transcribed.

Qualitative textual analysis followed a revised version of grounded theory (Glaser & Strauss, 1967). The analysis should be 'grounded' in interview data; resulting explanations draw from participant responses rather than relying on preexisting theoretical insights (Strauss & Corbin, 1990). This process was revised in the sense that the theoretical framework and interview questions played a role in my interpretation of the interview data. I read transcripts in their entirety, and then engaged in the process of open coding followed by axial and then selective coding. The themes that emerged from this process are presented in the results section. Quotations from interviewees serve to illustrate the results and give voice to the participants in the research (Creswell, 2007).

RESULTS

Why create a Tinder profile? Via the interviews, users' impression motivations were uncovered by how they spoke about their use of Tinder and what their ultimate goal was while using the dating app.

Tinder use as entertainment, ego-boost, or searching for a partner

When initially asked why they used Tinder, all but two of the interviewees described their use as entertainment or an ego-boost. Erwin, 34 years old, identified wholeheartedly with the notion that Tinder was just for entertainment: 'For me it's more like a game.' To illustrate, he pointed to the games folder on his iPhone. 'See, the app is right here, right next to Candy Crush.' Erwin added: 'It's so superficial. And as soon as you realize that and accept that, it starts to be fun.' Sergio, 46, said something similar: 'I started just for fun, you know, it was a thing to pass the time.' Others saw Tinder as an ego-boost. Colin, 21, said: 'Why do I use Tinder? In the first place, I think for attention.'

Other users expressed a different motivation for using the app. They joined Tinder as a way to recover from a breakup, with the motivation to quickly find a new partner. Ross, 26, said, 'I'd come [to the Netherlands] for [my ex-girlfriend] and I'd forgotten to make a balance. So then I needed Tinder instantly to work. I needed people to meet up, and hang out, immediately.' Susan, 34, was on Tinder for two months before meeting her current boyfriend on the app:

I used it every day. I'd just got dumped. I'd just turned 34, and I was set on not staying alone, just miserable and by myself, I'm going to do something ... I downloaded it, it was super easy. I think I was on it multiple times a day. I was lying in bed crying for my ex and then whenever I had a match I was like, 'Yes! There are still men out there that like me!'

Yet these motivations also changed over time. I heard from users who revealed a hope for finding love, after initially using it for entertainment or ego-boost. As the interviews progressed, there was a tendency for wishful thinking to emerge. Sergio put it this way: 'If I find someone I want to live with on Tinder it's ideal. But it's not something that I'm really looking for.' It worked the other way, too: Reinout, 27, spoke about his use of Tinder with a self-deprecating humor. Like others, he had an ambitious streak when downloading the app that later faded: 'At first I was really looking for a girlfriend. And now it's more like yeah, I just enjoy dating.' He had learned how to play the game, but when asked to articulate his 'ultimate goal,' he returned to a relationship mindset: 'The ultimate goal is to meet this one perfect match who will be my best friend for the rest of my life.'

Impression motivation in a stigmatized environment

Motivations for using Tinder are not clear-cut. Tinder's reputation as reported in the media varies from a 'hook-up' app[5] to a serious tool to finding relationships.[6] An emerging issue related to motivations was a still-present stigma for dating online. Aya, a 22-year-old student, refused any ambitions for love or a long-term relationship from Tinder. She said: 'I still feel it's strange to search for people online if there are so many real people out there … I know some people who have a relationship from Tinder but … I actually think I would feel ashamed.'

Colin agreed. He explained how the stigma of having a Tinder profile impacted his reluctance to initially pair his real Facebook account with the dating app: 'First I had a fake profile. Then I thought, why not? Everyone's doing it. It can't possibly hurt me.' Colin revealed his actual identity, but would not go so far as to wish for an actual relationship from a Tinder match: 'I have friends who have relationships from Tinder but I think it's more shameful than something to be proud of … Yes, you're a lovely couple but you met on an application on your phone.'

If a Tinder user has different motivations for using the app, and her motivations change over time, this can present challenges for impression construction. Blackwell et al. found something similar in their study on the dating app Grindr: Users 'can have a range of norms and expectations, with little evidence in terms of visible cues from others to confirm or refute these expectations' (2015, p. 1128). Despite such complications, a Tinder user must choose particular photos and text to present her-/himself in a desired way to potential partners. The next section will look at how Tinder users make these decisions.

CONSTRUCTING A DESIRED IMPRESSION

Impression construction is about choosing the kind of impression to create and deciding precisely how to go about doing so (Leary & Kowalski, 1990; Toma & Hancock, 2010). For my interviewees, impression construction on Tinder was a carefully chosen process. As with online dating websites, construction of the Tinder profile is of utmost importance: it represents 'the first and primary means of expressing one's self during the early stages of a correspondence and can therefore foreclose or create relationship opportunities' (Ellison et al., 2006, p. 423). On Tinder, whether one is using the app for entertainment, seeking an ego-boost, or an eventual relationship, success is defined by an attractive profile, validated through mutual right swipes.

How I want others to see me

All interviewees had at least two photos on their profile, and some chose not to include any accompanying text. All said they had chosen pictures they liked or pictures their friends liked. For some, friends' approval was communicated in person or via liking a photo on Facebook: The number of Facebook 'likes' pictures received played a big role in determining what they thought would get a positive response. Wildon, 43, stated: 'I chose these pictures because they have likes on Facebook. I don't know myself whether I look good in a picture.' Johanna, 32, put it like this: 'I chose these photos because that's the way I want men to see me. That's the way I see myself. The best way.' Ideal yet authentic was important to interviewees. Erwin agreed: 'Your Tinder profile should be realistic. There's nothing more annoying than someone saying, "I expected you to be different".' Johanna added, 'Yes I have pictures with my children. This is me, this is the total package, take it or leave it.'

Profiles were constructed with the aim to avoid certain appearances. According to one woman, seeming too sexual was one look to avoid. Christina, 40, claimed: 'I would never put up sexy pictures, or pics that don't look like me.' This also happened for men. Wildon said: 'I want women to see me as handsome. Serious. And I don't want to be the guy who is starting with questions about sex.' Aya referenced her photo choices by describing the kind of person she wanted to avoid looking like: 'I want guys to know I'm a student ... you can see that I'm not wearing that much makeup or excessive jewelry or those brands that different people wear.'

Profile experimentation

Every Tinder user must construct an impression. And yet, Tinder profiles are not static. Like other social media platforms, changing photos and text is possible and interviewees often took advantage of this opportunity. A number described their profile changes as experiments. Erwin put it this way: 'Yes, I change my Tinder profile. The second version I put up only two pictures and no text. I got amazing results.' Ross had two profiles on Tinder:

In the beginning I couldn't figure out which pictures to put up to get matches ... I just did it as an experiment, to see what would happen. One profile is completely normal ... the other profile I made it more sexual. More like topless pictures ... just to see if there's girls out there that would be [interested], because there are certain dating apps or websites that are sexual.

With perhaps the opposite intention, Johanna did a 'philosophical profile' experiment:

For a while, I had a Tinder account with only a philosophical quote for my profile photo. It was very interesting because the alpha males would match with me just to tell me that I'm a bitch and I don't understand how nature works. They couldn't understand it, couldn't get it, and that was a filter for me. That's not the kind of guy I want.

When asked about their profile creation, many interviewees referenced their potential Tinder matches as sources of inspiration. As their knowledge of how the app worked increased, they became expert on how they should present themselves, based on what they like and do not like about their potential matches. There was a constant comparison with others along with descriptions of what to embrace and what to avoid. This ranged from the type of photos and text to even the order of photos. Susan revealed, 'I saw lots of people having lots of text and that was not something I wanted to do ... the quote I had [on Tinder] represented something that I wanted to represent.' Reinout said: 'The first and last photo are important. There's a girl, first picture, beautiful. Last one? It's terrible. No, no, no.' Because he had repeatedly had this experience while searching for matches, he realized the importance of keeping up an attractive appearance from start to finish. Thomas, 52, put it like this: 'If I saw someone who had pictures thought-out like mine, I'd probably find that interesting. An interesting background would trigger me, so I'm putting on my profile what triggers me.'

All interviewees strove to make a positive impression, essentially aiming for a balance between an ideal and an authentic self-presentation. They hoped to demonstrate the kind of person they are, and, simultaneously, the sort of person they wanted to attract. Tinder users not only swiped to perfect their own profiles, but also swiped to find potential matches.

The swiping culture

In order to explore how Tinder users choose their matches, interviewees described their thought process in deciding whether to swipe left or right, thus rejecting or accepting a match. Interviewees swiped through potential matches in real time in order to demonstrate this process, but did not interact with past or current matches as this went beyond the scope of the inquiry.

When swiping, users first see the primary profile photo. Unsurprisingly, users do focus their evaluation on superficial characteristics present in this profile photo. As Colin stated: 'It's like sitting outside at a café and judging people as they walk by.' The main reason interviewees gave for rejecting potential matches was the rather subjective trait of unattractiveness. Johanna demonstrated this while swiping: 'This one? No. He has a round face and I really hate sunglasses.' Sergio knew what he was looking for, but was not sure how to articulate it: 'For me it's all about the pictures. I want to see a face and a face that I like ... someone must attract me. It can be the shape of the mouth, nose, hair, anything.'

But this superficial evaluation quickly veered into more specific characteristics. Johanna became more precise when asked to describe what constitutes a left swipe for her: 'There are the men with the fish and men posing with cars that aren't theirs and men drinking. No. Swipe swipe swipe.' Wildon had much to say about this topic: 'Stupid pictures, duck faces, and people with animals. Dolphins, elephants, tigers, all cliché. I screenshot them.' Later, while we swiped together, he paused on one potential match: 'Her profile says "Yoohoo! Are there still nice men?" I'm almost sure she's not my type. The yoohoo type.' Thomas covered a range of these:

This is going to sound awful ... I'm quite fit and sporty and I'm really not into fat people. Fat people, straight away. Also pictures with dogs. And also when the profile pictures don't show the person, like a landscape ... what are they hiding? And the other one is a picture of six or seven people. Who's the person?

Filtering on Tinder

What can a photo show? Appearance, of course. Attractiveness. Perhaps personality. But other aspects emerged in the process of discussing match selection, for example, more obvious traits such as age or race, and also less evident aspects such as perceived education disparities. These aspects became apparent when interviewees were asked to specify which matches they rejected.

As in past research, interviewees used a process known as filtering when choosing a match. However, here, filtering is examined in the pre-interpersonal communication phase, via profile assessment.

Photos reveal more obvious traits such as race and age. Colin was asked about who he swiped left on, and replied: 'Well, the non-Caucasian, and someone older than 30, that would be goodbye.' Colin and others also mentioned their Facebook likes helped signal the intelligence or general interests of a potential match. Christina revealed the following:

... For the most part I'm just attracted to white men, and they have to be fit ... I like very intellectual, nerdy guys, and when it's all like these pictures of them just partying with their friends, on the boat, at these techno parties, at the festival, it's like the same shit over and over.

I had more than one interviewee tell me that duck-face selfies signal low education. Erwin prefaced his growing pessimism for Tinder with the following: 'I consider eighty percent of the country to be of lesser intelligence. And that eighty percent is now taking over Tinder.' When asked for an example, he said: 'There are so many spelling mistakes. If I see one I'm gone. It's as simple as that.'

Attraction is subjective and laden with factors other than sex appeal. These results suggest a mirroring of self-presentation with one's potential matches, as users overwhelmingly reported searching for people like them. Optimists might say that Tinder could be the great leveler of matchmaking. Because Tinder users have to swipe through every potential match presented to them, filtered only by geographical proximity, age, and sex, people could perhaps discover they are attracted to those previously pre-filtered out. Interviewees here have revealed that other factors are just as important as looks, and the tendency for humans to seek out comparable others still emerges. There were few exceptions, from those who at least considered expanding their dating horizons. Wildon said: 'Sometimes I am curious about women who are not my type. But I don't think it would work in the long term.'

Thus, the process of choosing matches on Tinder is driven by physical attraction, but perceived similarity is also essential when selecting matches. Of course, this process may change when the need to connect outweighs the need to find a similar match. Reinout, 27, stated: 'There are days when I'm out of dates ... then I like more people on Tinder.'

CONCLUSION AND DISCUSSION

In this paper, I have examined Tinder users' pre-match impression management, looking at their motivations for using the app, the process they go through in choosing their profile photos and text, and how they swipe for potential matches. From entertainment to ego-boost to relationship seeking, users vary in their motivations for using the dating app, sometimes changing them over time. The still-present stigma of technologically mediated dating (Wildermuth, 2004) may impact users' willingness to view it as a tool for serious dating or relationship seeking: Using Tinder 'just for fun' is more acceptable. Yet, recent survey research on Tinder users found no differences in the motivations of Tinder users and dating website users, but did observe a slight difference in the sexual permissiveness of the groups (Gatter & Hodkinson, 2016). More research is needed to clarify this aspect of dating app use.

Photos are selected in an attempt to present an ideal yet authentic self, in line with past research (e.g., Ellison et al., 2006). Tinder users often search for potential matches to provide clues as to how to present themselves in order to attract others like them. This project did not examine reaction to particular matches, but rather focused on the general process of using

Tinder. It could be that users tweak profiles as a response to particularly attractive others. Future research, perhaps following dating app users over time, could examine these possibilities.

In a reduced cue environment, results show that Tinder users use these minimal cues to show who they are, primarily via photos. They also take advantage of the controlled environment to conduct profile experiments to see how change to their self-presentation may improve their approval from others on the app. Profile choices are contemplated and often changed, as users alter their profiles in order to experiment with how reactions vary.

In terms of selecting matches, interviewees demonstrate knowledge of a particular set of 'courting rules' (Hardey, 2008) explicit to the dating environment. Hardy describes this as follows: '... individuals have to learn how to "decode" the profiles displayed on these sites and make choices on the basis of these mediated interactions' (p. 1112). Such knowledge could facilitate the possibility of an off-line meeting. In terms of choosing who they want to interact with, findings here show that interviewees overwhelmingly search for similar others, though a few did use the opportunity to match with those they would not usually select. This points to another inclination predominate on dating sites: Homophily, or 'love of the same,' is the tendency people have to seek out others like themselves. People like those who are the same age, have the same race, and possess similar educational backgrounds (Harrison & Saeed, 1977; McPherson, Smith-Lovin, & Cook, 2001).

Tinder's swiping environment provides few technologically enabled filtering options. In this case, filtering went beyond appearance into other identity factors, such as perceived education level and indicators of socio-economic status. Unlike traditional dating websites that often ask for height, weight, race, or education level (Hancock et al., 2007; Skopek et al., 2011), there are often no other indicators beyond a few photographs and a few words of text. This allows a different perspective on Tinder: Impression management is focused on that brief moment where one decides whether to swipe right or left. At the same time, Tinder could also be viewed as a platform that includes and excludes as we do in real life.

This paper focused on the 'pre-match' phase of Tinder profile construction and match selection. In relation to self-presentation, Goffman (1959) imagined face-to-face communication and talked about the reciprocal influence on actions when in each other's immediate physical presence. Digital presentation is a fundamentally different context, and scholars have dealt with these distinctions. For example, Hogan (2010) takes Goffman's work and applies it specifically to online media. Hogan's thinking sees the Tinder user as curator, and curators 'filter on behalf of the audience … filtering implies that one can evaluate a *set* of things *before* they are presented for consumption' (p. 382).

At the same time, the promise of physical interaction plays a role here: Leary argues that 'people tend to be more concerned with how others view them when they anticipate future interaction with them' (1995Leary, M. R. (1995). Self-presentation: Impression management and interpersonal behavior. Madison, WI: Brown & Benchmark., p. 57). Even in this pre-match stage, with a lack of interpersonal or face-to-face interaction, it seems such influence occurs on Tinder. Users are imagining who will see them, both those they want to meet and those they do not want to meet, and their desired self-presentation is important to tweak and maintain even before chatting with a match on the app or meeting them in person. In her discussion about networked privacy, boyd (2012) points out how our data 'provides a probabilistic image of who we are based on comparisons to other people' (p. 348). Tinder's algorithm is not made public: though filtering criteria are limited, it is not entirely clear which profiles are presented to users, complicating knowing to whom users are comparing themselves.

Yet, dating 'profiles are essential for online daters because they constitute a gateway for future FtF dating' (Ellison et al., 2012, p. 2). This paper did not delve into what happens after a Tinder match, but this is an issue of great interest for future research: Face-to-face interaction with matches is an important continuation of impression management. Gershon (2010) describes this as 'media switching' and has explored its constraints on romantic relationships. This process may be complicated further in a relationship that began on a dating app. Hardey (2004) also argues that authenticity is key for successful communication between strangers attempting to develop a trusting relationship. The concept of authenticity is emerging in recent scholarship on Tinder use (Duguay, 2016). Still, '… physically copresent interaction still has to be managed in a manner which consolidates the dyadic encounter … such meetings may still be disappointing when physical copresence fails to match the expectation of one or both individuals' (Hardey, 2002, p. 582).

A note of caution is in order whenever studying a trendy technological innovation: Since this research was conducted, Tinder has changed its interface to allow users to include their education and work information on their profiles. Users can also 'Super Like' other profiles, thus bringing their interest and their own profile to the front of another user's queue. Such changes are a reminder of the importance of maintaining a broader theoretical lens and not focusing solely on technological features.

Beyond theoretical considerations, several issues related to the process of data collection are worth mentioning. The first is the difference in recruitment experiences between female and male Tinder users. Two weeks into my recruitment via the app, of the fourteen email responses I received, only three were from women. The three who eventually responded to my request also requested more information about the project and proof of my personal identity; male respondents did not do so. Such difficulty reminded me that for women on Tinder, I was presenting as a man – without an identifying picture – asking to meet and interview them. This issue highlights the importance of issues such as safety that women particularly face online (Spitzberg & Hoobler, 2002).

Further, it is important to keep in mind who actually responds to a request for academic research via a dating app. Though diverse in terms of age, my interviewees as a whole were almost all highly educated and white. Their stereotypes – and rejection – of those from other walks of life were in line with the past literature: This tendency has been found to be a common practice on dating sites when seeking potential partners (Fiore & Donath, 2005). But what about those Tinder users my interviewees rejected? What is their experience of self-presenting and selecting matches on Tinder? Even in past literature these individuals seem underrepresented, with an elite group giving voice to research findings. This is an important factor to remedy in future research.

This research has provided a look at the new phenomenon of mobile matchmaking apps and has helped discover similarities and differences with past research in terms of impression management, particularly in an environment of reduced cues and increased control, local proximity, and a reduced filtering process. It offers insight into user experiences and perceptions within a still under-researched area of inquiry and makes a case to continue researching mixed-mode relationships in the context of dating apps, where users anticipate a move from technologically mediated to face-to-face communication.

NOTES

1. See for example http://www.globalwebindex.net/blog/what-to-know-about-tinder-in-5-charts.
2. For a visual illustration of the app, see https://www.gotinder.com/. Tinder's features have changed since I conducted this research; this issue is dealt with in the discussion. Although most users create profiles with their official Facebook account, others do so with a

fake account, thus distorting their name, age, or other identifying features.

3. As of 14 March 2016, Tinder reports 1.4 billion swipes per day, 26 million matches per day, more than 10 billion total matches, and users in 196 countries. More here: https://www.gotinder.com/press.
4. Dutch data available here: http://www.cbs.nl/en-GB/menu/themas/bevolking/publicaties/artikelen/archief/2014/2014-4087-wm.htm?Languageswitch=on.
5. One of the most frequently cited being Nancy Jo Sales's 'Tinder and the Dawn of the "Dating Apocalypse"' published by Vanity Fair: http://www.vanityfair.com/culture/2015/08/tinder-hook-up-culture-end-of-dating.
6. See the Stuart Dredge piece 'Tinder is for more than just casual sex, says CEO Sean Rad' in *The Guardian*: http://www.theguardian.com/technology/2015/nov/05/tinder-app-users-casual-sex-long-term-relationships.

DISCLOSURE STATEMENT

No potential conflict of interest was reported by the author.

NOTES ON CONTRIBUTOR

JanelleWard is Assistant Professor in the Department of Media and Communication at Erasmus University, Rotterdam. Her research and consultancy expertise is in digital communication. She currently researches how people create and manage impressions on mobile dating apps. Janelle's past research looked at political contexts and particularly political consumerism: when people purchase products with ethical intentions. Her publications have primarily focused on how organizations use technology to mobilize citizens, particularly youth.

REFERENCES

1. Best, K., & Delmege, S. (2012). The filtered encounter: Online dating and the problem of filtering through excessive information. *Social Semiotics*, *22*(3), 237–258.

2. Bilton, N. (2014, October 29). Tinder, the fast-growing dating app, taps an age-old truth. *The New York Times.* Retrieved from http://www.nytimes.com/2014/10/30/fashion/tinder-the-fast-growing-dating-app-taps-an-age-old-truth.html

3. Birnholtz, J., Fitzpatrick, C., Handel, M., & Brubaker, J. (2014). *Identity, identification and identifiability: The language of self-presentation on a location-based mobile dating app.* Proceedings of mobile HCI, Toronto, Canada.

4. Blackwell, C., Birnholtz, J., & Abbott, C. (2015). Seeing and being seen: Co-situation and impression formation using Grindr, a location-aware gay dating app. *New Media & Society, 17*(7), 1117–1136.

5. boyd, d. (2012). Debate: Networked privacy. *Surveillance & Society, 10* (3/4), 348–350.

6. boyd, d., & Ellison, N. B. (2007). Social network sites: Definition, history, and scholarship. *Journal of Computer-Mediated Communication, 13*(1), 210–230.

7. Brubaker, J. R., Ananny, M., & Crawford, K. (2016). Departing glances: A sociotechnical account of 'leaving' Grindr. *New Media & Society, 18*(3), 373–390.

8. Couch, D., & Liamputtong, P. (2008). Online dating and mating: The use of the internet to meet sexual partners. *Qualitative Health Research, 18,* 268–279.

9. Creswell, J. (2007). *Qualitative inquiry and research design: Choosing among five traditions.* Thousand Oaks, CA: Sage.

10. Duguay, S. (2016). Dressing up Tinderella: Interrogating authenticity claims on the mobile dating app Tinder. *Information, Communication & Society,* 1–17.

11. Eigenraam, A., & Zandstra, P. (2014, October 9). 'Nog getinderd'? Dit is wat je als niet-single mist en niet mist aan Tinder. *NRC.* 2014. Retrieved from http://www.nrc.nl/nieuws /2014/10/09/nog-getinderd-dit-is-wat-je-als-niet-single-mist-en-niet-mist-aan-tinder/

12. Ellison, N., Heino, R., & Gibbs, J. (2006). Managing impressions online: Self-presentation processes in the online dating environment. *Journal of Computer-Mediated Communication, 11,* 415–441.

13. Ellison, N. B., Hancock, J. T., & Toma, C. L. (2012). Profile as promise: A framework for conceptualizing veracity in online dating self-presentations. *New Media & Society, 14*(1), 45–62.

14. Farnden, J., Martini, B., & Raymond Choo, K. K. (2015). *Privacy risks in mobile dating apps.* Proceedings of 21st Americas Conference on Information Systems (AMCIS 2015).

15. Feingold, A. (1990). Gender differences in effects of physical attractiveness on romantic attraction: A comparison across five research paradigms. *Journal of Personality and Social Psychology, 59*(5), 981–993.

16. Fiore, A., & Donath, J. S. (2005). *Homophily in online dating: When do you like someone like yourself?* Proceedings of CHI '05 extended abstracts on human factors in computing systems, Portland, OR, USA.

17. Gatter, K., & Hodkinson, K. (2016). On the differences between Tinder™ and online dating agencies: Questioning a myth. An exploratory study. *Cogent Psychology, 3*, 1–12.

18. Gershon, I. (2010). Breaking up is hard to do: Media switching and media ideologies. *Journal of Linguistic Anthropology, 20*(2), 389–405.

19. Gibbs, J. L., Ellison, N. B., & Heino, R. D. (2006). Self-presentation in online personals: The role of anticipated future interaction, self-disclosure, and perceived success in internet dating. *Communication Research, 33*(2), 152–177.

20. Glaser, B. G., & Strauss, A. (1967). *The discovery of grounded theory: Strategies for qualitative research.* Chicago: Aldine.

21. Goffman, E. (1959). *The presentation of self in everyday life.* New York, NY: Anchor Books.

22. Gonzales, M. H., & Meyers, S. A. (1993). 'Your mother would like me': Self-presentation in the personals ads of heterosexual and homosexual men and women. *Personality and Social Psychology Bulletin, 19*(2), 131–142.

23. Gudelunas, D. (2012). There's an app for that: The uses and gratifications of online social networks for gay men. *Sexuality & Culture, 16*(4), 347–365.

24. Hall, J. A., Park, N., Song, H., & Cody, M. J. (2010). Strategic misrepresentation in online dating: The effects of gender, self-

monitoring, and personality traits. *Journal of Social and Personal Relationships, 27*(1), 117–135.

25. Hamilton, N. F. (2016). Love and relationships online. In I. Connolly, M. Palmer, H. Barton, & G. Kirwan (Eds.), *An introduction to cyberpsychology* (pp. 71–85). New York, NY: Routledge.

26. Hamilton, R. J., & Bowers, B. J. (2006). Internet recruitment and e-mail interviews in qualitative studies. *Qualitative Health Research, 16*(6), 821–835.

27. Hancock, J. T., Toma, C., & Ellison, N. (2007). *The truth about lying in online dating profiles.* CHI 2007 Proceedings on online representation of self, 449–452.

28. Handel, M. J., & Shklovski, I. (2012). *Disclosure, ambiguity and risk reduction in real-time dating sites.* GROUP '12 Proceedings of the 17th ACM International Conference on Supporting Group Work, 175–178.

29. Hardey, M. (2002). Life beyond the screen: Embodiment and identity through the internet. *The Sociological Review, 50*(4), 570–585.

30. Hardey, M. (2004). Mediated relationships: Authenticity and the possibility of romance. *Information, Communication & Society, 7*(2), 207–222.

31. Hardey, M. (2008). The formation of social rules for digital interactions. *Information, Communication & Society, 11*(8), 1111–1131.

32. Harrison, A. A., & Saeed, L. (1977). Let's make a deal: An analysis of revelations and stipulations in lonely hearts advertisements. *Journal of Personality and Social Psychology, 35*(4), 257–264.

33. Hogan, B. (2010). The presentation of self in the age of social media: Distinguishing performances and exhibitions online. *Bulletin of Science, Technology & Society, 30*(6), 377–386.

34. Kazmer, M. M., & Xie, B. (2008). Qualitative interviewing in internet studies: Playing with the media, playing with the method. *Information, Communication & Society, 11*(2), 257–278.

35. Koestner, R., & Wheeler, L. (1988). Self-presentation in personal advertisements: The influence of implicit notions of attraction and role expectations. *Journal of Social and Personal Relationships, 5*(2), 149–160.

36. Kramer, N. C., & Winter, S. (2008). Impression management 2.0: The relationship of self-esteem, extraversion, self-efficacy, and self-presentation within social networking sites. *Journal of Media Psychology, 20*(3), 106–116.

37. Lampe, C., Ellison, N., & Steinfield, C. (2007). *A familiar face(book): Profile elements as signals in an online social network*. Presented at the CHI 2007, San Jose, California.

38. Leary, M. R. (1995). *Self-presentation: Impression management and interpersonal behavior*. Madison, WI: Brown & Benchmark.

39. Leary, M. R., & Kowalski, R. M. (1990). Impression management: A literature review and two-component model. *Psychological Bulletin, 107*(1), 34–47.

40. Lin, K., & Lundquist, J. (2013). Mate selection in cyberspace: The intersection of race, gender, and education. *American Journal of Sociology, 119*(1), 183–215.

41. Manning, J. (2014). Construction of values in online and offline dating discourses: Comparing presentational and articulated rhetorics of relationship seeking. *Journal of Computer-Mediated Communication, 19*(3), 309–324.

42. McPherson, M., Smith-Lovin, L., & Cook, J. M. (2001). Birds of a feather: Homophily in social networks. *Annual Review of Sociology, 27*(1), 415–444. doi: 10.1146/annurev.soc.27.1.415

43. Miller, H. (1995). *The presentation of self in electronic life: Goffman on the internet (pp. 1–8)*. Presented at the Embodied Knowledge and Virtual Space Conference, Goldsmiths' College, University of London.

44. Papacharissi, Z. (2002). The presentation of self in virtual life: Characteristics of personal home pages. *Journalism & Mass Communication Quarterly, 79*(3), 643–660.

45. Skopek, J., Schulz, F., & Blossfeld, H. (2011). Who contacts whom? Educational homophily in online mate selection. *European Sociological Review, 27*(2), 180–195.

46. Spitzberg, B. H., & Hoobler, G. (2002). Cyberstalking and the technologies of interpersonal terrorism. *New Media & Society, 4,* 71–92.

47. Stenson, C., Balcells, A., & Chen, M. (2015). *Burning up privacy on Tinder.* Presented at The Eleventh Symposium On Usable Privacy and Security, Ottawa, Canada.

48. Strauss, A., & Corbin, J. (1990). *Basics of qualitative research: Grounded theory procedures and techniques.* Newbury Park, CA: Sage.

49. Toma, C. L., & Hancock, J. T. (2010). Looks and lies: The role of physical attractiveness in online dating self-presentation and deception. *Communication Research, 37*(3), 335–351.

50. Toma, C. L., Hancock, J. T., & Ellison, N. B. (2008). Separating fact from fiction: An examination of deceptive self-presentation in online dating profiles. *Personality and Social Psychology Bulletin, 34*(8), 1023–1036.

51. Tufekci, Z. (2008). Grooming, gossip, Facebook and MySpace: What can we learn about these sites from those who won't assimilate? *Information, Communication & Society, 11*(4), 544–564.

52. Van de Wiele, C., & Tom Tong, S. (2014). *Breaking boundaries: The uses & gratifications of Grindr* (pp. 1–12). Presented at the UbiComp '14, Seattle, Washington.

53. Walther, J. B. (1996). Computer-mediated communication: Impersonal, interpersonal, and hyperpersonal interaction. *Communication Research, 23*(1), 3–43.

54. Wildermuth, S. M. (2004). The effects of stigmatizing discourse on the quality of on-line relationships. *CyberPsychology & Behavior, 7,* 73–84.

55. Woo, J. (2006). The right not to be identified: Privacy and anonymity in the interactive media environment. *New Media & Society, 8*(6), 949–967.

56. Zytko, D., Jones, Q., & Grandhi, S. A. (2014). *Impression management and formation in online dating systems.* Presented at the Twenty Second European Conference on Information Systems, Tel Aviv.

10

CHAPTER

Developed Democracies versus Emerging Autocracies: Arts, Democracy, and Innovation in Quadruple Helix Innovation Systems

Elias G Carayannis[1]* and David FJ Campbell[2,3]

1 European Union Research Center, Department of Information Systems & Technology Management, School of Business, The George Washington University, Duquès Hall, Funger Hall, Suite 515C, 2201 G Street, NW, Washington, D.C 20052, USA Full list of author information is available at the end of the article

ABSTRACT

Arts, democracy, and innovation co-evolve. While for the Triple Helix model the existence of a democracy is not necessary for knowledge production and innovation, the Quadruple Helix is here more explicit. The way how the Quadruple Helix is being engineered, designed, and 'architected' clearly shows that there cannot be a Quadruple Helix innovation system without democracy or a democratic context. The following attributes and components define the fourth helix in the Quadruple Helix: 'media-based and culture-based public,' 'civil society,' and 'arts, artistic research, and arts-based innovation'. By this, the fourth helix in the Quadruple Helix represents the perspective of the 'dimension of democracy' or the 'context of democracy' for knowledge, knowledge production, and innovation. This is particularly true when democracy is to be understood to transcend the narrow understanding of being primarily based on or being primarily rooted in government institutions (within Triple Helix). Civil society, culture-based public, quality of democracy, and sustainable development convincingly demonstrate what the rationales and requirements are for conceptualizing democracy broader. Political pluralism in a democracy co-evolves with the pluralism, diversity, and heterogeneity of knowledge, knowledge production, and innovation ('Democracy of Knowledge'). The Quintuple Helix extends the Quadruple Helix by aspects of the 'natural environments of society and economy,' 'social ecology,' and the 'socio-ecological transition'. Also, this environmental context of society can be better addressed in a democracy than in a non-democracy. The current world appears to be challenged by a race between developing democracies versus emerging autocracies over knowledge production and innovation. The contributions of arts, arts-based research, and arts-based innovation to knowledge production and innovation systems are manifold. Art helps and aids us in thinking 'beyond the box'. The traditional understanding of arts emphasizes the aesthetic dimension of arts. Art and arts can also be understood (and re-invented) as a manifestation of knowledge, knowledge production, and knowledge creation. Arts and artistic research are now being regarded as drivers for forming and pluralizing interdisciplinary and transdisciplinary configurations and networks with research in the sciences and the application and use of knowledge and innovation in the context of not only society and democracy, but also the economy. Art, arts-based research, and arts-based innovation contribute to creating (co-creating) the basis for new models of economic growth. This indicates opportunities for a creative design or creative design processes in the further co-evolution of knowledge economy, knowledge society, and knowledge democracy.

Keywords: AAA (arts, artistic research, and arts-based innovation) Co-evolution Cross-employment and multi-employmentDemocracy of knowledge Socio-ecological transition Multi-level innovation systems Innovation ecosystem Twenty-first century Fractal Research and Education and Innovation Ecosystem (FREIE) Networks and network governance Linear and non-linear innovation Quadruple Helix innovation systems Quadruple Helix Plus Quintuple Helix

BACKGROUND

The following is the vision and manifesto for knowledge production and innovation in Quadruple Helix innovation systems: arts, democracy, and innovation co-evolve. This is being stated explicitly by the concept and model of the Quadruple Helix innovation systems (Carayannis and Campbell [2006], [2009], [2010])[a]. How does the Quadruple Helix[b] differ from 'N-Tuple of Helices' (Leydesdorff [2012])? N-Tuple of Helices on the one hand, and the Quadruple Helix and the Quintuple Helix innovation systems (*Quadruple Helix Plus*) on the other, are not driven by the same emphasis, and their ramifications point into different directions (for a summary and overview, see also the interview in Park [2014]). While the N-Tuple of Helices can be regarded as an abstract meta-concept of helices, the Quadruple Helix is more concrete in its qualities by being interested in developing and proposing solutions and problem-solving approaches based on a broader conceptual and theoretical understanding. The Quadruple Helix focuses (as a final test) on reform, strategy, policy, and implementation processes and on transformation of economy, society, and democracy (economy in democracy). The Quintuple Helix could be interpreted as a Quadruple Helix Plus extension, thus representing a further development and evolution in the line of thinking that integrates the ecology (also social ecology) into our approaches for knowledge production and innovation (Carayannis and Campbell [2010]; Carayannis et al. [2012]).

The market should serve society, and the society should serve the people and individuals (and not the other way around). The role of the individual is paramount. Quadruple Helix emphasizes that 'people matter'. First of all, the Quadruple Helix is 'human-centered' and only secondarily 'institution-oriented'. The Quadruple Helix operates with a strong bottom-up momentum and encouragement. Therefore, for the Quadruple Helix, democracy matters and is key and crucial. Quadruple Helix addresses knowledge production and innovation in the context of democracy. There cannot be a Quadruple Helix (or Quintuple Helix) 'outside' (or without) democracy. Here we see a co-development and co-evolution of democracy and knowledge production and innovation for knowledge society and knowledge democracy. This addressed and asserted co-evolution of the Quadruple Helix (Quintuple Helix) with democracy is even more plausible, when democracy is not being understood in terms of a 'narrow' electoral democracy, but when democracy is being carried and driven by a 'broader' conceptual understanding, for which freedom, equality, sustainable development, and the idea of 'quality of democracy' are equally crucial (Campbell and Carayannis [2013a]; [2015]).

Triple Helix is possible within a democracy. However, Triple Helix is also possible without a democracy. The Triple Helix focuses on the knowledge economy, which may be approached by a non-democratic framework. Non-democratic (authoritarian) political regimes may be tempted to implement varieties of Triple Helix designs. Per definition, to already begin with a conceptual starting point, it is impossible for a non-democratic (authoritarian) political regime to try to implement a Quadruple Helix. *There is no Quadruple Helix without democracy.* In addition, evidence suggests that the ecological sensitivity of the Quintuple Helix (Carayannis et al. [2012]) can be more easily or realistically be implemented and promoted within a democratic context of knowledge production and innovation. For the Quadruple Helix, the 'democracy matters': this is in line with a view of a 'Neo-Renaissance' where democracy encourages development in action for smart, sustainable, and inclusive growth, by this advocating sustainable development. This should allow for 'happy accidents'. For discourses on knowledge and innovation, a *Democracy versus Technocracy* issue can be postulated, where technocratic (and bureaucratic) approaches to innovation in non-democratic regimes are being questioned and challenged by knowledge production and innovation in democracies. Also, for the developing countries and emerging markets, this has implications and ramifications, where there should be expectations that developments in knowledge and innovation are paralleled by progress in democratization (of course, this may not be always the case in empirical terms or empirically). Democracy acts as one of the levers that happy accidents in knowledge production and innovation are being transformed and translated into opportunities and benefits for society and to the people. Can there also be a 'democratic capitalism', and which attempts of realization can there be approached or tried out (Carayannis and Kaloudis [2010])?

The fourth helix in the Quadruple Helix consists of the following components or attributes: the 'media-based and culture-based public', 'civil society', and 'arts, artistic research, and arts-based innovation'. This fourth helix also could be paraphrased as the dimension of democracy (knowledge democracy) or the dimension of knowledge society in the context of democracy. Arts-based research offers opportunities for interdisciplinary and transdisciplinary configurations, re-configurations, and networks with the knowledge production in the sciences-based research. Arts-based innovation liberates the concept of innovation from primarily economically driven concerns and purposes, and encourages creativity for processes of knowledge production and innovation. Arts-based research and arts-based innovation possibly lead to a re-thinking and re-modeling (and 're-invention') of economic growth models not only in the advanced economies, but also in the emerging markets.

Metrics for a possible measurement of Quadruple Helix (Quintuple Helix) innovation systems could emphasize the following aspects and considerations: (1) the degree of pluralism, diversity, and heterogeneity of different forms and paradigms of knowledge, knowledge production, and innovation; and (2) the degree of networks and their dynamics and connectivity across the pluralistic architecture of knowledge and innovation. This would imply to look at the pluralism of knowledge and innovation and the pluralism of their structures and processes by which knowledge and innovation are interconnected and interwoven in networks and in 'networks in clouds and crowded crowding clouds'. Networking and 'clouds' ('crowds') implies to focus in greater detail on the *connecting synapses* in the architecture of knowledge and innovation systems. This notion of pluralism in knowledge and innovation is also being captured by the idea of 'Democracy of Knowledge', meaning that the pluralism in knowledge production communicates with pluralism (political pluralism) in a democracy, proposing here a co-evolution of democracy with systems of knowledge production and innovation (Carayannis and Campbell [2009], [2012], p. 55). Metrics of measurement of knowledge and innovation should therefore be sensitive for expressing, displaying, and mapping pluralism, diversity, and heterogeneity. Furthermore, the argument could be set up for discussion that forms or types of knowledge production may correspond with forms and types of innovation systems. Congruencies here could be as follows: Triple Helix innovation (Etzkowitz and Leydesdorff [2000]) interacts with 'Mode 1' and 'Mode 2' knowledge production (Gibbons et al. [1994]), while Quadruple Helix and Quintuple Helix innovation systems refer to a 'Mode 3' type of knowledge production. In terms of pluralism and diversity, the Quadruple Helix (Quintuple Helix) appears to be better designed and adapted for the social environments of democracy and advanced economies. In economic model building, the concept of 'isoquants' (*Isoquanten*) expresses relationships between indicators or factors and furthermore demonstrates what happens with regard to these patterns of interaction when there is a tendency or momentum of higher development. Higher levels of development could be accompanied by shifts in patterns of correspondence. In Figure 1, we suggest how an econometric model of isoquants could be applied to types of knowledge production and innovation systems. Figure 1 indicates opportunities for developing mature metrics of measurement for knowledge production and innovation systems. For further debate, there also should be a focus and sensitivity for how arts-based research and arts-based innovation influence and contribute to economic growth and thus may alter our models that underlie approaches to economic progress. There is an expectation that creativity is being crucially cross-fertilized by pluralism and heterogeneity in knowledge production and innovation. Without creativity, innovation is not viable in the long run.

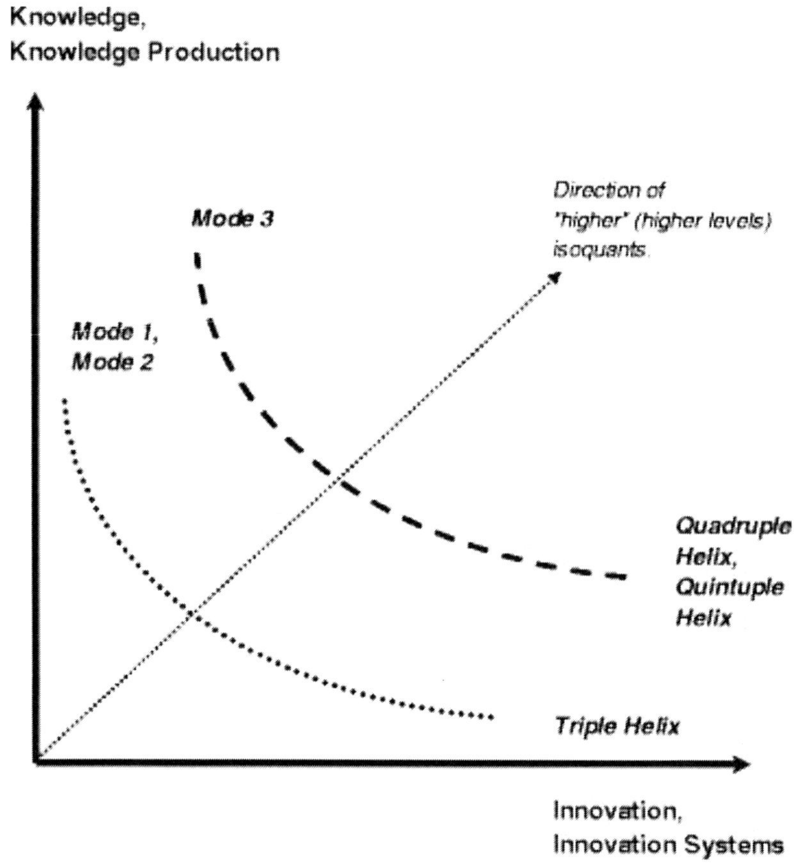

Figure 1 Possible metrics of measurement of knowledge and innovation (conceptual approach).

RESULTS AND DISCUSSION

Research questions on artistic research

This article contribution does not focus on the *arts as such*. The focus is on 'artistic research' and how artistic research relates to research, knowledge production (knowledge creation), innovation, and innovation systems. Artistic research may also have the potential to help us to better understand the arts themselves. By this, artistic research qualifies as an epistemic approach ('epistemic tool') that navigates to core meanings of arts and of 'art in the arts'. Artistic research, however, also bridges, cross-connects, and links the arts with knowledge production (research) and knowledge application (innovation) in the sciences or research and innovation that are based on the sciences[c]. Therefore, at least potentially, artistic research is also interdisciplinary in character. Artistic research adds to the development and formation of designs and architectures of interdisciplinary research platforms and research as well as innovation networks, where different disciplines in the sciences are interconnected with the disciplines in the arts through research and innovation activities.

In fact, interdisciplinary ambitions in the sciences are reinforced and excelled by bringing artistic research into play. Artistic research and also the arts in general help in creatively strengthening and unfolding the interdisciplinary drive in the sciences. In one understanding, interdisciplinarity does not happen 'automatically' in the organizational context of the sciences but requires an involvement of structures and processes that encourage a further development of interdisciplinarity (within the institutional framework of universities or of other higher education institutions). For example, academic careers often follow a disciplinary logic; therefore, inserting and introducing interdisciplinarity to organizations and networks requires to innovate and to re-invent the academic career logic. The organizational framing of transdisciplinarity creates even further challenges (on interdisciplinarity and transdisciplinarity; see for example Arnold [2013a], [2013b]). There exists and is the opportunity of configuring and re-configuring scientific research (research in the sciences) and artistic research, interwoven in arrangements of interdisciplinarity and transdisciplinarity. In fact, artistic research has all the potential to increase interdisciplinarity and transdisciplinarity also in research in the sciences, when scientific research and artistic research are being interlinked.

Our analysis is being driven by the following core research question: How does artistic research relate to research in the sciences, and how does artistic research relate to innovation and innovation systems? Our inclination is to engage further in formulating, developing, and designing propositions in reference to our research question. These propositions are more tentative in character; additional 'research about research' is necessary and may impact future research agendas. The analysis of our research question will be based on Carayannis and Campbell ([2013]) and will depart in iterations conceptually from there. We are motivated to inquire connections of artistic research to innovation and innovation systems by relying on and by applying consequently the concepts of the Quadruple and Quintuple Helix innovation systems. In fact, we believe that the Quadruple and Quintuple Helixes are designed (and driven) in a way and are carried by an understanding that emphasizes the importance of arts, arts universities, and artistic research for creativity, knowledge production, and innovation. Triple Helix represents a basic model of the innovation core and was developed by Etzkowitz and Leydesdorff ([2000]; Leydesdorff [2012]). Quadruple Helix (Carayannis and Campbell [2009]) and Quintuple Helix (Carayannis and Campbell [2010]) bring in additional perspectives and by this already 'contextualize the context' (see Figure 2). When we develop the importance and meaning of artistic research for research and innovation, we will follow in particular the conceptual logic of the models of the Quadruple and Quintuple Helix innovation systems. These models will serve as reference for artistic research. The Quadruple Helix and the Quintuple Helix express and emphasize why arts and artistic research are important for knowledge production and innovation.

Our following analysis is structured in the following sections. In the 'Arts and artistic research' section, we explore further the cross-connections and interconnection between arts and artistic research. This is based on an understanding of arts as a manifestation of knowledge. The 'Innovation systems in conceptual evolution: Mode 3 knowledge production in Quadruple and Quintuple Helix innovation systems' section embeds artistic research in the context of the concepts of Quadruple and Quintuple Helix innovation systems. We demonstrate how concepts of innovation and innovation systems have evolved and how knowledge production, innovation, and structures of organizations are intertwined in co-evolution. Finally, in the 'Conclusions' section, the conclusion to our analysis, we speculate on possible future scenarios of co-development of *arts, research, innovation, and society.* This co-development has all the potentials of becoming crucial for the further progress of innovation and innovation systems that drives knowledge economy, knowledge society, and knowledge democracy.

Arts and artistic research

What is art or what are the arts? This creates a challenge in the quest for finding or identifying answers. However, for our analysis presented here, a possible definition of arts is not of primary concern, because we will focus with greater emphasis on artistic research and its ramifications for knowledge production and innovation. Therefore, we reflect more briefly on the issue of what art is or what arts may be considered to be. There

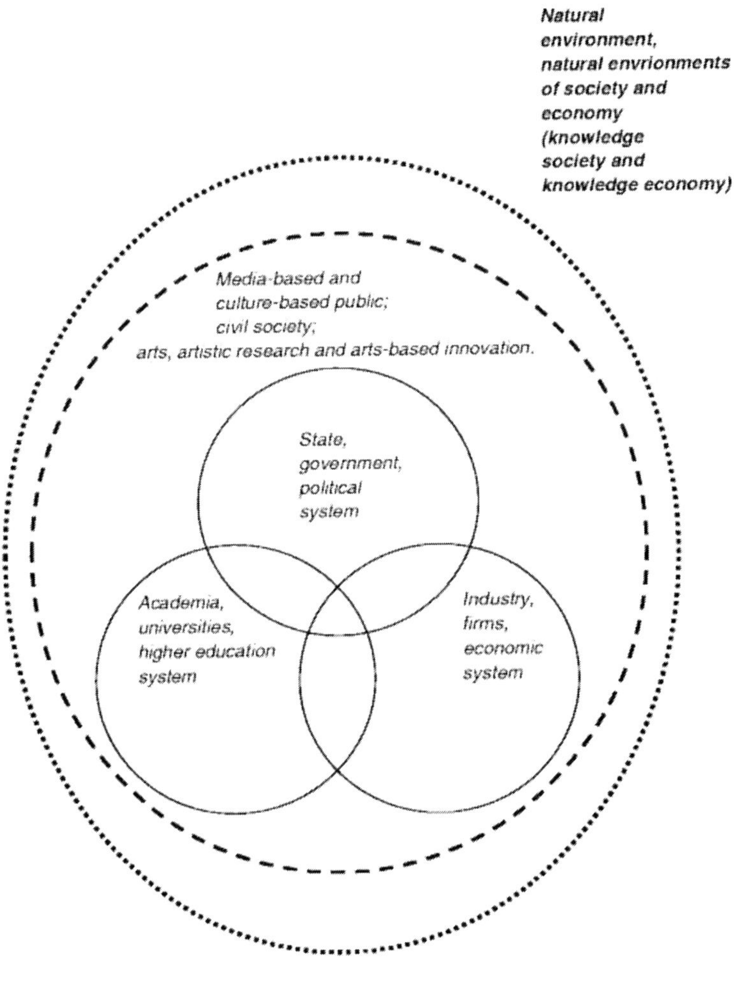

Source: Authors' own conceptualization based on Carayannis and Campbell (2010, p. 62; 2013).

Figure 2 The Quintuple Helix (five-helix model) innovation system.

exists not only one definition, but a pluralism of different definitions of arts (e.g., see Campbell [2013b]). We must recognize and should acknowledge a variety of different definitions of arts, also with competing, sometimes even conflicting meanings. There are contradictions between the available definitions of arts. Also, definitions of arts have changed over time and continuously will do so and are furthermore context-dependent. Are the arts older than the sciences? Art (as a concept and practice) exists now for several hundred, better several thousand, years (at least), so there was a sufficiently long time for a serious evolution of arts. Art (as a concept and practice) probably is even older than the sciences (modern sciences).[d] A pragmatic simplification could suggest that art is what artists are doing (and artists do very different things). Consequently, 'established art' is being represented by the established artists. But there exists no universal criterion or general standard as to who may qualify (or not) being an artist. We know that several of the most influential (and innovative) artists only had an impact later in their life, if not even for later generations. In these cases, the not-established (non-established) artists were even more important.

A 'traditional' understanding of arts frequently associates the arts with an 'aesthetic' dimension, which could be more abstract or more concrete. More concrete means to indicate a 'perceived beauty' also as an emotionally perceived beauty. Emotions imply then that beauty causes or is connected to emotions in the human 'observer' (also producer) of arts, when art is being perceived (created). The emotional spectrum can be comprehensive and diversified, and also controversial. Looking at art from a historical perspective, the aesthetic dimension of arts often was thought to express the beauty or perfect beauty (perfect order) of the world (the universe), of society, and also of individual people (for example, see the review and discussion in Öcal [2013], pp. 11 to 27).[e] This expression of beauty or perfection (*beautiful perfection*) could have religious connotations but was not necessarily linked to religious connotations. In several contexts, also cross-references were drawn between the beautiful, the perfect, and the good. Was this the case (the construction of meaning), then the beautiful was furthermore the morally or ethically good. The 'beautiful order' expressed the 'morally good order'. By associating art closely to an aesthetic dimension, arts can also fulfill 'aesthetic functions' or also those aesthetic functions, which society or specific communities want to assign to arts. Acknowledging modern (post-modern) reactions against traditional (too traditional) aesthetical concepts and conceptions of beauty, the aesthetic dimension may also be sub-clustered into a complex world of very different sub-dimensions. Can the absence of beauty also be expressed in the aesthetical dimension? Is there the 'beauty of the non-beauty' or the 'beauty of the ugly', and does the ugly also fulfill aesthetic functions? Does the aesthetic dimension capture equally the presence and also the absence of beauty? Can beauty (aesthetics) be measured without referring to a specific context? In a terminology of *measurement of beauty*, beauty could be represented on a dimension (a scale or multi-dimensional scale) of aesthetics that expresses the presence (presence and/or absence) of beauty, allowing for gradual degrees of beauty or also the expression of 'positive beauty' and 'negative beauty'

In addition to aesthetics, what are the possible alternative dimensions for conceptualizing and 'measuring' the arts that complement and expand the aesthetic dimension of arts? Beyond aesthetics, how can we conceptualize arts further? In fact, we are interested in also promoting an understanding of arts which drives arts *further and beyond an only-aesthetic-understanding of arts*, which does not deny the aesthetic dimension (dimensions) but intends to complement the aesthetic dimension of arts. Therefore, we want to suggest the following as a new reference point for debate:

Art and arts can also be understood (and re-invented) as a manifestation of knowledge, knowledge production and knowledge creation. Furthermore, knowledge production and knowledge creation extend to knowledge application and knowledge use.

The here presented approach to arts introduces knowledge as an additional dimension for defining and understanding arts. This additional dimension does not replace but extends the aesthetic dimension of arts, by this making the arts clearly multi-dimensional (see Figure 3).[f] Consequences of this are (when we follow the logic of that particular knowledge definition of arts) that the arts cannot be understood comprehensively and sufficiently only on the basis of aesthetics alone. Of course, we could speculate whether the knowledge involvement of arts implies a knowledge for which aesthetic considerations play frequently an important role (forms of beauty or non-beauty).[f]Aesthetics may interact with different forms of reasoning or intelligence, such as intuition or emotional intelligence. Research and progress in the sciences clearly are also driven by and benefit obviously from intuition or emotional intelligence. Clearly, there is furthermore more of a need to continuously reflect which dimensions (in addition to aesthetics and knowledge) may also be of further relevance for art and arts.

By introducing this additional knowledge dimension of arts, complementing the dimension of aesthetics, it is being acknowledged that also forms of arts-based research and arts-based innovation are existing and can emerge further. In fact, artistic research represents one crucial expression of arts as a manifestation of knowledge. What are the differences between arts-based research and artistic research? Boundaries here are obviously fluid and depend on specific positions of perspective. Connotations of artistic research imply that the research is not only arts-based, but that research and arts are actually being intertwined and interlinked with each other more directly (on 'artistic research', see furthermore Damianisch [2013] and Mateus-Berr [2013]). Artistic research is more immediately and straightforwardly connected to the arts than arts-based research, at least being seen from a conceptual understanding. Of course, also artistic research and arts-based research overlap. In fact, opportunities of interdisciplinary and transdisciplinary combinations of artistic research,

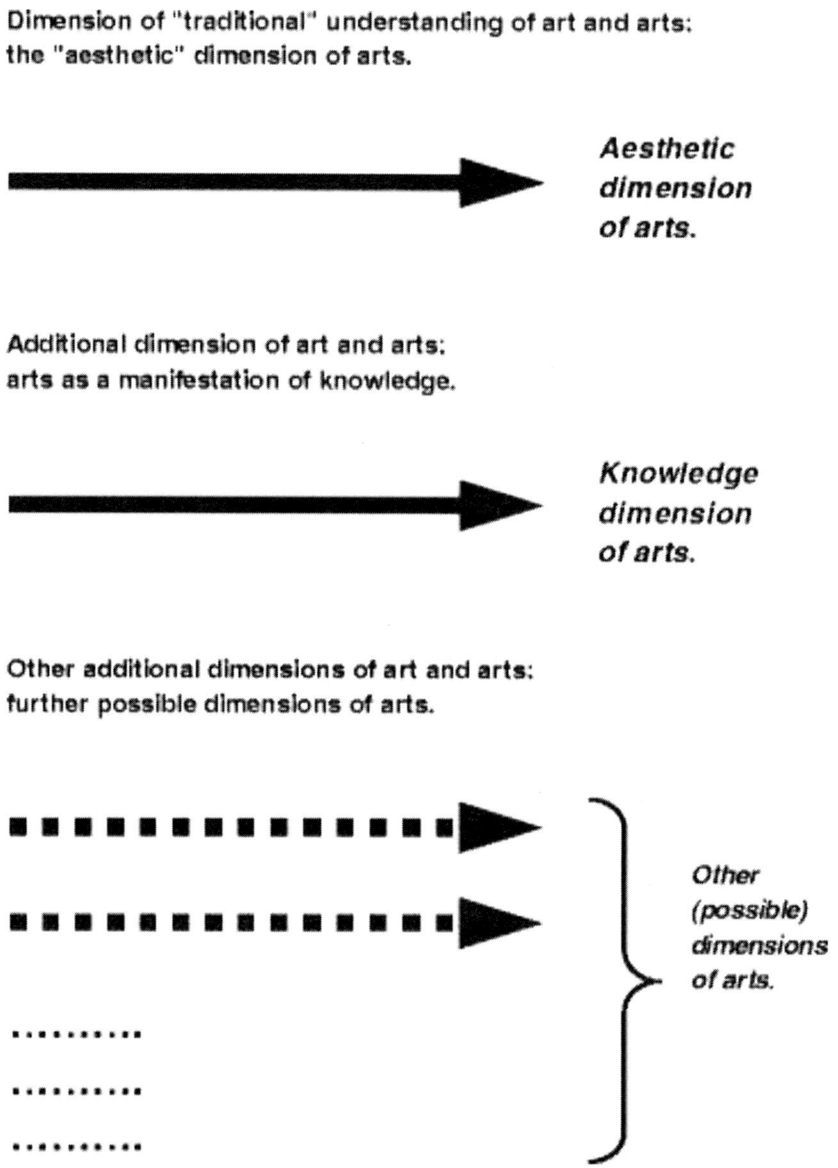

Figure 3 Dimensions of conceptualization and measurement of art and arts.

arts-based research, and academic research (in the sciences), extended by configurations of arts-based innovation and sciences-based innovation, are arising and can be utilized by institutions, organizations, communities, and networks. 'In a short-cut, *transdisciplinarity may be defined as the application of interdisciplinarity* (transdisciplinarity = application of interdisciplinarity?)' (Campbell and Carayannis [2013a], p. 34). Of course, we have to admit that there also can be the transdisciplinarity of a 'disciplinarity in application'. Transdisciplinarity usually (always) refers to forms of application. Often (but not always), interdisciplinarity is more application-friendly than disciplinarity (see again Arnold [2013b]).

The proposition (that knowledge, knowledge production, and knowledge creation qualify as an additional dimension for art and arts) has the implication that the arts and our understanding of arts are being opened to knowledge and the 'tree of ramifications' of knowledge. In fact, this bridges the arts with research and innovation. The arts are interconnected with research and innovation in the arts and with research and innovation in the sciences. 'Art as a manifestation of knowledge' draws interdisciplinary and transdisciplinary configurations in connectivity, where knowledge production and innovation in the arts are extended to knowledge production and innovation beyond the arts. Furthermore, we can argue that in the context of a knowledge understanding of the arts, the artistic research actually is important for comprehending arts and also for developing the arts further. (There is no comprehensive understanding of the arts without artistic research?) Artistic research also touches on epistemic implications for the arts. Already in the introduction, we asserted 'By this, artistic research qualifies as an epistemic approach ("epistemic tool") that navigates to core meanings of arts and of "art in the arts"'. Is aesthetics more than only-beauty, or what is the beauty of epistemology?

Having introduced knowledge, knowledge production, and knowledge creation as a second dimension for art and arts, in addition to the dimension of aesthetics, we want to speculate on some of the implications and ramifications of this intellectual endeavor. The following are the propositions we want to suggest for further discussion:

1. Art as a manifestation of knowledge: Our proposition is that art and arts can also be understood as a manifestation of knowledge. This knowledge-based definition refers to knowledge production and knowledge creation (research)[h] as well as to the application and use of knowledge (innovation). We introduced knowledge as a second (additional) dimension for defining arts that complements the first (and more traditional) dimension of arts, which is aesthetics. In our opinion, knowledge does not replace but complements aesthetics for a broader understanding of arts. Dimensions in addition to aesthetics and knowledge appear also to be possible and valid for arts. (In the context of our analysis here,

however, we do not engage further in speculating on dimensions of arts beyond aesthetics and knowledge.)

2. Arts, artistic research, interdisciplinary, and transdisciplinary combinations of research in the arts and research in the sciences, innovation, and innovation systems: Artistic research represents one outflow in the consequence of approaching arts as a manifestation of knowledge. In fact, it could be argued that artistic research helps in better understanding arts in all the possible ramifications. To turn the argument, without artistic research, our pictures of arts are incomplete, probably also too fragmented. Without artistic research, our visions of arts are insufficient. Based on this paradigm of knowledge, arts as a manifestation of knowledge and artistic research clearly enable to cross-connect and interconnect the arts with knowledge. Discourses in knowledge are being bridged with discourses in the arts. Research and innovation can spread from the domain of knowledge to the domain of arts. Arts and artistic research are now being regarded as drivers for forming and pluralizing interdisciplinary and transdisciplinary configurations and networks with research in the sciences and the application and use of knowledge and innovation in the context of not only society and democracy, but also the economy. Arts and artistic research aid and add in widening our horizons of knowledge production and knowledge creation within the sciences. With arts and artistic research, the domain (domains) of intelligence can be more fully leveraged for knowledge creation, which is also important for knowledge production in the sciences. What are forms or sources of knowledge production (intelligence) beyond language or the use of (written) texts? Arts may also be utilized as an unconventional strategy for preparing grounds for The New 'beyond horizon', for encouraging and experimenting with unconventional configurations of 'interdisciplinarity in transdisciplinarity'. Arts as a manifestation of knowledge and artistic research redefine the arts in a way, making it then obvious as to why the arts are crucial (at least potentially) for innovation and whole innovation systems (national systems of innovation or multi-level innovation systems; Carayannis and Campbell [2006]; Carayannis and Campbell [2012], pp. 32 to 35). One radical proposition would be that without strong and continuously evolving cross-references to arts, every comprehensive innovation system (national or multi-level architectured innovation system) is only premature, operating below possible capabilities. 'The arts excel as innovation systems'. Arts and artistic research also contribute to Quality of Democracy and innovation capabilities in

democracies and in processes of further democratization (Campbell and Carayannis [2013a]).

3. Epistemic implications of arts and artistic research: Arts as a manifestation of knowledge and artistic research also emphasize the epistemic implications of arts.[i] This is not being seen as being in conflict with the aesthetic dimension of arts, since beauty or non-beauty are per se neutral with regard to epistemic potentials. Epistemic ramifications of arts and artistic research are manifold, diverse, and heterogeneous. For example, artistic research engages the arts in interdisciplinary and transdisciplinary networks of research in the sciences (or in networks of research in the sciences and the arts). Arts and artistic research help us to think and imagine beyond the 'written text'. Imagination and science fiction are references for the powers of fantasy and may inspire processes and scenarios of knowledge creation and knowledge production, long before a particular knowledge application or technology implementation is realistic ('thinking in possibilities before possible uses'). 'Fiction or science fiction may serve as stimulators for creative ideas, with the potential of being later transformed, at least partially (and of course not always), into new knowledge creation and production. We can also call this the creativity of knowledge creation' (Carayannis and Campbell [2010], p. 48). Arts and artistic research can be used for designing 'virtual worlds' that could not exist 'outside in the world' or that contradict the 'outside world', leading to paradoxical phenomena and furthermore to questions of what reality is or 'what really exists?'. One example is the Dutch graphic artist M.C. Escher (Maurits Cornelis Escher), who lived from 1898 until 1972[j]. Escher engaged in drawing 'impossible constructions'[k] that actually represent 'logical contradictions' (seen from a certain perspective of perception). Two famous drawings of his are 'Ascending and Descending' (1960)[l] and 'Waterfall' (1961)[m]. We could speculate whether the arts help us in seeing 'impossible worlds'. Is there a contradiction between the picture (image) and a conventional logical explanation, which reality is truer (or is already this wording a contradiction in itself)? These briefly discussed examples illustrate only tentatively and fragmentarily (in a partial manner) the whole spectrum of epistemic implications of arts and artistic research that appear to be possible. Further research and further artistic work is necessary.[n]

Innovation systems in conceptual evolution: Mode 3 knowledge production in Quadruple and Quintuple Helix innovation systems

Universities, or higher education institutions (HEIs) in general, have three main functions: teaching and education, research (research and experimental development, R&D), and the 'third mission' activities, for example, innovation (Campbell and Carayannis [2013], p. 5). In reference to 'arts universities', now the question and challenge arise as to what extent and in what way the arts universities differ from the (more traditional) universities in the sciences. Arts universities obviously place an emphasis on the arts, and the arts are not identical with the sciences. However, also arts universities frequently make references to the sciences, and thus, also arts universities can express competences in teaching and in carrying out research in the sciences. The other major challenge of arts universities is to engage in artistic research and 'arts-based innovation'. By this, arts universities (and other higher education institutions in the arts) are also being linked to and are being interlinked with national innovation systems and multi-level innovation systems. This widens the whole interdisciplinary and transdisciplinary spectrum of higher education systems. Artistic research furthermore complements the 'teaching of arts' at arts universities (see also the propositions formulated by Bast [2013]). Hybrid and innovative combinations of universities of arts and universities of the sciences are possible and indicate organizational opportunities for promoting creativity.

University research, in a traditional understanding and in reference to universities in the sciences, focuses on basic research, often framed within a matrix of academic disciplines, and without a particular interest in the practical use of knowledge and innovation. This model of university-based knowledge production also is being called Mode 1 of knowledge production (Gibbons et al. [1994]). Mode 1 is also compatible with the linear model of innovation, which is often being referred to Bush ([1945]). The linear model of innovation asserts that first there is basic research in the university context: gradually, this university research will diffuse out into society and the economy. It is then the economy and the firms that pick up the lines of university research and develop these further into knowledge application and innovation, for the purpose of creating economic and commercial success in the markets outside of the higher education system. Within the frame of linear innovation, there is a sequential 'first-then' relationship between basic research (knowledge production) and innovation (knowledge application)

The Mode-1-based understanding of knowledge production has been challenged by the new concept of Mode 2 of knowledge production, which was developed and proposed by Gibbons et al. ([1994], pp. 3 to 8, 167). Mode 2 emphasizes a knowledge application and a knowledge-based problem solving that involves and encourages the following principles:

'knowledge produced in the context of application', 'transdisciplinarity', 'heterogeneity and organizational diversity', 'social accountability and reflexivity', and 'quality control' (see furthermore Nowotny et al. [2001], [2003], [2006]). Key in this setting is the focus on a knowledge production in contexts of application. Mode 2 expresses and encourages clear references to innovation and innovation models. The linear model of innovation also has become challenged by non-linear models of innovation, which are interested in drawing more direct connections between knowledge production and knowledge application, where basic research and innovation are being coupled together not in a first-then, but in an 'as well as' and 'parallel' (parallelized) relationship (Campbell and Carayannis [2012]). Mode 2 appears also to be compatible with non-linear innovation and its ramifications.

The Triple Helix model of knowledge, innovation, and university-industry-government relations, which was introduced and developed by Henry Etzkowitz and Loet Leydesdorff ([2000], pp. 111 to112), asserts a basic core model for knowledge production and innovation, where three 'helices' intertwine, by this creating a national innovation system. The three helices are identified by the following systems or sectors: academia (universities), industry (business), and state (government). Etzkowitz and Leydesdorff refer to 'university-industry-government relations' and networks, putting a particular emphasis on 'tri-lateral networks and hybrid organizations', where those helices overlap in a hybrid fashion. Etzkowitz and Leydesdorff ([2000], p. 118) also explain how, in their view, the Triple Helix model relates to Mode 2: the 'Triple Helix overlay provides a model at the level of social structure for the explanation of Mode 2 as an historically emerging structure for the production of scientific knowledge, and its relation to Mode 1'. More recently, Leydesdorff ([2012]) also introduced the notion of N-Tuple of Helices.

Mode 1 and Mode 2 may be characterized as 'knowledge paradigms' that underlie the knowledge production (to a certain extent also the knowledge application) of higher education institutions and university systems. Success or quality, in accordance with Mode 1, may be defined as: 'academic excellence, which is a comprehensive explanation of the world (and of society) on the basis of "basic principles" or "first principles", as is being judged by knowledge producer communities (academic communities structured according to a disciplinary framed peer review system)'. Consequently, success and quality, in accordance with Mode 2, can be defined as: 'problem-solving, which is a useful (efficient, effective) problem-solving for the world (and for society), as is being judged by knowledge producer and knowledge user communities' (Campbell and Carayannis [2013], p. 32). A Mode 3 university, higher education institution, or higher education system would represent a type of organization or system that seeks creative ways of combining and integrating different principles of knowledge production and knowledge application (for example, Mode 1 and Mode 2), by this encouraging diversity and heterogeneity and also creating creative and innovative organizational contexts for research and innovation. Mode 3 encourages the formation of 'creative knowledge environments' (Hemlin et al. [2004]). 'Mode 3 universities' and Mode 3 higher education institutions and systems are prepared to perform 'basic research in the context of application' (Campbell and Carayannis [2013], p. 34). This has furthermore qualities of non-linear innovation. Governance of higher education and governance in higher education must also be sensitive, whether a higher education institution operates on the basis of Mode 1, Mode 2, or a combination of these in Mode 3. The concept of 'epistemic governance' emphasizes that the underlying knowledge paradigms of knowledge production and knowledge application are being addressed by quality assurance and quality enhancement strategies, policies, and measures (Campbell and Carayannis [2013b], [2013c]).

Emphasizing again a more systemic perspective for the Mode 3 knowledge production, a focused conceptual definition may be as follows (Carayannis and Campbell [2012], p. 49): Mode 3 '... allows and emphasizes the co-existence and co-evolution of different knowledge and innovation paradigms. In fact, a key hypothesis is: The competitiveness and superiority of a knowledge system or the degree of advanced development of a knowledge system are highly determined by their adaptive capacity to combine and integrate different knowledge and innovation modes via co-evolution, co-specialization and co-opetition knowledge stock and flow dynamics' (see Carayannis and Campbell [2009]; on 'Co-Opetition', see Brandenburger and Nalebuff [1997]). Analogies are being drawn and a co-evolution is being suggested between diversity and heterogeneity in advanced knowledge society and knowledge economy, political pluralism in democracy (knowledge democracy), and the quality of a democracy. The 'Democracy of Knowledge' refers to this overlapping relationship. As is being asserted, 'The *Democracy of Knowledge*, as a concept and metaphor, highlights and underscores parallel processes between political pluralism in advanced democracy, and knowledge and innovation heterogeneity and diversity in advanced economy and society. Here, we may observe a hybrid overlapping between the *knowledge economy, knowledge society and knowledge democracy*' (Carayannis and Campbell [2012], p. 55). The Democracy of Knowledge, therefore, is further reaching than the earlier idea of the 'Republic of Science' (Polanyi [1962]).

The main focus of the Triple Helix innovation model concentrates on university-industry-government relations (Etzkowitz and Leydesdorff

[2000]). In that respect, Triple Helix represents a basic model or a core model for knowledge production and innovation application. The models of the Quadruple Helix and Quintuple Helix innovation systems are designed to comprehend already and to refer to an extended complexity in knowledge production and knowledge application (innovation); thus, the analytical architecture of these models is broader conceptualized. To use metaphoric terms, the Quadruple Helix embeds and contextualizes the Triple Helix, while the Quintuple Helix embeds and contextualizes the Quadruple Helix (and Triple Helix). The Quadruple Helix adds as a fourth helix the media-based and culture-based public, the civil society, and arts, artistic research, and arts-based innovation (Carayannis and Campbell [2009], [2012], p. 14; see also Danilda et al. [2009]). The Quadruple Helix also could be emphasized as the perspective that specifically brings in the dimension of democracy or the context of democracy for knowledge, knowledge production, and innovation. The Quintuple Helix innovation model even is more comprehensive in its analytical and explanatory stretch and approach, adding furthermore the fifth helix (and perspective) of the 'natural environments of society' (Carayannis and Campbell [2010], p. 62) (see Figures 4 and 2).

The Triple Helix is explicit in acknowledging the importance of higher education for innovation. However, it could be argued that the Triple Helix sees knowledge production and innovation in relation to economy; thus, the Triple Helix models first of all (primarily) the economy and economic activity. In that sense, the Triple Helix frames the knowledge economy. The Quadruple Helix brings in the additional perspective of society (knowledge society) and of democracy (knowledge democracy). The Quadruple Helix innovation system understanding emphasizes that sustainable development of and in economy (knowledge economy) requires that there is a co-evolution of knowledge economy and knowledge society and knowledge democracy. The Quadruple Helix even encourages *the perspectives of knowledge society and of knowledge democracy* for supporting, promoting, and advancing knowledge production (research) and knowledge application (innovation). Furthermore, the Quadruple Helix is also explicit that not only universities (higher education institutions) of the sciences, but also universities (higher education institutions) of the arts should be regarded as decisive and determining institutions for advancing next-stage innovation systems: the interdisciplinary and transdisciplinary connection of sciences and arts creates crucial and creative combinations for promoting and supporting innovation. Here, in fact, lies one of the keys for future success. The concept and term of 'social ecology' refers to 'society-nature interactions' between 'human society' and the 'material

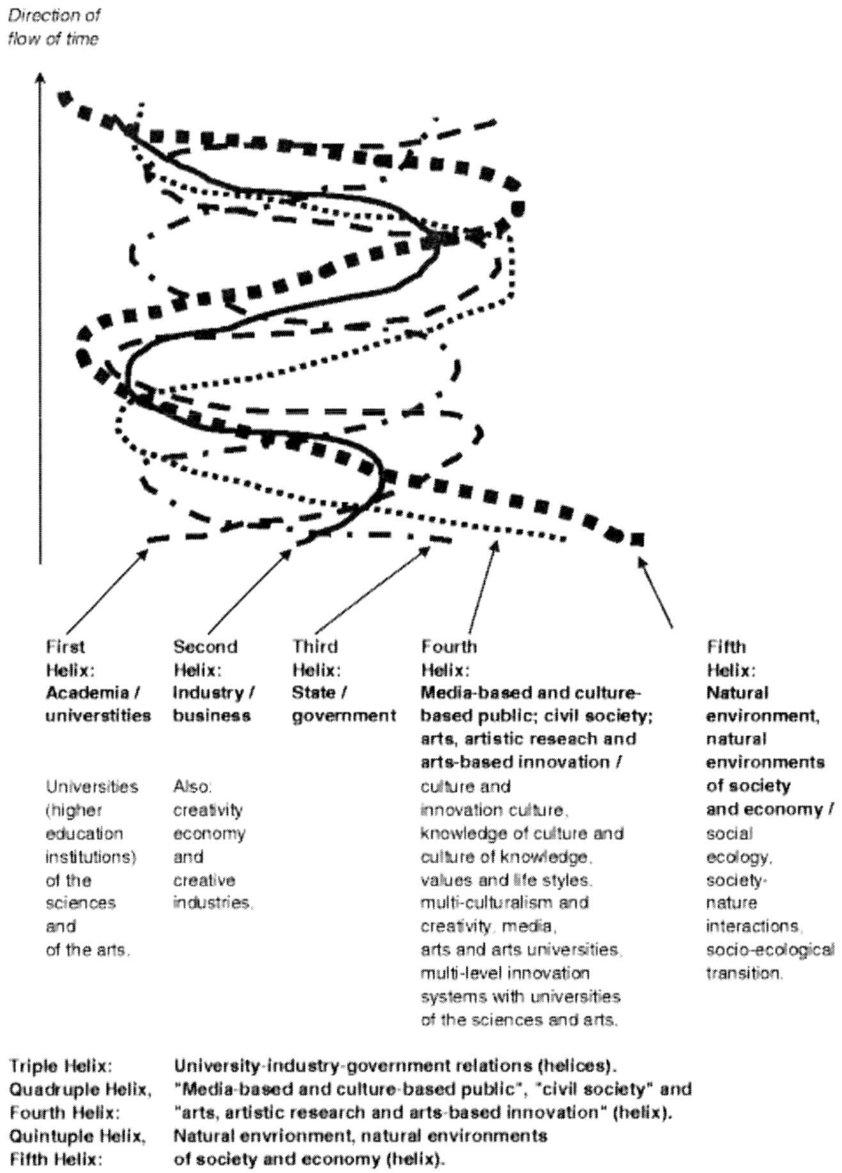

Source: Authors' own conceptualization based on Etzkowitz and Leydesdorff (2000, p. 112), Carayannis and Campbell (2009, p. 207; 2012, p. 14; 2013) and Danilda et al. (2009).

Figure 4 The Quadruple and Quintuple Helix innovation systems.

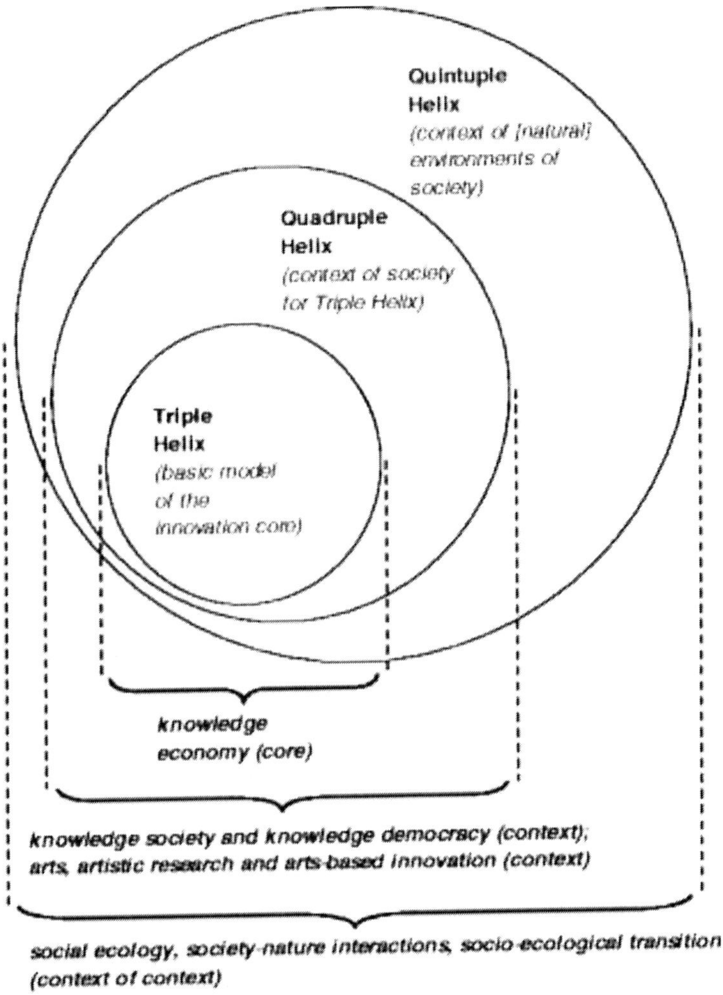

Source. Authors' own conceptualization based on Carayannis, Barth and Campbell (2012, p. 4) and Carayannis and Campbell (2013).

Figure 5 The Quadruple and Quintuple Helix innovation systems in relation to society, economy, democracy, and social ecology.

world' (see, for example, Fischer-Kowalski and Haberl [2007]). The European Commission ([2009]) identified the necessary socio-ecological transition of economy and society not only as one of the great next-phase challenges, but also as an opportunity, for the further progress and advancement of knowledge economy and knowledge society. The Quintuple Helix refers to this socio-ecological transition of society, economy, and democracy, and the Quintuple Helix innovation system is therefore ecologically sensitive. Quintuple Helix bases its understanding of knowledge production (research) and knowledge application (innovation) on social ecology (see Figure 5). Environmental issues (such as global warming) represent issues of concern and of survival for humanity and human civilization. But the Quintuple Helix translates environmental and ecological issues of concern also in potential opportunities, by identifying them as possible drivers for future knowledge production and innovation (Carayannis et al. [2012]). This, finally, defines also opportunities for the knowledge economy. "The Quintuple Helix supports here the formation of a win-win situation between ecology, knowledge and innovation, creating synergies between economy, society and democracy" (Carayannis et al. [2012], p. 1).

CONCLUSIONS

The terms and concepts of Mode 3 knowledge production and Quadruple Helix innovation systems were first introduced to international academic debate by Carayannis and Campbell ([2006], 2009) and were later developed further (Carayannis and Campbell [2012]). The same applies to the Quintuple Helix (Carayannis and Campbell [2010]). From the beginning, the media-based and culture-based public as well as universities and other higher education institutions of the arts were being regarded as crucial attributes and components of the Quadruple and Quintuple Helix innovation systems, implying that arts are essential for the progress and evolution of innovation systems (see again Figures 4 and 2). In our analysis here, we developed more specifically the Quadruple and Quintuple Helix innovation systems in terms and in favor of arts, artistic research, and arts-based innovation. We wanted to demonstrate the full momentum and flexibility of the Quadruple and Quintuple Helix for conceptually addressing and integrating art and arts.

More generally speaking, further ramifications of Mode 3 knowledge production in Quadruple Helix and Quintuple Helix innovation systems are as follows:

1. *Multi-level innovation systems, the global and the local (GloCal)*: Lundvall was pivotal for introducing the concept of the 'national innovation system'. Lundvall ([1992], pp. 1, 3) explicitly

acknowledges that national innovation systems are challenged in permanence (but are also extended) by regional as well as global innovation systems. Here, Kuhlmann ([2001], pp. 960 to 961) could be paraphrased, and the assertion that as long as nation states and nation state-based political systems exist, it is plausible to use the concept of the national innovation system. More comprehensive in its analytical architecture than the national innovation system is the concept of the 'multi-level innovation system' (Carayannis and Campbell [2012], pp. 32 to 35). In a spatial understanding, multi-level innovation systems compare the national not only with the sub-national (regional, local), but also with the transnational and global levels (see, for example, Kaiser and Prange [2004]; furthermore, see Pfeffer [2012]). However, it is also important to extend multi-level innovation systems to the challenges and potential benefits and opportunities of a non-spatial meaning, understanding, and 'mapping': 'Therefore, multi-level systems of knowledge as well as multi-level systems of innovation are based on spatial and non-spatial axes. A further advantage of this multi-level systems architecture is that it results in a more accurate and closer-to-reality description of processes of globalization and *gloCalization*' (Carayannis and Campbell [2012], p. 35).

2. *Linear and non-linear innovation*: Knowledge application and innovation are being challenged and driven out of an interest of combining and integrating linear and non-linear innovation. Keys to here are diversity, heterogeneity, and pluralism of different knowledge and innovation modes, and their linking together via an architecture of co-evolving networks. Firms, universities, and other organizations can engage (at the same) in varying and multiple technology life cycles at different levels of maturity. Another way, how to think non-linear innovation is being suggested by the concept of cross-employment (Campbell [2011]; Campbell [2013a]). As a form and type of multi-employment, cross-employment emphasizes that the same individual person may be employed by two (or more) organizations at the same time, where one organization could be located closer to knowledge production and the other to knowledge application (innovation): if those organizations are also rooted in different sectors, then cross-employment acts also as a trans-sectoral networking (Campbell and Caraynnis [2013], pp. 65, 68). Cross-employment can furthermore bridge different sectors and disciplines in the sciences with different disciplines in the arts. What results is a 'Mode 3 Innovation Ecosystem': 'This parallel as well as sequentially time-lagged unfolding of technology life cycles also expresses

characteristics of Mode 2 and of nonlinear innovation, because organizations (firms and universities) often must develop strategies of simultaneously cross-linking different technology life cycles. Universities and firms (commercial and academic firms) must balance the nontriviality of a fluid pluralism of technology life cycles' (Carayannis and Campbell [2012], p. 37; see furthermore Dubina et al. [2012]). The relationship between networks, 'cooperation and competition' (Co-Opetition), represents a challenge and sensitive issue and allows for different creative answers in organizational representation and manifestation.

3. *Twenty-first century Fractal Research, Education and Innovation Ecosystem (FREIE)*: Here, the understanding of FREIE is 'This is *a multilayered, multimodal, multinodal, and multilateral system*, encompassing mutually complementary and reinforcing innovation networks and knowledge clusters consisting of human and intellectual capital, shaped by social capital and underpinned by financial capital' (Carayannis and Campbell [2012], p. 3).
4. *Linear and non-linear innovation, and the causality of 'if-then' and of 'if-if' relations*: The hybrid overlapping of linear innovation and of non-linear innovation displays also possible ramifications and draws associations to models of causality and their re-modeling. 'We can speculate, whether this parallel integration of linearity and nonlinearity not also encourages a new approach of paralleling in our theorizing and viewing of causality: *in epistemic (epistemological) terms, the so-called if-then relationships could be complemented by (a thinking in) "if-if" relations*' (Carayannis and Campbell [2012], p. 24; see also Campbell [2009], p. 123).

At the beginning of our analysis (in the introduction), we formulated the following research questions: How does artistic research relate to research in the sciences, and how does artistic research relate to innovation and innovation systems? We were inclined to develop the interrelation and interlinkage between arts, research, and innovation on the basis of the concepts of the Quadruple and Quintuple Helix innovation systems. We wanted to address *art and artistic research* (AAA) in the context of the Quadruple Helix and Quintuple Helix. 'Arts, research, innovation, and society' may be regarded as a program with implications for theory, policy, and practice.[o] In the following, we develop further a few more propositions. These propositions should be regarded as input for discussion and discourse:

1. 'Arts as a manifestation of knowledge': By defining arts as a manifestation of knowledge (in complementary extension of a more traditional understanding of the aesthetic dimension of arts),

the artistic research and arts-based innovation then interflow directly with art and arts. Artistic research helps to explain the arts. Artistic research also contributes to the epistemic potential of the arts and in arts. Universities and other higher education institutions of the arts are challenged to respond to artistic research and to implement strategies for developing artistic research, which also informs and drives university teaching. Through knowledge creation, knowledge production, knowledge application, and knowledge use, research in the arts and arts-based innovation are being interconnected with research in the sciences and sciences-based innovation. Arts and artistic research add to the interdisciplinary and transdisciplinary spectrum of research organizations and of research networks and can assist the sciences (also science) in building interdisciplinary arrangements. Interdisciplinarity often qualifies as a good basis for transdisciplinarity. Arts and artistic research foster heterogeneous processes of diversification and pluralization within knowledge production and innovation. Arts and artistic research promote creativity, which are keys for knowledge creation, knowledge production, and innovation.

2. Art and artistic research p in Quadruple and Quintuple Helix innovation systems: The concepts of the Quadruple Helix and Quintuple Helix innovation systems are explicitly sensitive for the roles of arts and of artistic research for innovation. Within the context of that line of thinking, arts, artistic research, and arts-based innovation (AAA) are essential for the further evolution and progress of innovation systems. Universities and other higher education institutions of the arts represent crucial organizations for innovation systems (national and multi-level innovation systems). In multi-level innovation systems, the global, national, and local innovation systems co-evolve in parallel and in being mutually intertwined.

3. 'Arts, research, innovation, and society' and the quality in economy and the quality of democracy: Innovation may not be narrowed down to economic concerns and economic activities. Innovation is more than only economics. Arts, research, innovation, and society contribute to creating the basis for new models of economic growth, where 'growth in quality' challenges the traditional focus on 'quantitative growth' of selected economic benchmarks or indicators. Arts, artistic research, and arts-based innovation are keys for advanced economies as well as the emerging markets. Arts, research, innovation, and society furthermore interrelate and cross-link with Quality of Democracy. This indicates opportunities

for a creative design or processes of creatively designing the further co-evolution of knowledge economy, knowledge society, and knowledge democracy.

4. Arts, artistic research, arts-based innovation, and 'Creative Knowledge Environments': Creative Knowledge Environments (Hemlin et al. [2004]) are interested in contributing to capabilities of knowledge creation, knowledge production, and innovation in organizations and in networks. Cross-employment (Campbell [2013a]) defines one strategy for doing so (also for non-linear innovation). Arts, artistic research, and arts-based innovation furthermore represent approaches that add to the formation of Creative Knowledge Environments. Arts, artistic research, and arts-based innovation and innovations are at least 'potentially' relevant to all organizations and networks that are engaged in research (knowledge creation and knowledge production) and innovation (knowledge application and knowledge use). This clearly demonstrates the possibilities and opportunities of arts, artistic research, and arts-based innovation for the sustainable development and the 'innovative re-invention' of organizations and networks that are involved in and perform in knowledge production. There is a need for more creative organizational design. In aggregation, this is also of relevance for whole innovation systems.

Arts, democracy, and innovation co-evolve. At the beginning of our article analyses, we introduced the blueprint for a vision and manifesto on knowledge production and innovation in Quadruple Helix innovation systems. The Quadruple Helix regards itself to be human-centered oriented. While for the Triple Helix model the existence of a democracy is not (per se) necessary for knowledge production and innovation, the Quadruple Helix is here more explicit. The way how the Quadruple Helix is being engineered, designed, and architected clearly shows that there cannot be a Quadruple Helix innovation system without democracy or a democratic context. The following attributes and components define the fourth helix in the Quadruple Helix: media-based and culture-based public, civil society, and arts, artistic research, and arts-based innovation. By this, the fourth helix in the Quadruple Helix represents the perspective of the dimension of democracy or the context of democracy for knowledge, knowledge production, and innovation. This is particularly true when democracy is being understood to transcend the narrow understanding of being primarily based *on* or being primarily rooted *in* government institutions (within Triple Helix). Civil society, culture-based public, quality of democracy, and sustainable development convincingly demonstrate what the rationales and requirements are for conceptualizing democracy broader

(Campbell and Carayannis [2013a]).[q] Political pluralism in a democracy co-evolves with the pluralism, diversity, and heterogeneity of knowledge, knowledge production, and innovation (Democracy of Knowledge, see Carayannis and Campbell, [2009], [2012], p. 55). We postulate here a congruence of structures and processes in democracy and in innovation systems. The Quintuple Helix extends the Quadruple Helix by aspects of the natural environments of society and economy, the social ecology, and the socio-ecological transition. Also, this environmental context of society can be better addressed in a democracy than in a non-democracy. The current world appears to be challenged by a race between developing democracies versus emerging autocracies over knowledge production and innovation.

The contributions of arts, arts-based research, and arts-based innovation to knowledge production and innovation systems are manifold. Art also helps us to think beyond the box. Finally, we want to engage in an artistic 'turn of turns'. In an attempt of free association with the work of M.C. Escher[r], we close here the analysis with the following two words: *Ascending Waterfalls. This is the beginning of beginning.*

METHODS

The article focuses on creatively designing, redesigning, and discussing concepts of knowledge production and innovation. The points of departure are the Triple Helix, Quadruple Helix, and Quintuple Helix for innovation, and Mode 1, Mode 2, and Mode 3 for knowledge production. These concepts are all published and thus publicly accessible. Particularly, the concept of the Quadruple Helix innovation system is being discussed and conceptually further developed in a twofold assessment approach: (1) democracy and knowledge production and innovation, and (2) art, arts-based research, and arts-based innovation. This should contribute to a continuous discourse building of the Quadruple Helix. The emphasis is on a deepening reflection.

ENDNOTES

[a]Concepts and terms in reference to knowledge production and innovation will be explained in greater detail later in this article.

[b]For example, see on Google Scholar the tracking of references to the concept of the Quadruple Helix innovations systems: http://scholar.google.at/scholar?cites=2775322365902273481&as_sdt=2005&sciodt=0,5&hl=en.

[c]Within the context of our analysis, the plural term 'sciences' always includes the natural sciences, life sciences, and also the social sciences and

humanities (human sciences). For us, the *sciences* address the whole and complete disciplinary spectrum. Therefore, 'sciences' is not equivalent to science. When we use the shorter expression 'scientific research', we actually always mean the research in all of the sciences.

[d]Of course, depending on how the sciences or knowledge production in the sciences is defined or is being defined, we may arrive here at different conclusions. Perhaps, the sciences (or pre-forms of the sciences) are just as old as the arts (pre-forms of arts). *Reasoning and aesthetic sensitivity represent universal categories of humanity.*

[e]In her Master thesis, Derya Öcal also reflects on the question, *What is Art? Was ist Kunst?*

[f]We would like to add that art also reflects culture and the experiences on which culture is being based or which influence culture. This is also being expressed and recognized in the following: *The Knowledge of Culture and the Culture of Knowledge* (Carayannis and Pirzadeh [2014]).

[g]Perhaps it would be fairer to talk here about the 'beauties' or 'non-beauties' always in the plural form (in contrast to a 'singular' beauty or non-beauty). Depending on the specific context (time, era or society), the understanding of the beautiful often varied or varies. Also, within a specific context, quite different understandings of the beautiful can co-exist (partly in contradiction to each other). In that sense, beauty obviously depends upon culture.

[h]In the context of our analysis here, we use knowledge production and knowledge creation as interchangeable concepts. We could speculate whether 'knowledge creation' fits better for purposes of describing processes (knowledge-based processes) in the arts and in artistic research than the term 'knowledge production'.

[i]Epistemology or an 'epistemic base' may apply as a concept to very different fields. For example, also policy and policies can be discussed under aspects of 'epistemic governance' (Campbell and Carayannis [2013b], [c]). In fact, to utilize an epistemic base in unusual contexts has all the qualities of a potentially innovative approach or approaching.

[j]See http://en.wikipedia.org/wiki/M._C._Escher.

[k]For an overview, see http://www.mcescher.com/gallery/.

[l]See http://www.mcescher.com/gallery/impossible-constructions/ascending-and-descending/ and http://en.wikipedia.org/ wiki/Ascending_and_Descending.

[m]See http://www.mcescher.com/gallery/impossible-constructions/waterfall/ and http://en.wikipedia.org/wiki/Waterfall_(M._C._Escher).

[n]'I would add that arts and expressions of arts evolve with time and styles. And the beauty of it is that, unlike reality, arts make everything possible ("impossible world") by the simple fact that art is a matter of perception with limitless possibilities. There is more freedom in arts than in real life, and even more than for science fiction were some plausible technicality should be featured' (direct quote from an anonymous review and reviewer for this article).

[o]See also the announcement of the close launch of a whole new book series on the topic of 'arts, research, innovation, and society'. The title of this book series is exactly: *Arts, Research, Innovation and Society* (http://www.springer.com/series/11902).

[p]In the context of our analysis here, *art and artistic research* (also *arts, artistic research, and arts-based innovation*) refers to a conceptual 'Triple A' of the qualities of arts-based knowledge production and arts-based innovation.

[q]To turn this line of thinking, autocracies are not interested to allow the development of a free and mature civil society. On the contrary, autocracies want to control and suppress the rise of an independent civil society.

[r]See again our analysis in the 'Arts and Artistic Research' section.

COMPETING INTERESTS

The authors declare that they have no competing interests.

AUTHORS' CONTRIBUTIONS

EGC and DFJC jointly developed (co-created) the concepts of the Quadruple Helix and Quintuple Helix innovation systems. The competence background of EGC is in economics and of DFJC in political science. Both authors are fascinated by arts, arts-based research, and arts-based innovation. Both authors read and approved the final manuscript.

AUTHOR DETAILS

1 European Union Research Center, Department of Information Systems & Technology Management, School of Business, The George Washington University, Duquès Hall, Funger Hall, Suite 515C, 2201 G Street, NW,

Washington, D.C 20052, USA. 2 Unit for Quality Enhancement (UQE), University of Applied Arts Vienna, Oskar Kokoschka-Platz 2, Vienna 1010, Austria. 3 Faculty for Interdisciplinary Studies (iff), Institute of Science Communication and Higher Education Research (WIHO), Alpen-Adria-University Klagenfurt, Schottenfeldgasse 29, Vienna 1070, Austria.

REFERENCES

1. Arnold M: Interdisciplinary research (interdisciplinarity).In *Encyclopedia of creativity, invention, innovation and entrepreneurship* Edited by: Carayannis EG, Dubina N, Seel N, Campbell DFJ, Uzunidis D. Springer, New York, NY; 2013a, 1105–1113. [http://link.springer.com/referenceworkentry/ 10.1007/ 978–1-4614–3858–8_302][http://www.springerreference.com/ docs/html/chapterdbid/378730.html]

2. Arnold M: Transdisciplinary research (transdisciplinarity).In *Encyclopedia of creativity, invention, innovation and entrepreneurship* Edited by: Carayannis EG, Dubina IN, Seel N, Campbell DFJ, Uzunidis D. Springer, New York, NY; 2013b, 1819–1828. [http://link.springer.com/referenceworkentry/10.1007 /978–1-4614–3858–8_337][http://www.springerreference.com/ docs/html/chapterdbid/378740.html]

3. Bast G: Preparing a "creative revolution" – arts and universities of the arts in the creative knowledge economy.In *Encyclopedia of creativity, invention, innovation and entrepreneurship* Edited by: Carayannis EG, Dubina IN, Seel N, Campbell DFJ, Uzunidis D. Springer, New York, NY; 2013, 1471–1476. [http://link.springer.com/referenceworkentry/10.1007/978–1-4614–3858–8_442][http://www.springerreference.com /docs/html/chapterdbid/378818.html]

4. Brandenburger AM, Nalebuff BJ: *Co-Opetition*. Doubleday, New York, NY; 1997.

5. Bush V: *Science: the endless frontier*. United States Government Printing Office, Washington, D.C; 1945.

6. Campbell DFJ: "Externe Umwelten". Außensichten auf das iff. In *iff. Interdisziplinäre Wissenschaft im Wandel*. Edited by: Arnold M. LIT, Vienna; 2009:99–134.

7. Campbell DFJ: *Wissenschaftliche „Parallelkarrieren" als Chance. Wenn Wissenschaft immer öfter zur Halbtagsbeschäftigung wird,*

könnte eine Lösung im „Cross-Employment" liegen. Guest Commentary for DIE PRESSE (February 2, 2011). 2011.

8. Campbell DFJ: Cross-employment.In *Encyclopedia of creativity, invention, innovation and entrepreneurship* Edited by: Carayannis EG, Dubina IN, Seel N, Campbell DFJ, Uzunidis D. Springer, New York, NY; 2013a, 503–508. [http://link.springer.com/referenceworkentry/10.1007/978-1-4614-3858-8_254][http://www.springerreference.com/docs/html/chapterdbid/378719.html]

9. Campbell GS: Speaking pictures: innovation in fine arts.In *Encyclopedia of creativity, invention, innovation and entrepreneurship* Edited by: Carayannis EG, Dubina IN, Seel N, Campbell DFJ, Uzunidis D. Springer, New York, NY; 2013b, 1716–1722. [http://link.springer.com/ referenceworkentry/ 10.1007/978-1-4614-3858-8_484][http://www.springer reference.com/docs/html/chapterdbid/378853.html]

10. Campbell DFJ, Carayannis EG: Lineare und nicht-lineare knowledge production: innovative Herausforderungen für das Hochschulsystem. *Zeitschrift für Hochschulentwicklung* 2012, 7(2):64–72. [http://www.zfhe.at/index.php/zfhe/article/view/448]

11. Campbell DFJ, Carayannis EG: Quality of democracy and innovation.In *Encyclopedia of creativity, invention, innovation and entrepreneurship* Edited by: Carayannis EG, Dubina IN, Seel N, Campbell DFJ, Uzunidis D. Springer, New York, NY; 2013a, 1527–1534. [http://link.springer.com/ referenceworkentry/ 10.1007%2F978-1-4614-3858-8_509][http://www.springerreference.com/docs/html/chapterdbid/378878.html]

12. Campbell DFJ, Carayannis EG: *Epistemic governance in higher education. Quality enhancement of universities for development. SpringerBriefs in Business.* Springer, New York, NY; 2013b.

13. Campbell DFJ, Carayannis EG: Epistemic governance and epistemic innovation policy.In *Encyclopedia of creativity, invention, innovation and entrepreneurship* Edited by: Carayannis EG, Dubina IN, Seel N, Campbell DFJ, Uzunidis D. Springer, New York, NY; 2013c, 697–702. [http://link.springer.com/referenceworkentry/10.1007/978-1-4614-3858-8_271][http://www.springerreference.com/docs/html/chapterdbid/378723.html]

14. Campbell DFJ, Carayannis EG: Explaining and comparing quality of democracy in Quadruple Helix structures: the quality of democracy

in the United States and in Austria, challenges and opportunities for development.In *Cyber-development, cyber-democracy and cyber-defense. Challenges, opportunities and implications for theory, policy and practice* Edited by: Carayannis EG, Campbell DFJ, Efthymiopoulos MP. Springer, New York, NY; 2015. [http://www.springer.com/social+sciences/political+science/book/978-1-4939-1027-4]

15. Carayannis EG, Barth TD, Campbell DFJ: The Quintuple Helix innovation model: global warming as a challenge and driver for innovation. *Journal of Innovation and Entrepreneurship* 2012, 1(1):1–12. [http://www.innovation-entrepreneurship.com/content/pdf/2192-5372-1-2.pdf] 10.1186/2192-5372-1-1

16. Carayannis EG, Campbell DFJ: "Mode 3": meaning and implications from a knowledge systems perspective. In *Knowledge creation, diffusion, and use in innovation networks and knowledge clusters. A comparative systems approach across the United States, Europe and Asia*. Edited by: Carayannis EG, Campbell DFJ. Praeger, Westport, Connecticut; 2006:1–25.

17. Carayannis EG, Campbell DFJ: "Mode 3" and "Quadruple Helix": toward a 21st century fractal innovation ecosystem. *International Journal of Technology Management* 2009, 46(3/4):201–234. [http://www.inderscience.com/browse/index.php?journalID=27&year=2009&vol=46&issue=3/4][http://www.inderscience.com/search/index.php?action=record&rec_id=23374&prevQuery=&ps=10&m=or] 10.1504/IJTM.2009.023374

18. Carayannis EG, Campbell DFJ: Triple Helix, Quadruple Helix and Quintuple Helix and how do knowledge, innovation and the environment relate to each other? A proposed framework for a trans-disciplinary analysis of sustainable development and social ecology. *International Journal of Social Ecology and Sustainable Development* 2010, 1(1):41–69. [http://www.igi-global.com/free-content/41959][http://www.igi-global.com/article/triple-helix-quadruple-helix-quintuple/41959]

19. Carayannis EG, Campbell DFJ: *Mode 3 knowledge production in Quadruple Helix innovation systems. 21st-century democracy, innovation, and entrepreneurship for development. SpringerBriefs in Business*. Springer, New York, NY; 2012.

20. Carayannis EG, Campbell DFJ: Mode 3 knowledge production in quadruple helix innovation systems: Quintuple Helix and social ecology.In *Encyclopedia of creativity, invention, innovation and*

entrepreneurship Edited by: Carayannis EG, Dubina IN, Seel N, Campbell DFJ, Uzunidis D. Springer, New York, NY; 2013, 1293–1300. [http://link.springer.com/referenceworkentry /10.1007 /978-1-4614-3858-8_310][http://www.springerreference.com/docs/html/chapterdbid/378732.html]

21. Carayannis EG, Kaloudis A: A Time for action and a time to lead: democratic capitalism and a new "new deal" for the US and the world in the twenty-first century. *Journal of the Knowledge Economy* 2010, 1(1):4–17. [http://link.springer.com/article/10.1007/s13132-009-0002-y]10.1007/s13132-009-0002-y

22. Carayannis EG, Pirzadeh A: *The knowledge of culture and the culture of knowledge. Implications for theory, policy and practice.* Palgrave Macmillan, Houndmills; 2014.

23. Damianisch A: Artistic research.In *Encyclopedia of creativity, invention, innovation and entrepreneurship* Edited by: Carayannis EG, Dubina IN, Seel N, Campbell DFJ, Uzunidis D. Springer, New York, NY; 2013, 122–127. [http://link.springer.com/referenceworkentry/10.1007/978-1-4614-3858-8_473][http://www.springerreference.com/docs/html/chapterdbid/378843.html]

24. Danilda I, Lindberg M, Torstensson B-M: *Women resource centres. A quattro helix innovation system on the European agenda. Paper.* 2009.

25. Dubina IN, Carayannis EG, Campbell DFJ: Creativity economy and a crisis of the economy? Coevolution of knowledge, innovation, and creativity, and of the knowledge economy and knowledge society. *Journal of the Knowledge Economy* 2012, 3(1):1–24. [http://link.springer.com/article/10.1007/s13132–011–0042-y] 10.1007/s13132-011-0042-y

26. Etzkowitz H, Leydesdorff L: The dynamics of innovation: from national systems and "Mode 2" to a triple helix of university-industry-government relations. *Research Policy* 2000, 29: 109–123. 10.1016/S0048-7333(99)00055-4

27. *The world in 2025. Rising Asia and socio-ecological transition.* European Commission, Brussels; 2009.

28. *Socioecological transitions and global change. Trajectories of social metabolism and land use.* Edward Elgar, Cheltenham; 2007.

29. Gibbons M, Limoges C, Nowotny H, Schwartzman S, Scott P, Trow M: *The new production of knowledge. The dynamics of science and research in contemporary societies.* Sage, London; 1994.

30. Hemlin S, Allwood CM, Martin BR: *Creative knowledge environments. The influences on creativity in research and innovation.* Edward Elgar, Cheltenham; 2004.

31. Kaiser R, Prange H: The reconfiguration of national innovation systems – the example of German biotechnology. *Research Policy* 2004, 33: 395–408. 10.1016/j.respol.2003.09.001

32. Kuhlmann S: Future governance of innovation policy in Europe – three scenarios. *Research Policy* 2001, 30: 953–976. 10.1016/S0048-7333(00)00167-0

33. Leydesdorff L: The Triple Helix, Quadruple Helix, ..., and an N-Tuple of Helices: explanatory models for analyzing the knowledge-based economy? *Journal of the Knowledge Economy* 2012, 3(1):25–35. [http://link.springer.com/article/10.1007/s13132-011-0049-4] 10.1007/s13132-011-0049-4

34. *National systems of innovation. Towards a theory of innovation and interactive learning.* Pinter Publishers, London; 1992.

35. Mateus-Berr R: Applied design thinking lab and creative empowering of interdisciplinary teams.In *Encyclopedia of creativity, invention, innovation and entrepreneurship* Edited by: Carayannis EG, Dubina IN, Seel N, Campbell DFJ, Uzunidis D. Springer, New York, NY; 2013, 73–116. [http://link.springer.com/referenceworkentry/10.1007/978-1-4614-3858-8_437][http://www.springerreference.com/docs/html/chapterdbid/378815.html]

36. Nowotny H, Scott P, Gibbons M: *Re-thinking science. Knowledge and the public in an age of uncertainty.* Polity Press, Cambridge; 2001.

37. Nowotny H, Scott P, Gibbons M: Mode 2 revisited: the new production of knowledge. *Minerva* 2003, 41: 179–194. 10.1023/A:1025505528250

38. Nowotny H, Scott P, Gibbons M: Re-thinking science: Mode 2 in societal context. In *Knowledge creation, diffusion, and use in innovation networks and knowledge clusters. A comparative systems approach across the United States, Europe and Asia.* Edited by: Carayanni EG, Campbell DFJ. Praeger, Westport, Connecticut; 2006:39–51.

39. Öcal D: *Kunst und Politik. [Art and Politics.] Master Thesis.* University of Vienna, Vienna; 2013.

40. Park HW: Transition from the Triple Helix to N-Tuple Helices? An interview with Elias G. Carayannis and David F. J. Campbell. *Scientometrics* 2014, 99(1):203–207. [http://link.springer.com/article/10.1007%2Fs11192-013-1124-3][http://download.springer.com/static/pdf/907/art%253A10.1007%252Fs11192-013-1124-3.pdf?auth66=1397308723_4cb0003877af5305d5dc202280b9cd6d&ext=.pdf] 10.1007/s11192-013-1124-3

41. Pfeffer T: *Virtualization of universities. Digital media and the organization of higher education institutions.* Springer, New York, NY; 2012.

42. Polanyi M: The republic of science: its political and economic theory. *Minerva* 1962, 1: 54–74. [http://sciencepolicy.colorado.edu/students/envs_5100/polanyi_1967.pdf][http://fiesta.bren.ucsb.edu/~gsd/595e/docs/41.%20Polanyi_Republic_of_Science.pdf] 10.1007/BF01101453

11
CHAPTER

Beyond Gamification: Reconceptualizing Game-Based Learning in Early Childhood Environments

Jason Nolan[a]* and Melanie McBride[b]

[a]*Early Childhood Studies, Ryerson University, 350 Victoria Street, Toronto, Ontario, Canada M5B 2K3;*

[b]*Communication and Culture, York University, Toronto, Ontario, Canada*

ABSTRACT

The recent promotion and adoption of digital game-based learning (DGBL) in K-12 education presents compelling opportunities as well as challenges for early childhood educators who seek to critically, equitably and holistically support the learning and play of today's so-called digital natives. However, with most DGBL initiatives focused on the increasingly standardized 'accountability' models found in K-12 educational institutions, the authors ask whose priorities, identities and notions of play this model reinforces or neglects. Drawing on the literatures of early childhood studies, game-based learning, and game studies, they seek to illuminate the *informal* contexts of play within the 'hidden' and 'null' curricula of DGBL that do not fit within the efficiency models of mainstream education in North America. In the absence of a common critical or theoretical foundation for DGBL, they propose a conceptual framework that challenges what they regard to be the institutionally nullified dimensions of *autonomy, play, affinity* and *space* that are essential to DGBL. They contend that these dimensions are ideally situated within the inclusive and play-based curriculum early childhood learning environments, and that the early years constitute a critically significant, yet overlooked, location for more holistic and inclusive thinking on DGBL.

Keywords: DGBL, digital game-based learning, early childhood education, videogames, situated learning, hidden and null curriculum, critical pedagogy

INTRODUCTION

Children aged 3-10 represent the largest demographic of the virtual worlds and online games (Kzero, 2011) they themselves choose to engage with and identify as significant to them (de Castell & Jenson, 2004). The recognition that digital games are a critically important dimension of younger children's lives at home and in early years locations presents challenges as well as opportunities for early childhood education. Though there is a growing body of work on the use of technology in the early years, much of the research has focused on multimedia tools and popular culture as pathways to multi-modal literacies with relatively little focus on digital games (Marsh, 2010; Marsh, Brooks, Hughes, Ritchie, & Roberts, 2005; Plowman, McPake, & Stephen, 2008; Plowman, Stephen, & McPake, 2010; Wolfe, & Flewitt, 2010; Yelland, 2010).

Though funding and access to resources are often cited as the primary barrier to technology adoption and integration in early learning contexts, the role of negative dispositions towards technology and digital games has not been fully considered as a contributing factor in the relatively slow pace of technology adoption in early childhood education. These negative dispositions towards technology may also reflect more than a century of dominant values informed by the naturalist and romantic beliefs about children, childhood and children's learning. For example, in their study of technology integration in a UK preschool, Wolf and Flewitt (2010) report that practitioners believed children's lives were made 'toxic' by technology use, and that these practitioners only rarely encouraged or supported the use of computers or other technologies for learning.

Though many parents and practitioners alike still regard digital games to be less developmentally, socially and culturally beneficial to children than outdoor play (Agger & Shelton, 2007; Frost, 2010). a growing number of early childhood scholars (Marsh, 2010; Wohlwend, 2009; Yelland, 2010;

Zevenbergen, 2007; are advocating the inclusion of digital technology and culture in early childhood education. In addition to reinforcing the value of these technologies in the development of multi-modal and emergent literacies, these scholars suggest that early childhood learning environments can play a vital role in bridging the digital divide across socio-economic and cultural barriers to access (Zevenbergen, 2007). However, institutional preoccupations with data and accountability are primarily focused on operational, skills-based technology use rather than its meaning-making potential and role in the development of multi-modal literacies (Wolf & Flewitt, 2010).

In the absence of a critical pedagogy for digital game-based learning (DGBL), educators unfamiliar with games or gaming culture are faced with confusing and conflicting agendas which they are ill-equipped to interrogate. Miller, Robertson, Hudson, and Shimi (2012) have pointed out that most of the conceptualization and implementation of DGBL have focused primarily on the upper-years, far removed from the sociocultural literacies and experiences of children's 'at home' technology use (Plowman et al., 2010). As de Castell and Jenson (2003a) contend, the true pedagogical and ludic spirit of learning with games 'resides in the engagement itself and not in its extrinsically defined "learning outcomes"' (p. 52). Given the formative role of play in children's learning and development, we ask how any vision of game-based learning can be so far removed from the conditions and states of being that inspire players to engage in the 'lusory attitude' of game-play (Suits, 2005).

Drawing on the critical pedagogical theories of the 'hidden curriculum' (Jackson, 1968/1990) and 'null curriculum' (Eisner, 1985), we seek to identify whose priorities, identities or cultures are promoted or neglected in the current conceptualizations of DGBL and how they may be reconnected with the meaning, pleasure and engagement they hold for children in their out-of-school lives. We make use of Eisner's null curriculum, in particular, as a conceptual heuristic through which to illuminate dimensions of informal learning and play with digital games that are often inconvenient for mainstream K-12 education. Focusing attention on these inconvenient dimensions is important, because, as de Castell and Jenson (2003b) suggest, 'the ways in which tools are shaped and organized are often decontextualized, either hidden or abstracted from their particular sociocultural location' (p. 652). We propose that a critical inquiry into the institutionally nullified dimensions of *autonomy, play, affinity* and *space* may form the basis for developing a framework from which to reconceptualize a more inclusive and meaningful vision of DGBL and play. From this standpoint, we situate the early years as the ideal location in which the nullified dimensions of DGBL can be understood.

BACKGROUND: GAME-BASED LEARNING BEYOND THE 'DIGITAL NATIVE'

Currently among the most profitable sectors of cultural production, digital games have also captured the attention of institutions that wish to harness the engagement and motivational qualities of games for a variety of *non*-gaming purposes from consumer loyalty to advocacy and education (McGonigal, 2011). Among the most popular expressions of this paradigm is the trend of 'gamification', which emerged in 2008 as a behavioural marketing strategy that makes use of surface-level game features such as badges, achievements or rewards as an 'incentive' for consumer loyalty (Bogost, 2011). In industry parlance, the addition of these features to non-gaming appliances or services is referred to as a 'game layer' - such as points systems or rewards (Deterding, 2011).

Within educational domains, another variant of the non-gaming use of games is DGBL. Proponents of DGBL such as Prensky (2007) claim that digital games can promote collaboration, problem-solving, and communication, experimentation and the exploration of identities (Gee, 2007; Gee & Hayes, 2009; Prensky, 2012), which are resonant with the use of other technologies in the early years: 'extending knowledge of the world, acquiring operational skills, and developing dispositions to learn' (Plowman et al., 2010). At issue is the problem that proponents of DGBL often make causal links between between their claims and learning theories drawn from behaviourism, constructivism, narrative and cognitive psychology, and various other educational theories (Van Eck, 2006) leading to assumptions regarding observations that are better characterized as coincidences.

Among the most contentious arguments made by DGBL proponents is the reliance on the 'digital natives' argument (Bennett, Maton, & Kervin, 2008), which has rhetorically constructed a 'new' generation of learners who are supposed to be 'fluent' in digital technologies (Prensky, 2007, 2012, Tapscott, 1999).The most established proponents of DGBL (Gee, 2007; Gee & Hayes, 2009; Prensky, 2007) argue that there is some inherent, though unsubstantiated, interest and identification with digital games that appeals to the digital native learner regardless the social, cultural, and environmental contexts. However, as de Castell and Jenson (2003a) point out:

technology's principal use in curriculum development has had little to do with transformation and far more to do with its principal appeal to educational administrators: its unprecedented capabilities for surveillance, control, and documentation – all basically forms of record-keeping – and so of 'educational *accountability*'. (p. 48)

It comes as no surprise that many models of DGBL more closely resemble the extrinsic, reward-driven structures of gamification in exchange for a *pedagogical capital* divorced from intrinsic meaning or motivation. By pedagogical capital, we mean the acquisition of outcomes, practices and dispositions valued and rewarded within formal education rather than those of the player. For many teachers, the idea of adding, what is referred to as a 'game layer' to existing curriculum seems a logical, if not desirable, goal. As with gamification, there is a question of whether these school-approved games will actually engage children meaningfully in curricula and activities that hold no intrinsic value.

We suggest that autonomy, in the form of learner-initiated choices, is a core dimension of informal game-based learning. Furthermore, we argue that the highly 'incidental' learning that occurs as a result of these autonomous choices reflects a lived experience of digital gaming culture that is often counter-institutional. Ignoring the importance of autonomy, many models, particularly those emerging in American and Canadian K-12 education are more gamified than gameful.

The gamified vision of DGBL that fits best with standardized curriculum is one of competition, achievement and reward structures that reflects what Jackson (1968/1990), and many since, have referred to as a 'hidden curriculum' of values and norms in keeping with existing power relations and institutional structures. Within the resistance pedagogy of Freire (2004), hooks (1994) and others, the hidden curriculum refers to the hegemonic values and social relations that underlie traditional education (Flinders, Noddings, & Thornton, 1986) that are unexamined in relation to current educational technology trends. In the context of DGBL, the hidden curriculum pertains not only to the operationalization of games for schooling, but the use of games and game-based structures that serve to reinforce and emphasize the competitive and hierarchical values of institutional education. As Miller et al. (2012) note, the freedom, cooperation and open-endedness of early years learning and play are put aside in favour of competition and a reduced set of options for engagement in K-12 education. We contend that the authentic motivation for game-play is not simply an affordance of engaging game features, but resides within the player's lived and embodied sociocultural identity and her own situated contexts for playing a game.

Unlike traditional institutional schooling, the everyday learning and play that occur with games is organized and directed by the child rather than a teacher (Jonassen & Strobel, 2006; Meyer, 2002; Silvia, 2008; Stevens,

Satwicz, & McCarthy, 2008). While proponents of DGBL such as Gee and Hayes (2009) and Prensky (2007) characterize game-based learning as 'situated' within videogame communities and cultures, their use of the term most often refers to the 'semiotic domains' of the symbolic culture of the game as a text (i.e. affinity for the game and culture of *World of Warcraft*) rather than the lived and sociocultural life-world of the player. For example, a child who plays a game because it is meaningful to a friend with whom they want to spend more time is not playing the game because of a specific investment in the semiotic domain of that game but their feelings and interest in their friend. Given the more 'parallel' play styles of today's videogame players, it is easy to presume that their interest resides in the game and not with the friend whose co-location in room or online may otherwise appear secondary.

An awareness of the need to give children a voice in their learning choices, and a recognition that there are many children's voices to listen to, are on the rise, in part due to the influence of the Reggio Emilia approach (Edwards, Gandini, & Forman, 1998; Gandini, 2008). Over the past decade, post-foundationalist and reconceptualist movements in early childhood education, particularly in Australia, New Zealand and Canada, have interrogated the grand narratives of developmental and behaviourist traditions in early childhood education according to critical intersectionalities of race, class, gender and sexuality (Canella, 1997; Dahlberg, Moss, & Pence, 2006; Grieshaber & Cannella, 2001; Pacini-Ketchibaw & Pence, 2005). Similarly to the Reggio Emilia approach, reconceptualist and post-foundationalist scholars have largely neglected technology or digital culture as an important site of struggle for more emergent questions of theory and praxis, just as, by and large, most thinking on games for learning misses the opportunity to explore cultural contexts (de Castell & Jenson, 2003b) as well as the socio-spatial dimensions of the player's space (de Kort & Isselsteijn, 2008). And there is room for movement in both directions.

Reconceptualizing DGBL through a more holistic and critical lens must also critically interrogate the essentialism of digital native narratives in order to address the learning and play needs and interests of children who do *not* enjoy or respond to the extrinsic, procedural logic of games or those who do not regard games as play. For as problematic as it may be for educators to undervalue or neglect digital games, it is equally as irresponsible to promote the idea that today's children, by virtue of their immersion and access to digital culture, love and are engaged by, digital games. This presents educators with the apparent double-bind of two conflicting agendas for DGBL: On the one hand, proponents of DGBL, who may never have had direct or lived experiences of classroom teaching, are advocating on behalf of the learning and literacy offered by games without having to take into account the real and varied challenges faced by today's diverse learners. Conversely, classroom practitioners who do the daily work of addressing these issues have scant opportunities to engage in the personal and professional learning required for developing the expertise necessary for incorporating DGBL into their professional practice, even if they have an intrinsic personal interest in digital gaming. We are left with a situation where both sides expect the other to have the expertise and experience to provide the solutions necessary to make DGBL work in the classroom, not realizing that they have, in effect, nullified or excluded the very dimensions required for *potentially* successful implementations of DGBL.

DIMENSIONS FRAMEWORK: AUTONOMY, PLAY, AFFINITY AND SPACE

Our proposed framework identifies four key dimensions of informal learning with digital games that appear to be most inconvenient or incompatible with the formal and increasingly standardardized curriculum models of K-12. The first and arguably most important of these dimensions is 'autonomy', which is the basis for anything we might call truly 'self-directed' play. The second of these dimensions is 'play', which, for our purposes, describes an open-ended, autonomous and intrinsically motivated activity that is not easily instrumentalized within performance-oriented goals. In the context of informal gaming, as opposed to organized institutional activities such as sport, play is voluntary, chosen freely, and chosen because it brings joy or pleasure (Schwartzman, 1978). The third dimension is 'affinity', which reflects the role of interest in particular objects, processes, experiences, and locations that draw or attract us beyond extrinsically defined values of what we should like or wish to engage. The final and fourth dimension is 'space', which is the environment or ecological location in which informal game-play occurs. In the context of autonomy, play and affinity, we are considering the significance of the space of game-play as a factor that is particularly inconvenient in relation to relocalizing the more socioculturally situated aspects of game-play within the school. The following sections serve to illuminate the interrelationship of our framework of autonomy, play, affinity and space as a means of reconceptualizing DGBL from the living room to the classroom.

Dimensions of *autonomy*

Autonomy is essential to young children's socio-emotional development, critical thinking and decision-making (boyd & Jenkins, 2006; Helwig, 2006;

Kamii, 1991; Marx & Steeves, 2010; Nolan, Raynes-Goldie, & McBride, 2011). As an open-ended, child-initiated activity with unknowable outcomes, children's *at-home*digital game-play provides many opportunities for autonomous learning through explorations that promote cycles of theory-building, testing, and reflection, in ever increasing levels of complexity. These explorations are rarely repetitive and static, but rather unfoldings and bifurcations predicated on their own goals and interests at the moment, at least when unmediated by adult interventions. A child's motivation to play a game, her choice of game, selection of playmates and the location, time and duration of her play are all mediated by the possibility and degree of autonomy. Though younger children's digital activities are subject to a higher degree of restrictive mediation, surveillance, and adult intervention (Nolan & McBride, submitted; Nolan et al., 2011), children's access, explorations and engagement with technologies in the home is more meaningfully situated than the often inauthentic and limited range of technological activity that takes place in the classroom. While there are legitimate legal and ethical reasons why younger children's activities are more limited in early learning environments, children's development of critical literacies (Macnaughton, 2005; Vasquez, 2004 is contingent on a reasonable provision of autonomy in relation to their inquiry and exploration of selected texts, which includes their play with digital games.

Some aspects of children's play with games may be relocalized in early learning environments insofar as children have similar opportunities for autonomous engagement. Jenkins (2006) articulates the problem of heteronomous, rather than autonomous, limits in relation to the relocalization of children's informal literacies within the predetermined hierarchies classroom space: 'Schools impose a fixed leadership hierarchy .... Even the most progressive schools set limits on what students can write compared to the freedom they enjoy on their own' and children are actually 'deskilled' when they then attempt to relocate their own literacies within the classroom (p. 193). Similarly, the spirit of inquiry that guides children's everyday play with digital games resides in the 'player, not the teacher or program, having autonomy over the interaction (the degree, kind, with whom, etc.)' (de Castell & Jenson, 2003a, p. 50). The contradiction between what children are able to do and learn on their own, versus what is prescribed in learning institutions, is at issue when institutions seek to appropriate play and game-play as instruments for the achievement of institutional goals or aims, while nullifying autonomous play.

The meaningful, playful and autonomous exploration of limits and self-regulation through DGBL would require an equally engaged and open-ended pedagogical orientation. As we are suggesting (Nolan &

McBride, submitted). in the context of game-based play, teachers and parents can develop more authentic and negotiated rules through a mediation strategy that involves more parallel and co-play than restrictive or heteronomic approaches, which are more common among parents with negative dispositions towards videogames and among non-gaming parents of younger children. Studies of parental mediation have shown that parents with a negative disposition towards the child's chosen cultural object are more likely to engage in restrictive, rather than collaborative, mediation practices (Nathanson, 2002; Nikken & Jansz, 2006; Oosting, IJsselsteijn, & de Kort, 2008). By extension, the challenge for non-gaming teachers in relation to their approach to children's digital gaming cultures is to develop experience in helping children explore their own goals and interests in relation to digital game-play, to avoid imposing a 'learning layer' and to look for emergent learning situated in the child's intrinsic interests and motivations.

From their socio-anthropological study *Children's Games in Streets and Playgrounds*, Opie and Opie (1984) report that children's informal play, unlike the play that is directed or supervised by adults or external structures, arises in the child (the one who is playing) and is mediated by the child's immediate context and her own needs, goals and curiosity. An outcome of child-initiated play is the child's own creation of rules that guide further play, sometimes resulting in the codification of these rules. When children are playing, they need freedom to not only choose the form that the play will take, but to choose everything about it, with minimal interference from others; this extends to the tools and technologies they use in their play as well: 'Play is unrestricted, games have rules' (Opie & Opie, 1984 p. 3).

Suits (2005) refers to the player's adoption of the rule-set of a game as a *lusory attitude*. When children participate in the creation of their own rule-sets through play, they engage in a meta-cognitive and autonomous inquiry into the nature of good or bad rules. Games encourage children to examine rule-sets in ways that schools cannot. However, teachers can use games, if they choose to permit these explorations, to support the child's autonomy in relation to the support and provision of the conditions that contribute to the child's experience of choice.

Miller et al. (2012), in their study of the off-the-shelf game *Nintendogs* among 5–7-year-old Scottish school children, found that successful learning outcomes were linked to the level of children's autonomy when they were playing. The importance of using games that were 'familiar' and 'from home' resulted in greater general motivation on learning tasks, group processing and peer interaction. Though they make no claims to generalizability of their findings, this example reinforces the importance of the role of choice, autonomy and 'familiarity' in younger children's game-play, which is otherwise nullified by more top-down and prescriptive approaches to digital content selection, implementation and pedagogy.

Dimensions of *play*

Along with autonomy, play is among the most ambiguous and nullified dimensions of DGBL. Play is both a central priority in early childhood education and the site of struggle between conflicting beliefs about the nature, value and purpose of DGBL. Despite much discipline-driven debate about the meaning of play, there is some general agreement about its core characteristics. In her comprehensive overview of the anthropology of play, *Transformations,* Schwartzman (1978) identifies several 'assumed' features of play:

Play is first of all assumed to be pleasurable and enjoyable, to be characterized by freedom and spontaneity, and to elicit active (as opposed to passive) engagement by players. Second, it is generally assumed that play is unproductive and without 'real' consequence in life – its motivations are said to be intrinsic as opposed to extrinsic. (p. 327)

Schwartzman's definition, like many others, emphasizes the *non-purposive* (i.e. 'unproductive') aspects of pleasure and freedom of play that sets it apart from other kinds of activities. This emphasis on play as a voluntary and freely chosen activity is important in the context of education, particularly in relation to the needs of marginalized children and children with disabilities. Goodley and Runswick-Cole (2010) extend this emphasis on freedom in a definition that reflects new sociology of childhood and critical disability perspectives:

play [is] freely chosen, personally directed and intrinsically motivated. Freely chosen play means that children choose when and what play they undertake, play is not part of a curriculum or a programme and does not have steps that need to be completed. When play is personally directed, it is children themselves who agree the roles or rules of the activity, as well as the outcomes, if any. (p. 506)

In *The ambiguity of play,* Sutton-Smith (1997) catalogues hundreds of 'rhetorics' of play that he breaks down into seven categories or overarching rhetorics, including: power, fate, identity, frivolity, self, progress, and the imaginary, each of which 'can be examined as a representation of the way people value some kind of play' (p. 204). For Sutton-Smith, *play as progress* is applied to ideas of a child's growth, development and adaptation through play, whereas *play as fate* is associated with games of chance such as gambling, and is situated in adults. *Play as power*describes sports, athletic activities and competition, while *play as identity* relates to community, festival and belonging to, and promoting, a shared social sense of self. His notion of *play as the imaginary* is where he locates creativity and innovation, along with flexibility and even irrationality, and the rhetoric of the *self* contains solitary and private interests, hobbies and pursuits that require sustained attention. The final rhetoric of *play as frivolous* highlights the importance of the traditional fool or trickster who is a playful critic of the existing dominant social power structures.

Sutton-Smith suggests that the rhetorics of play as 'progress' or 'growth', promoted within developmental psychology and education, are among the most hegemonic means of inscribing the irrational aspects of play with 'rational' attributes. However, the developmental basis of these play rhetorics promoted in the National Association for the Education of Young People's *Developmentally appropriate practice in early childhood programs serving children from birth through age 8* (Chung & Walsh, 2000; National Association for the Education of Young People, 2009) are now widely contested by more recent reconceptualist and post-foundationalist movements in early childhood education, which critique Developmentally Appropriate Practice for lacking inclusivity or cultural 'intersubjectivity' (Jordan, 2003; Langford, 2010; Lewin-Benham, 2008; Pacini-Ketchabaw & Pence, 2005). These critiques, which are gaining acceptance primarily among new sociology of childhood movements also challenge the universality of developmental stages as naturally occurring, *a priori* events (Goodley & Runswick-Cole, 2010).

Considering the seven rhetorics of play – *progress, fate, power, identity, imaginary, self* and *frivolity* – in the context of the three curriculum forms – *explicit, hidden* and *null* – provides a heuristic for understanding the differences between the rhetorics of play valued by institutions, and the rhetorics of play nullified by those institutions that survive in the player's out-of-school world. Eisner describes three locations for different learning opportunities that not only demarcate what can or cannot be undertaken in a schooling environment but also point to the varying levels of awareness of what is actually taking place. As mentioned above, the explicit curriculum is acknowledged in the teacher's curriculum, learning materials, textbooks, and institutional policy, while the hidden is present in the unwritten social practices of the school or learning environment, and the null can be ascertained by reflecting on what is missing, left out or occluded under the influence of the explicit and hidden curricula (Flinders et al., 1986). The play rhetorics (*progress, fate, power, identity, imaginary, self* and *frivolity*) can be tabled along with the three forms of curriculum (*explicit, hidden* and *null*) to reveal the clusters of the rhetorics of play that find a place in the explicit, hidden and null curricula of schools.

Explicit	Hidden	Null	
Progress	√	√	
Power	√	√	
Identity	√	√	
Imaginary			√
Self			√
Frivolity			√
Fate			√

The play rhetorics of power and progress fall under the purview of the explicit and hidden curricula, which are arguably most compatible with the hegemonic structures of competition, winning and achievement that are present in most education, sports and commercial game environments. The play rhetorics of imaginary, self, frivolity and fate are relegated to the null curriculum, to be excluded as much as possible in most education sports and commercial game environments. Though the explicit and hidden curricula are not always seen as working in tandem, the triumvirate rhetorics of progress, power and identity are inscribed across both locations. In Sutton-Smith's conceptualization of the rhetorics, progress, power and identity are *played* out wherever the curriculum engages issues of adaptation, growth, socialization, status and success, community and cooperation. Conversely, play rhetorics of the imaginary, self, frivolity and fate constitute the null dimension of play because they are less purposive, and more difficult to operationalize.

The above chart shows what is excluded *from* formal learning environments, and promoted *in* formal learning environments, as well as what is visible and what is hidden in formal learning environments. The play rhetorics of imaginary, self, frivolity and fate are excluded or nullified. However, these notions of imaginary, self, frivolity and fate at the same time represent what is most valued by players at the core of play. This exclusion brings into question both what institutions think the value of play is and what aspects of play can find a home in these institutions.

If we want the aspects of the progress, power and identity rhetorics of play to be the core of the DGBL in schools and learning institutions, we are on the right path, but is this what we want? The imaginary, self, frivolity and fate, are not only excluded in this scenario, but they are at the same time the elements of play for which most players have the greatest affinity.

Dimensions of *affinity*

Among the most considerable challenges for institutional integration of DGBL in schools is the problem of intrinsic interest or 'affinity' that normally occurs in informal game-play contexts. Gee (2006) refers to gaming cultures as 'affinity spaces' where players locate their interests in a personally meaningful and intrinsic 'semiotic domain'. This is not particular to gaming, but any culture or community where interest drives participation. Both the educator's disposition as well as the institutional attitude towards games and play are critical factors in the context of integrating DGBL in early childhood learning contexts, and sustaining their value throughout education. It is important to note that holistic learning approaches such as Waldorf, Montessori and Reggio Emilia sometimes serve to reinforce negative attitudes towards children's choices to engage in technology-based learning and autonomous/unstructured play due to pre-conceived notions of what tools and technologies are appropriate for children, as well as their notions of ages and stages of development. Many early childhood learning programmes have over-determined the role of an unproblematized romantic notion of 'natural' play at the expense of critical and meaningful inquiry with technology. The result of this reticence to engage in both technology-based learning and autonomous/unstructured play is that, though there is much scholarship and popular discourse devoted to problems and risks associated with children's increasing use of technology, such as obesity or online safety issues, there is an absence of critical or holistic pedagogies of digital learning and play. Waldorf, Montessori and Reggio Emilia approaches, as well as reconceptualist and post-foundationalist positions, have a great deal to offer in terms of holistic and critically situated discourses on children's DGBL, provided there is a concomitant willingness to recognize the child's autonomy when it comes to their affinity spaces and play.

Zevenbergen (2007) suggests that theoretical, pedagogical and dispositional alignments influence teachers perceptions and assumptions about the 'purpose of play within these settings ... whether play is solitary, parallel or interactive are also considerations in the field' that need to be explicitly taken up (p. 25). Gee (2006) argues that videogames function as 'affinity' spaces with particular features that constitute unique 'routes of participation' and social organization linking an individual's interests for the content within the play space. Gee points out that the features of 'affinity' spaces also include unique motivations on the part of players, and also unique ways in which content and activities are organized, how knowledge is distributed, and how the power relations between 'experts' and 'noobs' and distributed 'leadership' structures are manifest.

Central to notions of affinity are interest and emotion, both of which have long been identified as 'sites of struggle' within the classroom and formal education (Knight-Diop & Oesterreich, 2009). Emotion is among the most nullified aspects of formal education, however, research on emotion in education has shown that when emotion is engaged, it is often addressed in behavioural terms related to regulation and management (Boler, 1999; Knight-Diop & Oesterreich, 2009). In theorizing the null curriculum, Flinders et al. (1986) suggest that 'affect' (i.e. emotion, values, disposition) represents the most important core element of the null curriculum. Starting with their own affinities, children make more autonomous choices and decisions that situate their observations and theory-building within a familiar domain. When affinity links the child's learning and play with a community of practice, there is an opportunity to connect the child's learning to larger world of experiences. For the teacher, this is an opportunity to explore shared meanings and co-exploration that encourages learning as a shared experience of meaning-making.

Though interest and meaning are increasingly stated priorities within education, Eisner (1985) argues that the explicit curriculum in and of itself rarely holds any intrinsic meaning for learners. The heteronomous structures of institutional learning are often at odds with the complex cognitive, cultural, social or emotional needs of any given learner or the types of social arrangements or conditions that support authentically 'meaningful learning' (Novak, 2002). According to Shuell (1990), meaningful learning 'is an active, constructive, and cumulative process that occurs gradually over a period of time .... It is a goal oriented process best characterized in terms of problem solving' (p. 540). Research on meaningful learning also suggests the central role of affinity to communities of practice in supporting and cultivating learning, skills and knowledge:

meaningful learning occurs when learners are active, constructive, intentional, cooperative, and working on authentic tasks. Human learning is a naturally active mental social process. ... Through formal and informal apprenticeships in communities of play and work, learners develop skills and knowledge that they then share with other members of those communities with whom they learned and practiced those skills. (Jonassen & Strobel, 2006 pp. 1–2)

The use of children's self-selected affinity spaces, such as games they are familiar with and enjoy (Miller et al., 2012), offer a context for the kind of open-ended, autonomous and situated meaningful learning that should be a key requirement for the use of DGBL. The environment in which this situated learning occurs influences and is influenced by the other dimensions of autonomy, affinity and play (Flinders et al., 1986; Meyer, 2002; Silvia, 2008; Silvia, Henson, & Templin, 2009). In order to support 'freedom' to play or learn in an open-ended manner, we must first be able to recognize what it looks and feels like in informal learning contexts and consider what aspects may or may not be easily re-localizable schools and other institutional learning spaces governed by non-negotiable regulations, policies and norms.

Dimensions of *space*

Unlike a grade-segregated classroom, out-of-school 'in-room' game-play can consist of single or multi-aged player groupings. As well, children of the same age and demographic are free to approach the same game in radically different ways free of grade-, age- or developmentally determined expectations and limits imposed by institutional learning environments. The features or conditions of the location of interaction have as important a role as adult mediation in how children choose to interact. Unlike school, however, the physical conditions of 'in-room' gaming are as changeable and liquid as the configurations of the participants (Stevens et al., 2008). As Eisner (1985) notes, the absence of the 'soft surface' in school architecture and furniture represents a hidden curricula 'that restrict the ways in which one can sit and that yield to no form of body pressure. Rooms seldom have a soft relief; there are few places for enclosure or for privacy' (pp. 96–97). While children's rights to privacy and play are both identified in the *United Nations Convention on the Rights of the Child* (UNICEF, 2011), the possibility that play might be contingent on a feeling of privacy and freedom from surveillance is an inconvenient question for formal educational spaces where play-based learning takes place without any choice as to who the child is playing with or around. Not only are other children present, but adults are continually engaged in observation, assessment and interference in children's activities.

Invoking Bourdieu's theory of habitus, Zevenbergen (2007) notes that the unique internal and situated nature of digital natives has 'profound' implications for early childhood practice. The internal, virtual and physical location of children's digital learning and play is a critical, yet under-researched, phenomena. In one of the few ethnographic studies of children's at-home game-play, Stevens et al. (2008) articulate the everyday cognition 'in-room' game-play:

in the living and family rooms where the players are sitting, crouching, or reclining. Phones ring, parents come in (or argue in the next room), and players interact in a variety of ways with friends, siblings, and material resources–other than the game–that are in room. (p. 43)

Game-play at home requires no external set of rules, planned agenda or pre-existing pedagogy. It can be contingent on negotiations with the social and cognitive resources available and the interests and dispositions of those whom they wish to play with, where they wish to play and how. However, the experience of privacy, a limited experience for many children even at home, is a nullified condition in institutional spaces where children are supervised, monitored and assessed, and socialized into accepting this level of surveillance as normal.

Opie and Opie (1984) conclude that children play differently when they do not feel they are being observed, noting that privacy is a fundamental aspect of autonomous play and the informal learning that ensues. But what happens when play and the players are subject to the gaze of outsiders, parents or adults? And what happens when we institutionalize this lived experience of play, and bring it under the institutional gaze? From Bentham's Panopticon and Joseph Lancaster and Andrew Bell's monitorial methods in the early Nineteenth Century, up to contemporary early learning environments and elementary school classroom practices of surveillance and observation, the modern educational system situates the student within a heteronomous environment (Crain, 2003; Hager, 1959; Hogan, 1989; Lepper & Greene, 1975). If we are to introduce a meaningful and inclusive curricula of DGBL, a curricula that is arguably play-based, environmental considerations such as privacy cannot be overlooked.

Unlike the institutional spaces that children must be socialized into, out-of-school and in-room gaming environments are flexible and familiar spaces that are often repurposed and organized by children (Stevens et al., 2008). Game spaces are created by children themselves in a context in which 'each can get different things out of the space – based on their own choices, purposes, and identities' (Gee, 2006 p. 20). Bogost (2008) reminds us, video game players have 'their own culture and values. Video game players often self-identify as "gamers" and devote a major part of their leisure time to video games ... video game play could be understood as a 'community of practice" (p. 119). At the same time, Jonassen and Strobel (2006) point out that through 'formal and informal apprenticeships in communities of play and work, learners develop skills and knowledge that they then share with other members of those communities with whom they learned and practiced those skills' (pp. 1–2). The situated 'learning arrangements' characteristic of both convergence and informal game-play also serve to destabilize traditional structures of authority in important and meaningful ways (de Castell & Jenson, 2004). Through this 'in-room' gaming context, another child takes the role of teacher, which is redefined as a 'just-in-time' resource (Stevens et al., 2008) rather than an arbitrary credentialed 'professional' in an institution the learner had no hand in selecting. The teacher emerges almost spontaneously as whoever has the knowledge, ideas, or inspiration in the moment and the willingness to share.

Whereas the professional educator must maintain control and struggle with directing attention and classroom management in order to execute mandated learning policy, especially with rote or repetitive tasks, the practice principle in play and gaming environments forces us to acknowledge that children are more than happy to engage in rote or repetitive tasks when they choose to (Meyer, 2002). Children will engage if activities are attractive or interesting to them, and if they are able to choose how and when, and for how long, they engage in an activity, particularly if they are experiencing either success or meaningful failure in the activity. In such situations, the game-play is the curriculum content, and the content itself draws children together into shared exploration and learning. As such, informal learning in participant-directed environments is predicated on a sense of social engagement, feelings of comfort, trust, respect, happiness, and being valued. These affordances are not unique to gaming but characteristic of what Henning (1998) describes as the 'everyday cognition' that 'takes place at the juncture of everyday interactions' (p. 143). Early childhood researcher Vasquez (2004; Vasquez & Branigan Felderman, 2013) describes the 'negotiated' space for inquiry into the 'everyday' texts of children's own experience as a 'curriculum' for critical literacies. Though the situated and informal learning characteristic of gamers is often framed as an emergent phenomena, much of the learning and play that occurs in situated informal gaming contexts resembles reconceptualist and constructivist movements in early childhood learning, such as Reggio Emilia inspired dialogue and listening. The emphasis on 'listening' to children's '100 languages', which may be non-verbal, sensory or performative (Edwards, Gandini, & Forman, 1998), is pedagogically critical in relation to the multi-model literacies particular to digital games.

CONCLUSION

Without sustained critical inquiry, institutional approaches to educational technology integration and DGBL will only serve to reinforce, rather than disrupt, the behaviourist, rationalized and purposive frameworks that continue to transform learning and play into joyless performance and productivity outcomes. Though institutions may attempt to resist all that is not sufficiently 'practical' and not 'easily measured' within competitive learning outcomes (de Castell & Jenson, 2003b; Firlik, 1996), it is possible that reconceptualist and post-foundationalist pedagogies of early childhood education may offer the most direct route to the otherwise nullified head, heart and hands of ludic pedagogy.

Though policy and funding constraints are often cited as barriers to technology integration in early childhood learning centres, we argue that these problems are also dispositional. Digital games, and technology in general, remain a peripheral priority for many early childhood educators whose aversions to digital culture are often articulated according to outdated pedagogical and developmental standpoints. Reconceptualist pedagogies, however, may constitute a unique opportunity to re-cast 'play-based' learning priorities within a more inclusive digital approach that has implications far beyond early years. Zevenbergen (2007) notes that the use of technology in the early years is a matter of equity as much as pedagogical significance:

The provision of quality learning environments within the preschool has considerable potential to add digital capital to those students for whom the home experience is digitally impoverished in comparison with their more affluent peers. By addressing such differences as early as possible, the gap between the haves and have-nots can be reduced. (p. 27)

In the meantime, as early childhood educators debate the merits of technology integration, there is a need for more research of the 'everyday' conditions and features of children's digital gaming (Stevens et al., 2008).

If educators fail to develop a meaningful and inclusive critical pedagogy for DGBL, it will be left to a new class of educational consultants and developers to define learning and play. Rather than advancing education, we risk turning back the clock on decades of critical pedagogy and social research to replace the sandbox of play with a Skinner box of behavioural conditioning. Echoing Freire, Malaguzzi and others, de Castell and Jenson (2003b) conclude that 'without play, education becomes a force of compliance, not intelligence, and in this sense what we most urgently require of schooling today is that it can once again teach us to play, not to obey' (p. 49).

NOTES ON CONTRIBUTORS

Jason Nolan is director of the Experiential Design and Gaming Environments (EDGE) lab, and a professor in Early Childhood Studies at Ryerson University, and teaches in the joint Ryerson/York University graduate programme in Communication and Culture. His research focuses on adaptive design for children with disabilities, gaming/play, privacy/autonomy, sensory play, informal learning environments, virtual worlds, and the voices of 'digital natives'. [email: jnolan@ryerson.ca]

Melanie McBride is research associate in the EDGE lab at Ryerson University, and a MA candidate in Communication and Culture at York University. Her research is focused on children's informal learning and sensory play in digitally mediated social and gaming environments. [email: melanie1.mcbride@ryerson.ca]

REFERENCES

1. Agger, B., & Shelton, B. (2007). *Fast families, virtual children: A critical sociology of families and schooling*. Boulder, CO: Paradigm.
2. Bennett, S., Maton, K., & Kervin, L. (2008). The "digital natives" debate: A critical review of the evidence. *British Journal of Educational Technology*, 39(5), 775–786.
3. Bogost, I. (2008). The rhetoric of video games. In K. Salen (Ed.), *The ecology of games* (pp.117–139). Cambridge, MA: MIT Press.
4. Bogost, I. (2011). *Persuasive games: Exploitationware*. Gamasutra. Retrieved fromhttp://www.gamasutra.com/view/feature/6366/persuasive_games_exploitationware.php
5. Boler, M. (1999). *Feeling power: Emotions and education*. New York, NY: Routledge.
6. boyd, d., & Jenkins, H. (2006). *Myspace and Deleting Online Predators Act (DOPA)*. MIT Tech Talk. Retrieved from http://www.danah.org/papers/MySpaceDOPA.html
7. Canella, G. S. (1997). *Deconstructing early childhood education: Social justice and revolution*.New York, NY: Peter Lang.
8. de Castell, S., & Jenson, J. (2003a). Serious play: Curriculum for a post-talk era. *Journal of Curriculum Studies*, 35(6), 47–52.
9. de Castell, S., & Jenson, J. (2003b). OP-ED serious play. *Journal of Curriculum Studies*, 35(6),649–665.
10. de Castell, S., & Jenson, J. (2004). Paying attention to attention: New economies for learning. *Educational Theory*, 54(4), 381–397.

11. de Kort, Y. A. W., & IJsselsteijn, W.A. (2004). People, places, and play: Player experience in a socio-spatial context. *Computers in Entertainment*, 6(2), 18:1–18:11.
12. Chung, S., & Walsh, D. (2000). Unpacking child-centredness: A history of meanings.*Journal of Curriculum Studies*, 32(2), 215–234.
13. Crain, P. (2003). Children of media, children as media: Optical telegraphs, Indian pupils, and Joseph Lancaster's system for cultural replication. In L. Gitelman & G. Pingree (Eds.),*New media: 1740–1915* (pp. 61–80). Cambridge: MIT Press.
14. Dahlberg, G., Moss, P., & Pence, A. (2006). *Beyond quality in early childhood and care: Languages of evaluation*. London: Routledge.
15. Deterding, S. (2011). Gamification: Toward a definition. In D. Tan & B. Begole (Eds.),*Design, ACM CHI 2011* (pp. 12–15). Retrieved from http://gamification-research.org/wp-content/uploads/2011/04/02-Deterding-Khaled-Nacke-Dixon.pdf
16. Edwards, C., Gandini, L., & Forman, G. (Eds.). (1998). *The hundred languages of children: The Reggio Emilia approach to early childhood education*. Norwood, NJ: Ablex.
17. Eisner, E. (1985). *The educational imagination: On the design and evaluation of school programs*. New York, NY: Macmillan.
18. Firlik, R. (1996). Can we adapt the philosophies and practices of Reggio Emilia, Italy, for use in American schools? *Early Childhood Education Journal*, 23(4), 217–220.
19. Flinders, D., Noddings, N., & Thornton, S. (1986). The null curriculum: Its theoretical basis and practical implications. *Curriculum Inquiry*, 15(1), 33–42.
20. Freire, P. (2004). *The pedagogy of indignation*. Bolder, CO: Paradigm.
21. Frost, J. L. (2010). *A history of children's play and play environments*. New York, NY:Routledge.
22. Gandini, L. (2008). Introduction to the schools of Reggio Emilia. In S. Etheredge & L. Hill(Eds.), *Insights and inspirations from Reggio Emilia: Stories of teachers and children from North America* (pp. 24–28). Worcester, MA: Davis.
23. Gee, J. P. (2006). Semiotic social spaces and affinity spaces: From the age of mythology to today's schools. In D. Barton & K. Tusting (Eds.), *Beyond communities of practice: Language, power and social context* (pp. 214–232). Cambridge: Cambridge University Press.
24. Gee, J. P. (2007). *What videogames have to teach us about learning and literacy*. New York,NY: Palgrave.

25. Gee, J., & Hayes, E. (2009). *Public pedagogy through video games: Design, resources & affinity spaces.* Game Based Learning. Retrieved fromhttp://www.gamebasedlearning.org.uk/content/view/59/
26. Goodley, D., & Runswick-Cole, K. (2010). Emancipating play: Dis/abled children, development and deconstruction. *Disability & Society*, 25(4), 499–512.
27. Grieshaber, S., & Cannella, G. S. (2001). *Embracing identities in early childhood education: Diversity and possibilities.* New York, NY: Teachers College.
28. Hager, P. E. (1959). Nineteenth century experiments with monitorial teaching. *Phi Delta Kappan*, 40(4), 164–167.
29. Helwig, C. (2006). The development of personal autonomy throughout cultures. *Cognitive Development*, 21(4), 458–473.
30. Henning, P. (1998). Everyday cognition and situated learning. In D. Jonassen (Ed.),*Handbook of research on educational communications and technology* (2nd ed, pp.143–168). New York, NY: Simon & Schuster.
31. Hogan, D. (1989). The market revolution and disciplinary power: Joseph Lancaster and the psychology of the early classroom system. *History of Education Quarterly*, 29(3),381–417.
32. hooks, b. (1994). *Teaching to transgress. Education as the practice of freedom.* London:Routledge.
33. Jackson, P. (1968/1990). *Life in classrooms.* New York, NY: Teacher's College Press.
34. Jenkins, H. (2006). *Convergence culture.* New York, NY: New York University Press.
35. Jonassen, D., & Strobel, J. (2006). Modeling for meaningful learning. In D. Hung & M. S.Khine (Eds.), *Engaged learning with emerging technologies* (pp. 1–27). Dordrecht: Springer.
36. Jordan, R. (2003). Social play and autistic spectrum disorders: A perspective on theory, implications and educational approaches. *Autism*, 7(4), 347–360.
37. Kamii, C. (1991). Toward autonomy: The importance of critical thinking and choice making. *School Psychology Review*, 20(3), 382–388.
38. Knight-Diop, M., & Oesterreich, H. (2009). Pedagogical possibilities: Engaging cultural rules of emotion. *Teachers College Record*, 111(11), 2678–2704.
39. Kzero. (2011). *Virtual world registered accounts reach 1.7 bn in Q4.* Retrieved fromhttp://www.kzero.co.uk/blog/virtual-world-registered-accounts-reach-1-7bn-q4-2011/
40. Langford, R. (2010). Critiquing child-centred pedagogy to bring children and early childhood educators into the centre of a

democratic pedagogy. *Contemporary Issues in Early Childhood*, 11(1), 113–127.

41. Lepper, M. R., & Greene, D. (1975). Turning play into work: Effects of adult surveillance and extrinsic rewards on children's intrinsic motivation. *Journal of Personality and Social Psychology*, 31(3), 479–486.
42. Lewin-Benham, A. (2008). *Powerful children: Understanding how to think and learn using the Reggio approach*. New York, NY: Teachers College Press.
43. Macnaughton, G. (2005). *Doing Foucault in early childhood studies: Applying post-structural ideas*. London: Routledge.
44. Marsh, J. (2010). Young children's play in online virtual worlds. *Journal of Early Childhood Research*, 8(1), 23–29.
45. Marsh, J., Brooks, G., Hughes, J., Ritchie, L., & Roberts, S. (2005). *Digital beginnings: Young children's use of popular culture, media and new technologies*. Sheffield: University of Sheffield.
46. Marx, G., & Steeves, V. (2010). From the beginning: Children as subjects and agents of surveillance. *Surveillance & Society*, 7(3/4), 192–230.
47. McGonigal, J. (2011). *Reality is broken: Why games make us better and how they can change the world*. New York, NY: Penguin Press.
48. Meyer, R. (2002). Rote versus meaningful learning. *Theory into Practice*, 41(4), 226–232.
49. Miller, D., Robertson, D., Hudson, A., & Shimi, J. (2012). Signature pedagogy in early years education: A role for cots game-based learning. *Computers in the Schools*, 29(1–2),227–247.
50. Nathanson, A. (2002). The unintended effects of parental mediation of television on adolescents. *Media Psychology*, 4(3), 207–230.
51. National Association for the Education of Young People. (2009). *Developmentally appropriate practice in early childhood programs serving children from birth through age 8*. Retrieved from http://www.naeyc.org/files/naeyc/file/positions/PSDAP.pdf
52. Nikken, P., & Jansz, J. (2006). Parental mediation of children's videogame playing: A comparison of the reports by parents and children. *Learning, Media and Technology*, 31(2),181–202.
53. Nolan, J., & McBride, M. (submitted). Beyond ages and stages: Mixed-age mediation of young children's play in informal learning environments.
54. Nolan, J., Raynes-Goldie, K., & McBride, M. (2011). The stranger danger: Exploring surveillance, autonomy, and privacy in children's use of social media. *Canadian Children Journal*, 36(2), 24–32.

55. Novak, J. (2002). Meaningful learning: The essential factor for conceptual change in limited or inappropriate propositional hierarchies leading to empowerment of learners.*Science Education*, 86(4), 548–571.

56. Oosting, W., IJsselsteijn, W. A., & de Kort, Y. A. W. (2008, November). *Parental perceptions and mediation of children's digital game play at home: A qualitative study.* Short paper presented at the workshop on Designing for families, CSCW2008 conference, San Diego, CA.

57. Opie, P., & Opie, I. (1984). *Children's games in streets and playgrounds.* Oxford: Oxford University Press.

58. Pacini-Ketchabaw, V., & Pence, A. (2005). Contextualizing the reconceptualist movement in Canadian early childhood education. In V. Pacini-Ketchabaw & A. Pence (Eds.), *Early childhood education in motion: The reconceptualist movement in Canada* (pp. 5–20). Ottawa:Canadian Child Care Federation.

59. Plowman, L., McPake, J., & Stephen, C. (2008). The technologisation of childhood? Young children and technology in the home. *Children & Society*, 24(1), 63–74.

60. Plowman, L., Stephen, C., & McPake, J. (2010). Supporting young children's learning with technology at home and in pre-school. *Research Papers in Education*, 25(1), 93–113.

61. Prensky, M. (2007). *Digital game based learning.* Saint Paul, MN: Paragon House.

62. Prensky, M. (2012). *Digital natives to digital wisdom: Hopeful essays for 21st century learning.*Thousand Oaks, CA: Corwin.

63. Schwartzman, H. (1978). *Transformations: The anthropology of children's play.* New York,NY: Plenium Press.

64. Shuell, T. (1990). Phases of meaningful learning. *Review of Educational Research*, 60(4),531–547.

65. Silvia, P. (2008). Interest: The curious emotion. *Interest*, 17(1), 57–60.

66. Silvia, P., Henson, R., & Templin, J. (2009). Are the sources of interest the same for everyone? Using multilevel mixture models to explore individual differences in appraisal structures. *Cognition and Emotion*, 23(7), 1349–1406.

67. Stevens, R., Satwicz, T., & McCarthy, L. (2008). In-game, in-room, in-world: Reconnecting video game play to the rest of kids' lives. In K. Salen (Ed.), *The ecology of games: Connecting youth, games and learning* (pp. 41–68). Cambridge, MA: MIT.

68. Suits, B. (2005). *The grasshopper: Games, life and utopia.* Peterborough: Broadview Press.

69. Sutton-Smith, B. (1997). *The ambiguity of play*. Cambridge, MA: Harvard.
70. Tapscott, D. (1999). Educating the net generation. *Educational Leadership*, 56(5), 6–11.
71. UNICEF. (2011). *UN Convention on the Rights of the Child*. Retrieved fromhttp://www2.ohchr.org/english/law/crc.htm
72. Van Eck, R. (2006). Digital game-based learning: It's not just the digital natives who are restless. *Educause*, 41(2), 16–30.
73. Vasquez, V. M. (2004). *Negotiating critical literacies with young children*. Mahwah, NJ:Lawrence Erlbaum Associates.
74. Vasquez, V. M., & Branigan Felderman, C. (2013). *Technology and critical literacy in early childhood*. New York, NY: Routledge.
75. Wohlwend, K. (2009). Early adopters: Playing new literacies and pretending new technologies in print-centric classrooms. *Journal of Early Childhood Literacy*, 9(2), 117–140.
76. Wolfe, S., & Flewitt, R. (2010). New technologies, new multimodal practices and young children's metacognitive development. *Cambridge Journal of Education*, 40(4), 387–399.
77. Yelland, N. (2010). New technologies, playful experiences, and multimodal learning. In I. R. Berson & R. J. Berson (Eds.), *High-tech tots: Childhood in a digital world Charlotte* (pp.5–22). Charlotte, NC: Information Age Publishing.
78. Zevenbergen, R. (2007). Digital natives come to playschool: Implications for early childhood practice. *Contemporary Issues in Early Childhood*, 8(1), 19–29.

12 CHAPTER

Organization in the Crowd: Peer Production in Large-Scale Networked Protests

W. Lance Bennett[a]*, Alexandra Segerberg[b] and Shawn Walker[c]

[a] *Political Science and Communication, University of Washington, Seattle, WA 98195, USA;*

[b] *Political Science, Stockholm University, Stockholm 106 91, Sweden;*[c] *Information School, University of Washington, Seattle, WA, USA*

ABSTRACT

How is crowd organization produced? How are crowd-enabled networks activated, structured, and maintained in the absence of recognized leaders, common goals, or conventional organization, issue framing, and action coordination? We develop an analytical framework for examining the organizational processes of crowd-enabled connective action such as was found in the Arab Spring, the 15-M in Spain, and Occupy Wall Street. The analysis points to three elemental modes of peer production that operate together to create organization in crowds: the production, curation, and dynamic integration of various types of information content and other resources that become distributed and utilized across the crowd. Whereas other peer-production communities such as open-source software developers or Wikipedia typically evolve more highly structured participation environments, crowds create organization through packaging these elemental peer-production mechanisms to achieve various kinds of work. The workings of these 'production packages' are illustrated with a theory-driven analysis of Twitter data from the 2011–2012 US Occupy movement, using an archive of some 60 million tweets. This analysis shows how the Occupy crowd produced various organizational routines, and how the different production mechanisms were nested in each other to create relatively complex organizational results.

Keywords: connective action, crowd organization, Occupy, peer production, communication as organization

A remarkable protest wave swept the world in the first decade of the twenty-first century, beginning with a popular uprising in Iceland following the financial crisis in 2008, and moving through Tunisia, Egypt, Spain, the United States, and dozens of other countries (Castells, 2012; Gerbaudo, 2012). Although local issues and political contexts differed dramatically, all of the uprisings were characterized by local physical occupations and protest activities in which core participants were linked by digital media to each other and to more peripheral participants and bystander publics. The images, videos, websites, and other media artifacts also provided models for successive protests to appropriate, enabling very different mobilizations to learn and share practices and action repertoires despite differences in time, place, opportunity structures, culture, and issues. In addition, news of these events signaled a global integration of different types and systems of media – personal and mass, national and international – as the activation of vast social networks carried iconic videos and other participant content from cell phones and laptops into the mass media. As those images and direct citizen reports circulated within and beyond national borders, they engaged larger publics, both near and far, who, unlike the often passive audiences of mass media, had opportunities to also become participants by using the many affordances that interactive media offer.

Such mobilizations are characterized by what we have termed the logic of connective action in which participants engage with issues largely on individual terms by finding common ground in easy-to-personalize action frames that allow for diverse understandings of common problems to be shared broadly through digital media networks (Bennett & Segerberg, 2013). Among the strengths of connective movements are the speed of mobilizations, the flexibility of mediated crowds to shift among issue foci and action tactics, and the capacity to reach large-scale publics both directly through digital media and indirectly via conventional mass media channels. The outcomes of these mobilizations have often been impressive. The uprisings in Iceland forced new elections, led to the arrest of bankers and corrupt politicians, the nationalization of banks, and the creation of a citizen assembly to write a new constitution. Governments were toppled in Tunisia and Egypt. Popular challenges to the democratic legitimacy of governments arose in Spain and elsewhere in Europe. And a long ignored public discussion of inequality and political influence was initiated by the Occupy Wall Street protests in the United States (Bennett, 2012; Castells, 2012). Beyond these impressive outcomes, the limits of mediated crowds are also evident, as efforts to form parties, and other, more bureaucratic forms of political organization often meet with limited or undesirable results, as indicated by the aftermath of the Egyptian uprisings. Moreover, political transition efforts often encounter resistance from within as conflicts develop among participants who join the action with conflicting understandings, commitments, and goals (Bennett & Segerberg, 2013).

Despite the evident differences from conventional political organization based on more familiar principles of bureaucratic organization and collective action (Bimber, 2003; Bimber, Flanagin, & Stohl, 2012; Bennett & Segerberg, 2012, 2013; Olson, 1965), these dense networks of networks behave as hybrid organizations (Chadwick, 2013) that can produce remarkable levels of coordination. Many connective networks have been able to regulate their ranks to promote particular ideas and frames and to filter out or limit the dominance of others. While bureaucratic organizations still play a decisive role in some networks, established political groups and organizations have been marginalized in others because they were perceived as politically incompatible with the ethos of the crowd. Thus, despite its well-developed organization, the Muslim Brotherhood was not elevated to a position of leadership during the Tahrir Square mobilization in Egypt in 2011 (Kirkpatrick, 2011). Similarly, labor unions in Spain were kept on the periphery of the 15-M (or *indignados*) uprising in 2011 (Ancelovici, 2012; Anduiza, Cristancho, & Sabucedo, 2013). Meanwhile, the Occupy protests in the United States did not take up the overtures of various established political organizations such as Moveon.org (Harkinson, 2012). Examples of other kinds of work that crowds can do include: establishing various participation norms (e.g. non-violence, deliberative decision-making); shaping and shifting issue focus (inequality, corruption, real democracy); and developing action frames and participatory ethos (e.g. 'we are the 99%').

This paper focuses on large-scale networks of public action in which bureaucratic organizations are marginalized. The logic of connective action broadly suggests why organizational properties can occur among individuated populations using media networks to bridge and transcend their face-to-face interactions. Earlier studies show that highly personalized participation can occur within inclusive organizations such as the social forum movement that invite 'flexible political identities' (della Porta, 2005). We also know that personalized and loosely coordinated action can occur within technology-enabled networks such as consumer campaigns against corporations (Bennett, 2003). However, it is less clear how fine-grained organization develops in large-scale protest in the relative absence of structured or commonly bounded organizations with established norms, targets, demands, and broadly shared values. This paper offers preliminary answers to the question of how crowd organization is produced: how crowd-enabled networks are activated, structured, and maintained in the absence of recognized leaders, common goals, or conventional organization.

The puzzle is how large-scale crowd-enabled networks achieve coherent organization. The key empirical question is: What finer-grained associative mechanisms create and support that organization? We develop an analytical framework for examining basic organizational processes of crowd-enabled connective action, and, by extension, for understanding the role of communication as central to the organization of such networks. We argue for the significance of particular communication technologies and practices that serve as *stitching mechanisms* that connect different networks into coherent organization. We further identify a set of key peer-production mechanisms that, together, accomplish this network stitching, and show how crowd-enabled networks can grapple with the myriad inputs, demands, and events that shape the qualities and the fates of movements. These elements of peer production include: the *production, curation*, and *dynamic integration* of various types of information content and other resources that become distributed and utilized across the crowd. The workings of these mechanisms are illustrated with a theory-driven analysis of Twitter data from the 2011–2012 US Occupy movement, using an archive of some 60 million tweets. The analysis focuses on the large-scale public Occupy network over time (not the organization of a specific action or camp), and shows how the Occupy crowd produced various organizational routines by nesting or packaging different production mechanisms to create relatively complex organizational results. Identifying these peer-production mechanisms illuminates the role of communication as organization in crowd-enabled action and offers a more fine-grained understanding of how large-scale digitally networked protest crowds work.

The puzzle of organization in crowd-enabled action

It is easy to find many particular instances of coordination and organization in large-scale protests. The wave of protests mentioned earlier all involved physical sites of concerted local actions such as encampments, marches, and meetings. Such physical and often ritualized enactments provide the symbolic vocabularies, the dramatic narratives, and the provocations for reactions from authorities and other citizens that sustain action and focus public attention. However, it is challenging to identify the overarching coordinating processes that give the large-scale public media networks around these sporadic dispersed local activities varying measures of coordination, thematic coherence, and public recognition – in short, *coherent organization*.

The search for overarching organization begs a closer look at just what we mean by that term. We begin with a broad definition of organization aimed at bridging bureaucratic and less-conventional post-bureaucratic (Bimber, 2003) variants that require continuous peer production in order to function. There are many distinct forms of non-bureaucratic organization, ranging from networked organization, in which interdependent actors engage in reciprocal relationships in pursuit of common goals (Mueller, 2010; Powell, 1990), to virtual organizations, in which distributed and diverse entities collaborate by electronic means (De Sanctis & Monge, 1999). As we have discussed elsewhere (Agarwal, Bennett, Johnson, & Walker, forthcoming; Bennett & Segerberg, 2013), what these variations on conventional organization have in common with more bureaucratic forms are several fundamental capacities: (a) *Resource mobilization*: the capacity to produce, allocate, and utilize material and symbolic goods that enable recurring patterns of collective action to occur; (b) *Responsiveness to external conditions*: the capacity to react to near-term threats and opportunities, resulting in asserting different repertoires of action, and assessing their results; (c) *Long-term adaptation, change, or decline*: the capacity to develop new patterns of association internally as the organization grows, suffers resource loss, or receives less recognition by relevant publics. Such basic standards of coherent organization can be applied to a range of networked organization types, from highly bureaucratic coalitions (organizationally brokered) to loose bureaucratic networking (organizationally enabled), to crowd-enabled protests.[1] Some protest networks, of course, do not display these elements, but others do. We have shown elsewhere that crowds can display these qualities (Agarwal et al., forthcoming; Bennett & Segerberg, 2013), but it is less clear how they do it.

A first clue about how such organization works comes from earlier theory and research on peer-production communities. As large-scale crowds organize, we often see in the early stages the cooperative development of websites, the customization of social media platforms and channels, and the creation and sharing of content through these and other means of communication such as phones, SMS, and email. The processes and results of such communication resemble the self-organization of open peer production (Benkler, 2006) and open collaboration (Forte & Lampe, 2013). This is to be expected, as the logic at the heart of connective action, self-motivated sharing, is also the logic at the heart of much peer production (Benkler, 2006; Bennett & Segerberg, 2012, 2013).[2] Classic examples of peer production include information and knowledge commons such as Wikipedia, collaborative software such as Linux, and online news and discussion groups such as the political blog site the Daily Kos. They also include collaborative activist projects such as the network of Independent Media Centers (IMCs) of the global justice movement. Such projects may involve vast numbers of dispersed and differently engaged individuals that come together to create a common good – be it protest or software – around which further collective action will revolve (Bruns, 2008; Calderaro, 2011; Schweik & English, 2007). Despite the open-ended nature of such participation, peer-produced projects involve self-motivated production and self-organization: participants ideally contribute to the project in modular and granular ways and help shape the conditions of the action so that the projects build on self-selection and decentralization rather than coercion and hierarchically assigned tasks (Benkler, 2006, p. 59).

Some critics of mediated protests regard such participation as inherently weak, intermittent, and generally lacking in the leadership, commitment, and organization required to create effective movements (e.g. Gladwell, 2010). It is true that many attempts at scaling up protests fail, much as most production communities end up with few participants, most videos never go viral, and most blogs are seldom read. This said, successful peer production also often involves highly uneven levels of participation, experience an ebb and flow of participants, and rely on relatively small ratio of core participants even as smaller input from the broader crowd plays also plays a role (González-Bailón, 2013; Hill & Monroy-Hernández, 2013). The unfortunate critical equation of weak involvement with 'clicktivism' overshadows the point that large-scale and sustained (if uneven) involvement is achieved in some crowd-enabled networks. Our larger aim is to understand how this works.

Impressive headway has been made in understanding how mechanisms built into platforms and developed by participants help them navigate the many challenges of concerted action in conventional peer-production contexts. For example, filtering, recommendation, and reputation systems enable crowds to discern valuable or disruptive contributions (de Alfaro, Kulshreshtha, Pye, & Adler, 2011; Kriplean, Beschastnikh, & McDonald, 2008). New York, NY: Association for Computing Machinery.; Lampe, Johnston, & Resnick, 2007; Lampe & Resnick, 2004). Harmonizing horizontal participation with stratified access and roles helps intentional communities clarify goals, solve disputes, and encourage quality contribution (Crowston, Wei, Howison, & Wiggins, 2012; Hill, Shaw, & Benkler, n.d.; Kittur & Kraut, 2008; Luther, Fiesler, & Bruckman, 2013; Reagle, 2007; Weber, 2004; Welser et al., 2011; Zhu, Kraut, & Kittur, 2012). It also matters how community policies and norms are developed and upheld, and how they are coded into platforms (Butler, Joyce, & Pike, 2008). New York, NY: Association for Computing Machinery.; Forte, Larco, & Bruckman, 2009; Shaw, 2012).

Yet such mechanisms for self-organization have been studied mainly in production communities working under structuring conditions that are generally absent in large-scale crowd protests. Many of the commonly studied peer-production environments display some combination of common goals or outputs, bounded arenas of action, participant identification or reputation systems, moderation routines, software code or platforms that structure participation, clearly defined goals, and mechanisms beyond face-to-face processes for rewarding or discouraging particular kinds of contributions. Such features in turn echo the classic design principles that support self-governance in physical communities seeking to manage common pool resources, which include identifying a clearly bounded good and a bounded set of participants (Ostrom, 1990). Fred Turner observes that online commons-based peer-production communities, like their physical counterparts, depend on a specific 'structural and ideological scaffolding':

Structurally, such work requires a commons, a shared space that in most internet-driven accounts consists of digital messages, but which could be located as easily in some single geographical space. In these arenas, members of diverse social worlds can gather and collaborate toward some end. The commons in turn affords them visibility. Being able to be seen by one another makes it possible for workers to find one another, select projects, and build and maintain reputations. (Turner, 2009, p. 76)

These principles apply to diverse production communities, including Wikipedia, collaborative news collectives, open-source software developers, some technology businesses such as Google, the IMCs of the global justice movement, and even the annual Burning Man festival.

Structuring conditions may of course vary. Different communities emphasize some conditions over others, achieve them with better results, or substitute digital processes for physical locations and routines. For example, digital contexts that support stable user identities produce some of the effects of bounded participation without requiring formal 'membership', by enabling the reputation, recommendation, or stratified access systems that facilitate the peer organization of contributions. Further, in systems in which it is easy to change identities, core participants concerned with boundary work may concentrate their critical attention on those who remain anonymous or do not have a history (Crowston et al., 2012; De Alfaro et al., 2011; Forte et al., 2009). Similarly, gathering around a functional or clear common goal or self-awareness as an intentional community helps bring additional direction and focus to an arena or a project within it (Crowston et al., 2012; Haythornthwaite, 2009; Hill & Monroy-Hernández, 2013; Kittur, Pendleton, & Kraut, 2009). Highly stable crowd organizations such as Wikipedia may display organizational elements such as policies, rules, registered contributors, clear mission, stratified leadership, and a bureaucracy of sorts.[3] In short, peer-production projects may also change dramatically over time from their initial conception into more stable forms (Butler et al., 2008).

Such structuring conditions are less clearly present in the kind of large-scale public protest networks we focus on here. The face-to-face assemblies of the Occupy movement, which were critical to its core identity, did partly function as structuring arenas. However, there were also many of them, they were not bureaucratically coordinated, and they surely did not all share the same local concerns. Perhaps most importantly, the activists at the center of these dispersed physical arenas were a minority of total participants in the larger crowd. Both in numbers and impact, the whole crowd was larger than the sum of its parts, which magnified the movement's voice, impact, and palpable sense of public presence. This does not mean that peer production cannot occur in chaotic crowded contexts. However, without such contextual structuring, it is less clear how crowd production can manage more or less coherent organization. And yet it can.

The puzzle of organization in crowd-enabled action, then, is how large-scale protest networks achieve coherent organization in the absence of clear, codified structuring conditions that characterize the most commonly studied peer projects. We suggest one part of the answer lies in identifying the micro-sequences of different peer-production routines that constitute the organizing 'communication packages' in crowd-enabled networks. Through the resulting networks of crosscutting relations, distributed resources, and action plans, the broader crowd produces the discursive and technological routines that animate connective action – along with the noise that often disrupts and destroys it. We now turn to locating and identifying the finer-grained mechanisms involved in the production of coherent organization in the crowd. The analytical framework proceeds in two steps: (a) deciding where to look for higher-level organization in crowd-enabled networks and then (b) identifying the key peer-production mechanisms that produce order in the crowd.

Stitching mechanisms in crowd-enabled networks

Given their diffuse, densely layered, and dynamic character, crowds offer many arenas and technologies to study. However, not all of these myriad elements of crowds help understand the overarching organization. For example, some technologies may connect hyper-local networks (e.g. the email list for local assembly members), while others may serve as news and discussion sites at various levels of the crowd (e.g. Facebook pages for city protests). Our first aim is to identify the mechanisms for stitching such different networks together, and to understand how these network-management technologies work under different conditions.

Crowd-enabled protest networks often involve networks of many kinds and layers. A movement such as Occupy encompassed many kinds of actors, including Adbusters, participants in local camps and protests, Anonymous, supporters, and opponents who shared information and other resources mainly online, and broader publics who followed the protests as they emerged in daily life or in the media. It is also clear that many important elements that may be crucial to understanding the development of the networks are never digitally recorded or archived for public consumption. For example, many crucial elements of face-to-face assemblies have been lost despite being important at particular moments in time. However, our focus is on the ways in which crowds 'go public' through the digital production, sharing, preservation, and strategic routing of different types of resources. The aim here is to explain how crowds rise above local and face-to-face action to achieve large-scale, patterned organization. The diverse and dispersed layers in a crowd-enabled movement form a dynamic network of networks in which linking and switching activity directs the flow of information, frames, and other resources throughout the movement (Castells, 2009).

Figure 1 models the digital communication in a complex and multi-layered crowd-enabled network (Bennett & Segerberg, 2013). The three views of the same simple model show distinct networks (such as campaign and action sites, Facebook groups, Livestream, or YouTube channels) connecting via the dynamic networks that thread the layers together through stitching technologies (such as Twitter or SMS trees). This does not imply that everyone uses any particular platform or medium, but that linkages among different ones become crucial in the large-scale network for engaging participants and coordinating their relationships more broadly. Indeed, as Figure 1 suggests, the layers of networks may involve separate micro-networks that only partially overlap. These often disjointed networks of networks highlight the importance of the multiple points of connection via crosscutting agents and technologies that cut beyond reciprocal ties to traverse the different network levels, and circulate content and other resources through them. Similar to the idea that content that fails to spread is dead (Jenkins, Ford, & Green, 2013), the large-scale crowd will disintegrate if there is no connection between and circulation across its networks. This in turn suggests that stitching technologies (and the ways in which they are used) constitute important points of focus for understanding the role of communication as organization in the crowd-enabled network. The stitching technology is itself a key organizing mechanism in this context.

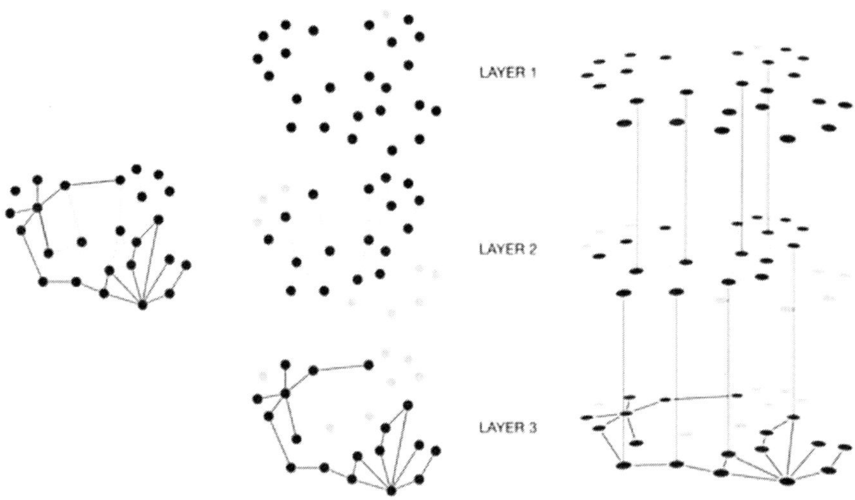

Figure 1. Layered crowd networks connected by stitching technologies. Used with permission of Martin Krzywinski, University of British Columbia Cancer Research Center (www.hiveplot.com).

Network stitching technologies – and the roles of particular platforms such as Twitter – may vary from one situation to another and change over time as old technologies decay and new ones emerge. For example, IMCs and other activist media hubs played a kind of stitching role in the global justice movement in ways that were partially echoed in various media work groups and bridging resource sites in Occupy (Juris, 2005; Lievrouw, 2011). Our first task was to identify key stitching or integrating mechanisms in Occupy. At first glance this may seem a dizzying task, as the protests involved numerous digital media platforms with millions of users interconnecting and feeding information in and out of dispersed physical communities over a period of many months. Multiple groups and networks surrounded the physical camps, as well as commercial platforms (such as

Facebook, Twitter, Livestream, and YouTube) and custom technologies (such as city websites, phone conference schedulers, or apps providing assistance in case of arrest). Some of these networks involved quite distinct networks or sub-crowds. Yet analyses slicing this digital traffic in different ways all revealed Twitter to be structurally central to the workings of this network of networks. Network crawls by the authors and members of the Social Media Lab at the University of Washington using different starting points and methods of analysis all revealed Twitter to be the most linked to node (see also Hemsley, 2011).[4] Various methods of mapping Occupy networks show Twitter to provide the dynamic network stitching mechanisms represented by the threads connecting different network layers on the right side of Figure 1.

Twitter is certainly not the only conceivable stitching mechanism in a crowd-enabled network, and there is likely to be more than one stitching mechanism in play in any given case. Moreover, it is important to note that Twitter can have quite different roles in other kinds of protest networks, such as those that are more tightly managed by conventional organizations (Bennett & Segerberg, 2013). This said, the technology's public, fine-grained routing, and quasi-broadcast qualities may make it particularly well suited to thread together the network of other networks that characterize crowds. Thus, in the case of Occupy, Twitter is central not simply because the data are 'there', or because Twitter data grant insight into all parts of the movement (which they do not), but because of their importance for dynamically connecting or stitching the multiple sub-networks into a large-scale movement. The next step for our purposes is to examine the micro-organizing processes that accomplish this stitching.

Processes of peer organization in the crowd

Understanding how large-scale crowd-enabled networks achieve coherent organization requires considering the organizing processes at work within the stitching mechanism(s). We propose a set of elemental processes that explain how stitching technologies such as Twitter can lend organization to the broader crowd. The mechanisms we propose are analytically distinct and may occur in nested sequences to build more complex organization. The set is basic, but not exhaustive. It consists in three basic categories of peer-production mechanisms:

Production: This category of mechanisms involves creating and publicizing (sharing) various kinds of resources within an action network. Basic production involves content created or distributed through one platform and shared across its network of users (be it Twitter, Facebook, Livestream, news sites, or blogs).

Curation: This family of mechanisms entails the preservation, maintenance, and sorting of digital assets created in the production process. Empirical measures of curation may focus on the negotiation and selection of content to be preserved; the affordances for accessing and sharing that content; archival distribution across sites that share a common resource pool such as music files or activist software development. Curation can thereby also entail the creation of norms and boundaries in particular user communities and their platforms.

Dynamic integration: This category of mechanisms enables contact, transmission, and switching among different actors, networks, platforms, and technologies. This mechanism is evident in switching and linking affordances such as patterns of links found over time in directed hashtags in tweets and in hashtag co-occurrence.

The first two mechanisms are familiar from the work on peer production discussed earlier: crowd-enabled networks share the challenges of ensuring the contribution and processing of content, and participants employ some of the same mechanisms to negotiate these challenges. While other familiar processes such as moderating and editing (important in projects focused on deliberation or the definitive development of content) seem less important in this stitching context, the third mechanism may be more particular to the needs of a multi-centered movement: the crowd-enabled network depends fundamentally on mechanisms that dynamically connect different parts of the crowd and circulate contributions across the different layers.

We distinguish the three processes analytically, but in practice they may nest into more complex operations. For example, a tweet is a unit of production that may be broken down into several sub-elements such as message content, links, retweets, hashtags, mentions, and other aspects of syntax which enable users to combine production with curation and dynamic integration. The logic of nesting in our framework is hierarchical. We can empirically demonstrate that production may occur without higher-level operations such as curation. By contrast, curation always implies production (as when messages are archived) and dynamic integration implies both production and curation (as when particular hashtags or numbers of hashtags are attached to tweets containing particular types of links). The important issue theoretically is how different production mechanisms occur (or do not occur) at critical junctures in the history of a crowd. For example, when the volume of crowd activity tails off, indicating that protests may be dying out, how do actors respond? Does the pattern of production (who produces what) change? Do patterns of curation (who retweets or follows whom) change? Do strategies for dynamic (re)integration change (how many of what hashtags are attached to particular patterns of production and curation)? In this analysis we track separable nested elements of these production processes. There are clearly mechanisms that blur these conceptual boundaries, as might be expected when actors see ways of performing more complex operations in fewer moves. For example, 'favoriting' a message on Twitter is used both to endorse and to bookmark. It archives the message, publicly signals that it is worth attention, connects two account holders (the source account is alerted), and is a message-specific public display of connection (Donath & boyd, 2004), since the message shows the number of times and by whom it has been favorited. While observing such operations may be useful for some analyses, we begin with more elemental units and their sequences.

Further, although the three elements of the peer-production package are integral to the organization of crowd-enabled networks, what is important is whether the crowd as a whole produces the package, not that all participants produce all parts. As in other peer-produced projects, participants in large-scale protest networks do not necessarily engage the different mechanisms equally, and many may contribute in seemingly insignificant ways. They also contribute for varying reasons and undertake different roles, with some preferring more (or less) vigorous roles and some shifting in their activity over time (Crowston et al., 2012; Fang & Neufeld, 2009; González-Bailón, Borge-Holthoefer, & Moreno, 2013; Hill et al., n.d.; Welser et al., 2011; Wilkinson, 2008). Even if participants do not always contribute to the larger cause wittingly, the broader crowd may identify particular contributions as valuable and therefore work to highlight some contributions and contributors more than others. Stable patterns of production and utilization emerge as people or sub-networks take particular actions, share responses to external events, or seek to transform the network's organizational form. It is the capacity of the broader crowd to produce the package of organizing processes at critical junctures that accounts for the organizational capacities of the larger network.

The following sections illustrate how the organizational mechanisms work in the crowd-enabled network of US Occupy 2011–2012. As should be clear, focusing on the production package highlights the action going into the organizing structure rather than the structural properties of the network per se. This analysis therefore complements but would also be enriched by closer analysis of diffusion mechanisms (Vasi & Suh, 2013) and other elements developed within social network analysis (Diani, 2011; Pavan, 2012). In what follows, however, we discuss each category of mechanism in turn. In each case we concentrate on the digital mechanism that most cleanly captures the gist of that category (respectively the link, the retweet, the multiple hashtag), but also successively trace how the categories (and their most associated mechanisms) nest in each other. The final empirical section spotlights how different parts of the crowd contribute to the whole.

Crowd-organizing mechanisms in action

The following study builds on an earlier project involving two of the authors that established that the Twitter networks in the Occupy movement in the United States displayed the general markers of organizational coherence outlined above (i.e. resource mobilization, responsiveness to external conditions, and long-term adaptation or change). That study identified empirical indicators of these organizational patterns in the Occupy-related Twitter traffic over time (see Agarwal et al., forthcoming). This section reviews the Twitter database developed in the earlier study, which we here re-analyze to examine how such coherence was achieved.

As already noted, numerous analyses using different network crawl starting points and methods showed unambiguously that Twitter was the overarching stitching mechanism of the Occupy movement. The Social Media Lab gathered a nearly complete set of Occupy Twitter data from mid-October, 2011, when a national network of occupations had been established and protest activity was at its peak, through the summer of 2012, when the national protests tailed off into scattered local actions.[5] (In addition to the data collection methods sketched below, the reader may wish to consult a more complete review of methods in a Social Media Lab working paper by Eckert, Walker, Hemsley, Nahone, & Mason, 2013). The peak data captured from 19 October 2011 to 31 December 2011 contained roughly 20 million Tweets (20,645,921). The archive collected tweets using Twitter's Streaming API track method, which returns tweets matching any of the search keywords occurring in the text, hashtags, @mentions, or URLs within a tweet. A curated list of 109 popular hashtags, keywords, and Occupy city accounts related to the Occupy movement was created by a panel of faculty and graduate students. The data stream was examined at regular intervals for emerging hashtags and keywords as the panel of experts tracked events related to Occupy and followed the burgeoning lists of hashtags (e.g. #occupywallstreet, #ows, #occupyoakland, and growing numbers of cities), places such as Zuccotti Park, key concepts such as '99%', political actions such as 'Occupy the ports', and 'Occupy Obama', and events such as 'pepper spraying cop'. The terms were added to the keyword list after review by the research team, resulting in a dynamic archive based on a list of 355 keywords as the collection continued through the summer of 2012, by which time the database included more than 200 million tweets.[6] A new data file was created each day and the file from the previous day was backed-up and processed. Processing included expanding shortened URLs, adding metadata to each tweet in order to make searching and sampling easier, and inserting it into a NoSQL database for storage and querying.[7]

The earlier study then coded links in tweets from the archive as proxies for various kinds of resources being produced and shared in Occupy networks. Random samples were drawn from three Occupy hashtags.[8] Two were keyed to local city protests (#occupyoakland and #occupyseattle), and samples were coded for their link content during the period of greatest protest activity from mid-October through the end of December 2011. The third hashtag, #ows, was the one that was most actively used over the course of the protests following its early emergence as a simpler version of #occupywallstreet. The hashtag #ows provided a common channel for actors at all levels – the stitching within the stitching mechanism – as well as serving as a feed for events specifically surrounding the New York City encampment in Zuccotti Park.[9] This tag remained active even after the camps were evicted and the protesters scattered in the early months of 2012.

The study coded the content of links in the three hashtag streams, and coded #ows over a longer period (October 2011 through March 2012) in order to capture behavior in the tail. A total of 30,675 links in tweets were coded for the three hashtags, and particular links and link types were tagged and tracked back into the entire data archive to examine broader flow patterns. For the present purposes, since the links represent different kinds of resources produced by the crowd, identifying the production, curation, and integrative engagement with various kinds of link resources offers a direct look at crowd-production processes. Analyzing these processes in three different hashtags enables comparison of different layers in the crowd and their reactions to events, and the possibility to monitor change over time. This analysis is supported and contextualized by ethnographic study of different Occupy networks conducted by other teams from the larger project.

Linking in the dataset was impressively diverse. Links included conventional news sites; political organizations; Occupy organizations; specialty and hybrid sites such as Kickstarter or Boing Boing; personal blogs and photo pages; and links to assistance with food, shelter, medical care, financial help, and legal services. The links were hand coded to distinguish the following categories:[10] (A) conventional news reports; (B) political commentary and activist media; (C) specialty sites beyond daily news and politics (e.g. Boing Boing, Tumblr pages, best of the web features, business reviews, Kickstarter fundraising initiatives.); (D) conventional government-linked organizations, such as parties, government institutions, and websites of officials; (E) issue-oriented, ideological, activist, and NGO organizations not specifically identified as Occupy groups; (F) clearly identified Occupy websites, campaigns, events, or organizations; (G) individuals' personal sites and postings; (H) music or individual celebrity websites; (I) other (including non-English-language sites), and (J) broken links. The coding also included sub-categories for E and F to capture whether the link pointed to sites based on branded commercial or more customized platforms: (E1) NGO and activist organizations not identified as Occupy and built on commercial media sites such as Facebook or YouTube, (E2) custom-built activist platforms and sites; (F1) Occupy sites based on commercial media platforms and (F2) custom-built Occupy sites.

Analyzing the broader patterns in the three hashtags, Agarwal et al. (forthcoming) found coherent organization in the form of resource allocation, responsiveness to external conditions, and long-term adaptation over time. In the analysis going forward, we focus on the finer-grained processes involved in achieving such coherence. At the time of Occupy, Twitter users could post messages of up to 140 characters, but could also insert shortened URL links into the messages. Account holders could remain anonymous behind pseudonyms, but whether anonymous or named, the account profiles and production patterns remained consistent over time. Users could retweet (resend) or mark as favorite messages to which they wanted to call attention, and could address other accounts publically by using the @mention function. The community-engendered convention of adding a hashtag (e.g. #keyword) to collate messages on a topic and make them publically searchable was widely used. While tweeting with a hashtag was a public, network-creating move, messages only appeared in public searches for a brief period of time (Bruns & Burgess, 2011; Murthy, 2013). Such specifications play into the organizing mechanisms and patterns that users develop within the streams to deal with such issues as the enforced brevity of the messages, anonymous contributors, and the ephemerality of a hashtag stream.

Production

The first category of mechanisms concerns the production and provision of content and resources. Forms of production may vary greatly. Resource provision in crowd-enabled protest networks requires contribution, but it does not require the contribution to be particularly original, created expressly for this purpose, or even offered knowingly with a larger crowd in mind. It can nevertheless provide raw material for organizing around different kinds of resources, including useful information, symbolic themes and identity frames, and resources for taking and coordinating action either immediately or in the future.

In Twitter, the main ways to introduce content resources into a network are the message text itself and links to content on other platforms (which also invokes curation). Our analysis of linking in the Occupy dataset indicates the variety of resources circulating at different levels of the larger movement. Figure 2 presents the resource linking distribution in the most-used hashtag, #ows, and the two local hashtags #occupyseattle and #occupyoakland during the peak period of protest activity in 2011. As the pie charts show, the #ows tweets linked fairly evenly to news (link coding category A) and to personal content (G) such as blogs, YouTube videos, and Flickr pages (27%). By contrast, contributors to the two local hashtags linked disproportionately to personal content (36% for Seattle and 44% for Oakland), reflecting the prominence of networks of interpersonal connections and on-the-ground organizational resources in the more local settings.

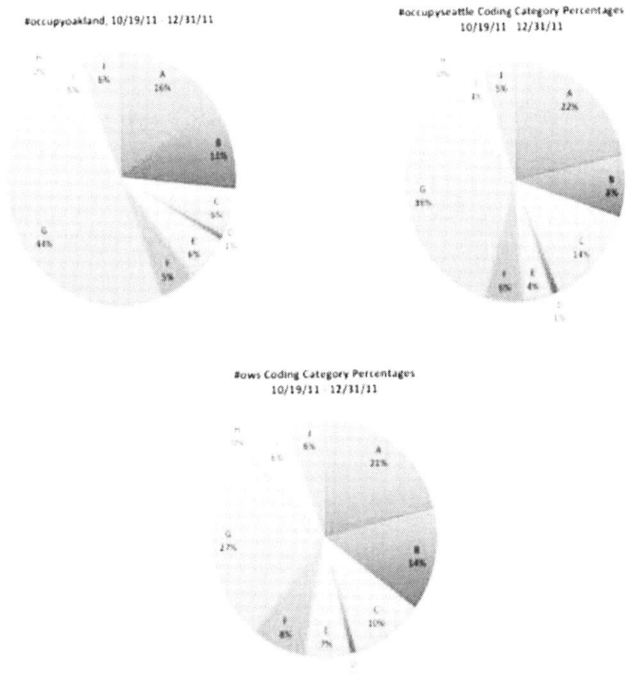

Figure 2. The proportions of each link resource category in each hashtag from 19 October 2011 through 31 December 2011. A = news; B = commentary/opinion; C = specialty sites; D = government; E = general political organization; F = Occupy site; G = personal content; H = music/celebrity; I = other; J = broken link. Source: Agarwal et al. (forthcoming).

These overall patterns result from complex divisions of production within each stream. The coded sample of non-retweeted tweets in the #ows stream revealed that 90% of contributors consistently linked to one type of resource only. As Figure 3 shows, the single-link producers linked disproportionately to news sites (A) or to personal site postings (G). The colored edges that connect different link-type clusters reveal that roughly 10% of user accounts in the coding sample produced more than one resource type. Figure 4 shows how this multiple resource production was spread more evenly across a variety of resource categories. Mapping the coded links back into the larger dataset enabled us to track a more dynamic and complete picture of resource production, picking up members of the crowd who also retweeted each coded link type or who produced the same links independently of each other. Adding up the volume of actors who produced more than one type of resource via non-retweeted links, retweeted links, and independent production of the same links showed that fully 27.7% of users in the dataset for the 2011 peak phase of the protests sent multiple link types. The curation and dynamic integration processes involving these multiple-resource producers will become important later in our analysis. For now, the production pattern shows that the dominant numbers of news links and personal site postings in #ows came from a large number of 'specialist' resource contributors, whereas a smaller but still impressive number of multiple-resource contributors ensured a diverse flow of link types. Both specialists and generalist producers are in this sense important in the stream.

Figures 2–4 thus suggest interesting divisions of labor in the production of resources. Not only are different layers of the crowd characterized by production of different resource types (Figure 2), but within a given network, some users produce fairly specialized types of resources (e.g. their own blogs or YouTube videos) while others generate more diverse content, as shown in Figures 3 and 4. Moreover, these production dynamics shift over time, both in response to specific events and as the crowd tails off. For example, during police raids the general production of links declined in all three hashtags, as people used Twitter more for direct communication about what to do in the crisis. Within such critical episodes of crisis or confrontation, the types of links that were produced also shifted. Confrontational events were often accompanied by spikes in news links to share impressions of how news media and public officials were engaging with events. As the camp-centered movement tailed off, production again changed: links to news (A) and personal postings (G) that dominated the #ows in 2011 (Figure 2) gradually receded, and the proportion of total links to Occupy (F) sites rose to constitute some 21% of total links by April 2012. This suggests that resource production shifted in efforts to point activists to Occupy sites where regrouping and reorganization might take place. A critical question about these large and shifting resource-production patterns is whether the crowd can make sense of the flood of resources coming in, and keep track of them for future use. This is where curation comes in.

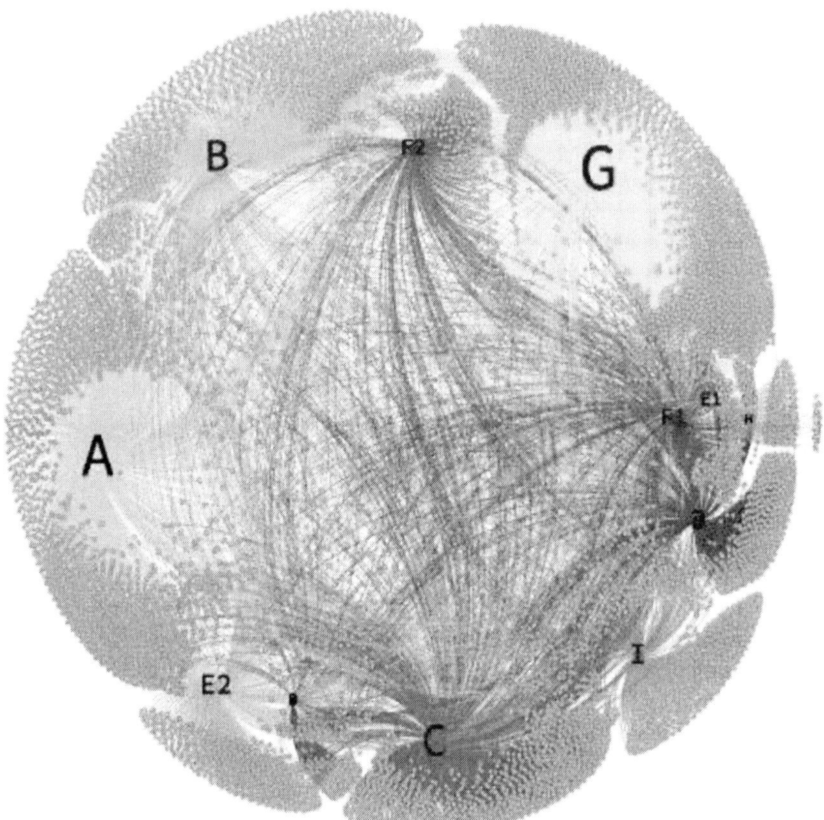

Figure 3. Individual tweets containing different resource links. Nodes represent user accounts linking to particular kinds of resources. Edges connecting different link categories represent users producing multiple kinds of link resources. Data come from the #ows link coding samples of 22,407 tweets with links drawn randomly every other day from 19 October 2011 to 30 April 2012. A = news; B = commentary/opinion; C = specialty sites; D = government; E = general political organization; F = Occupy site; G = personal content; H = music/celebrity; I = other; J = broken link.

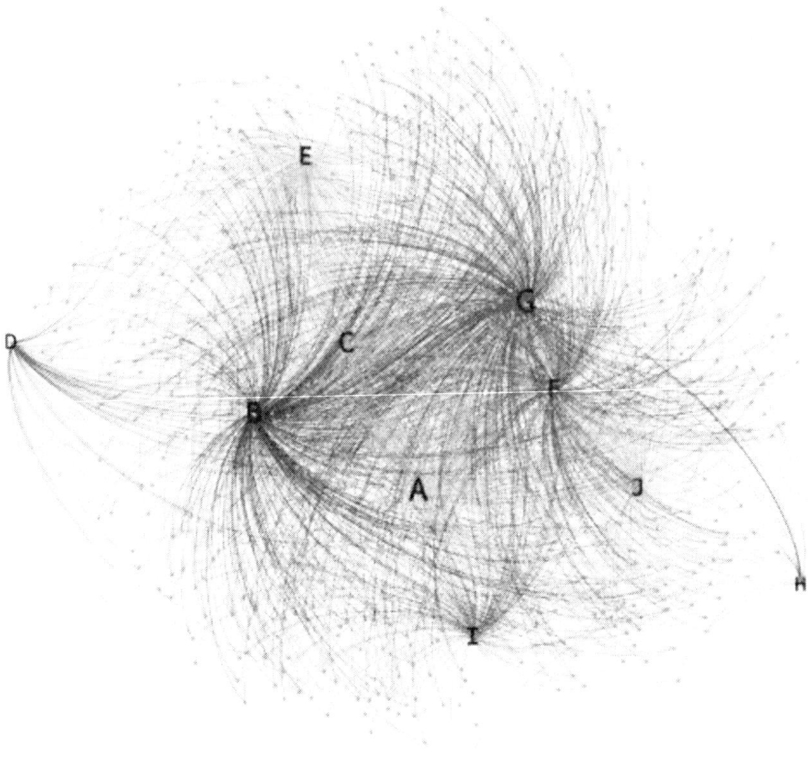

Figure 4. User accounts producing more than one link type in the #ows coding sample, representing about 10% of the sample of coded accounts shown in Figure 3. (This figure becomes 27.7% when the coded links are mapped back into the whole dataset, picking up retweets and multiple independent linking to the same resource site.) A = news; B = commentary/opinion; C = specialty sites; D = government; E = general political organization; F = Occupy site; G = personal content; H = music/celebrity; I = other; J = broken link.

CURATION

Our second category of peer-production mechanisms involves the preservation and maintenance of digital assets. Faced with the noise created by millions of producers, the crowd must sort, distribute, and draw timely attention to resources and sources. It is not so much viral spread that is significant from this perspective, but rather the on-going work to process what is of interest, and to distinguish it from less valuable or counterproductive contributions (including rumors and spam). Users monitor, filter, preserve, amplify, and endorse relevant, useful and desirable content in various networked curation processes (Barzilai-Nahon, 2008; Jenkins, Clinton, Purushotma, Robison, & Weigel, 2006; Lampe & Resnick, 2004; Meraz & Papacharissi, 2013; Shaw, 2012). As we discuss in later sections, complex forms involve cross-platform curation and promoting the contribution of specific participants.

Platforms such as Twitter are associated with relatively intense curation practices: contexts in which posts are only visible temporarily put pressure on users to preserve and highlight potentially important contributions before they disappear (Bernstein et al., 2011). We here focus on the most basic form of preserving and amplifying content on Twitter, which is to retweet, or forward, a message so that it is seen by one's followers or by those following the hashtags that may be attached to the message. Retweeting draws attention to a message (and its resource link), while suggesting possible endorsement,[11] adding longevity, and circulating it beyond the original network for which it was posted (boyd, Golder, & Lotan, 2010; Starbird & Palen, 2012). Other forms of Twitter curation include following, @mention, and favoriting.

As we discuss further below, there was a fair amount of noise in the Occupy Twitter traffic in the form of malignant spam and commercial promotion. However, various curation processes in the crowd such as selective retweeting and following helped dampen this noise. Indeed, the retweeting activity of the overall crowd was impressive. We matched the coded link sample of 13,424 codes in #ows for the period October–December 2011 with tweets in the entire dataset for the key link types A (news), B (commentary), C (specialty), F1 (commercial Occupy site), F2 (custom-built Occupy site), and G (personal site). This procedure identified 228,254 tweets with the same URLs, and 109,048 of them were retweets. This suggests a robust production flow (the same links being tweeted or retweeted by different accounts), resulting in broad crowd curation of the resources regarded as interesting or important to share. Occupy activists have reported retweeting deliberately to 'boost signals' and spread resources to people within but also beyond the movement (Penney & Dadas, 2013, p. 9). Emerging studies indicate the discernment with which this was done. Thorson et al. (2013) show that while almost half of the YouTube material that was linked in Occupy-related tweets was retweeted at least once, videos tagged with Occupy but actually promoting something else were not retweeted at all. As can be seen in Table 1, roughly half of the activity in the streams #ows and #occupyoakland consisted in retweeting, with some interesting differences between the streams as circumstances change.

Table 1. Tweets with links and retweets with and without links in hashtags #ows and #occupyoakland during peak activity periods in 2011.

Tweets				**Retweets**			
Date	**Total**	**With URLs**		**Total**	**% of all tweets**	**With URLs**	
#ows							
19 Oct. – 14 Nov.	1,504,643	816,276	54%	669,897	44%	351,443	52%
15 Nov. – 18 Nov.	792,623	312,676	39%	428,671	54%	161,389	38%
19 Nov. – 31 Dec.	1,496,575	849,264	57%	643,762	43%	343,555	53%
Total tweets	3,793,841	1,978,216		1,742,330		856,387	
#occupyoakland							

Tweets				Retweets			
Date	Total	With URLs		Total	% of all tweets	With URLs	
19 Oct. - 14 Nov.	345,381	149,343	43%	202,800	59%	81,126	40%
15 Nov. - 18 Nov.	21,025	9,193	44%	11,192	53%	4,446	40%
19 Nov. - 31 Dec.	85,099	39,464	46%	45,536	53%	18,999	42%
Total tweets	451,505	198,000		259,528		104,571	

Source: Adapted from Agarwal et al. (forthcoming).

Our earlier studies of Twitter as a stitching mechanism in crowd-enabled protest networks indicate that periods of relative normalcy favor the production and retweeting of links to mobilize resources across networks. The volume of links (especially to particular types of resources such as links to potentially stabilizing political organizations) also increases dramatically as the crowd tails off (Bennett & Segerberg, 2013). Times of crisis or confrontation, by contrast, typically involve less linking and higher volumes of retweets (also with fewer links), suggesting more real-time communication about the actions and needs of protesters. This pattern was also echoed in the #ows stream as the crowd shifted to respond to external crises. Table 1 shows the rise in retweets corresponding with a drop in tweets with links in the #ows stream from 15 to 18 November 2011, three days that saw police raids and camp evictions in New York City, which was both a local and national crisis for the movement. The period stretching to the end of the year marked a return to relative calm for #ows as indicated in the increase in both tweets and retweets with resource links. By contrast, Oakland was among the most active local protest sites, with numerous, nearly continuous confrontational activities such as camp evictions, strikes, port closures, and other clashes with police and city authorities scattered across the entire period of time shown in Table 1. Displaying something of a permanent crisis pattern, #occupyoakland contained more direct communication with fewer links and higher levels of retweets across the entire period. Tightly managed local camps with a focus on staging protest events may in general be expected to generate more direct personal communication focused on messaging and curation of important messages via retweeting rather than focusing on less action-oriented resource linking.

The different production and curation patterns in the two streams illustrate how different networks within a larger movement may develop their own production cultures and direct participant focus on different events. This is not surprising: as noted earlier, highly adaptive and flexible technologies such as Twitter may fulfill different functions in the different contexts. Our data further suggest that the #ows stream was both a local and national stitching mechanism, making it all the more interesting from an overarching organizational perspective. While curation helps explain how particular actors in the crowd keep resources in play, other mechanisms are involved in helping people channel those resources as the crowd itself changes in composition. The final category in the production package thus concerns the work to selectively integrate networks and resources over time.

Dynamic integration

Where the first two organizing processes involve the provision and ordering of resources in crowd-enabled networks, the third category of mechanisms concerns the work to integrate different parts of the crowd in response to events and states of organizational growth or decline. Since large-scale protests involve networks of networks, a continuing challenge is to connect the networks in meaningful ways. Fuller integration would likely involve a transition to a different organizational form. The crowd-enabled network depends more on partial integration and circulation, which in turn depends on the capacity of participants to create and recreate multiple connections between different network parts. This may include decisive brokerage or 'switching' connections between two significant networks (Castells, 2009), but must also involve multiple mundane connections that stitch the different networks in the crowd together. The capacity to connect networks is crucial for coherent organization in large-scale crowds.

One basic means of network integration involves circulating content across platforms (Baym & Shah, 2011). Elemental ways of achieving such cross-platform integration in Twitter is by linking to particular platforms and retweeting those links. What distinguishes this integration operation from basic production and curation is the type of platforms involved in the cross-linking. For example, tweeting and retweeting links to Occupy web sites, Livestream channels, or city Facebook groups better accomplish cross-platform organizational integration than retweeting a news story in *The New York Times* or *The Guardian,* or simply linking to an individual's blog or YouTube channel. We illustrate differences in these link patterns below. These kinds of nested and directed flows show how actors can use simple combinations of communication moves to create important shifts in crowd attention and engagement.

Cross-platform linkages become important for connecting layers or levels of crowds. Many local Occupy networks faced distinct challenges (cf. the discussion of Oakland earlier) and focused on different concerns (Jensen & Bang, 2013; Thorson et al., 2013). By archiving and linking videos and local sites to broadly used platforms such as Livestream or national campaign sites, activists in one sector of the crowd directed broader attention to critical local events such as camp evictions or successful actions that actors in other sectors might emulate. These cross-platform linkages provided more chances for local networks and broader publics to engage with each other. For example, the proliferation of cross-platform links to the trove of personal stories on the original 'we are the 99%' Tumblr blog surely aided the popular adoption of that idea as the master frame for the Occupy protests (Gaby & Caren, 2012). The empirical burden of deciding when cross-platform linking becomes integrative depends on determining which platforms enable high-volume interaction enabling coordination, commitment, or planning among participants in the crowd.

A more clearly distinguished form of dynamic integration involves Twitter users forging together whole network streams. The most direct way of doing this in Twitter is to insert one or more hashtags into a message to send it simultaneously into different sectors of the crowd and make them aware of each other. As noted, a hashtag makes a message searchable under that #keyword and allows people interested in a topic to gather all messages related to this. Adding multiple hashtags utilizes the stitching potential of a platform such as Twitter to the fullest by momentarily assembling complex networks that may be attentive to different streams. Mark Tremayne notes that some users seem to specialize in such action, and offers an excellent example of when a single instance of brokerage was decisive for the development of an early Occupy-related network, namely when a retired high-school teacher tweeted '#fuckyouwashington & #operationwallstreet' in July 2011 (Tremayne, 2013, p. 10). A large-scale network relies on the calibration and dynamic effected in multiple such instances.

Our Occupy dataset, which was compiled from multiple request terms (not just hashtags), reflected diligent hashtag use throughout the entire period. What is noteworthy is that overall integration strategies of appending multiple hashtags shifted in phase with events. This becomes clear if we examine the trend for messages containing particular link types (suggesting different message types) over time. For example, in #ows multiple hashtag use remained fairly constant amongst those linking to personal sites (coded as G), but, as the protests tailed off, the number of hashtags increased in tweets linking to Occupy sites, particularly to custom-built sites (coded F2) that tended to be more central nodes in most national level networks. Participants using #ows and linking to personal sites and postings (G) consistently included a high number of other hashtags throughout the entire period of analysis, suggesting that individuals worked hard to get their contributions distributed as widely as possible. As Figure 5 shows, the hashtag inclusion rate for links to personal sites hovers steadily around 3 and 4.[12] Given the personalized engagement that characterized the protests, it is not surprising that links to personal content were both the most common kind of production (shown in Figure 1) and the type of production that tailed off most slowly as the protests died off (top graph in Figure 5).

By contrast, as noted above, the use and number of multiple hashtags rose in messages linking to Occupy custom-built sites (F2; e.g. city websites or campaigns such as Occupy Our Homes). Figure 6 shows an increase from around 2 tags per message to between 3 and 4 by the end of the protests. Recall that the proportion of total Tweets containing links to Occupy organization sites also rose across time, suggesting that the crowd sought to stabilize and reorganize itself in its dying days. Other crowd-enabled movements that we have studied have displayed similar patterns of increase in hashtags and links to particular types of organizational resources in phases of rupture, decline, and transition. Such integration efforts seem to intensify precisely as other organizational aids (e.g. geographical arenas) disappear (Bennett & Segerberg, 2013).

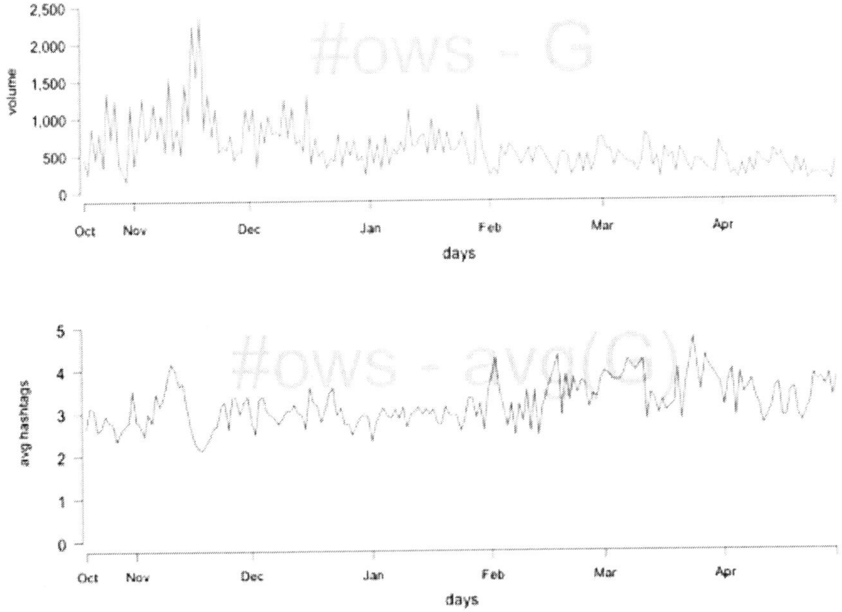

Figure 5. Tweet volumes (top) and average number of hashtags in tweets (bottom) that included links to personal content (G) and that used the #ows hashtag. Based on coded samples matched back into the entire database. Daily from 19 October 2011 to 30 April 2012.

Up to this point we have concentrated on the three distinct production types involved in explaining the role of communication in the organizing processes. We now approach the Occupy crowd from a more dynamic perspective showing how actors in the crowd worked together to put various organizing communication packages into play.

Producing the organizational package: core and co-producers

Beyond identifying the mechanisms of peer production, it is important to consider differences in how participants engage them and what this means for the development of the broader networks of networks that constitute the crowd. The capacity for coherent organization of crowd-enabled protest networks hinges on how core contributors and more peripheral participants work together to produce the package of elemental organizing processes.[13]

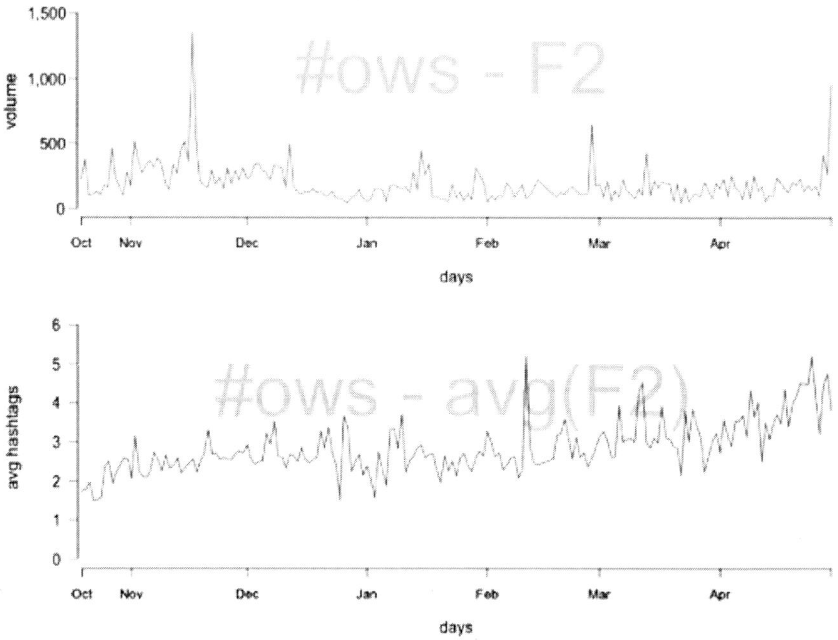

Figure 6. Tweet volumes (top) and average number of hashtags in tweets (bottom) with links to Occupy sites not built with branded commercial technology, such as city, tech developer, and campaign sites (coding category F2) and that used the #ows hashtag. Based on coded samples matched back into the entire database. Daily from 19 October 2011 to 30 April 2012.

As noted, not all participants in peer production will produce all or many parts of the organizational package. Contributors will undertake different roles and participate to varying degrees, and most are not mindful of the needs of the whole. Elite or early movers may trigger broad activation and information flows (Barzilai-Nahon, Hemsely, Walker, & Hussain, 2011; Margetts, John, Escher, & Reissfelder, 2012) or create an important 'attention backbone' to highlight other contributions (Benkler, Roberts, Faris, Solow-Niederman, & Etling, 2013, p. 1). Yet protest crowds may also work judiciously, if not with deliberate oversight, to enhance the network's capacity for coherent organization (González-Bailón, 2013; González-Bailón et al., 2013). The many small and fitful contributions of the crowd, whether in production, curation, or dynamic integration, are all potentially important in this process. As noted, participants can make a difference by highlighting contributions, actors, and streams that are deemed valuable. What is interesting is that while crowds often promote elites or those with prior credentials (González-Bailón et al., 2013; Suh, Hong, Pirolli, & Chi, 2010), in-network authority may also shift over time as recognition grows or the needs of the crowds change: curation processes may 'crowdsource to prominence' non-elite actors on the basis of their in-network contributions (Meraz & Papacharissi, 2013, p. 21; cf. Kwak, Lee, Park, & Moon, 2010; Starbird & Palen, 2012). Such activity helps shape the network and its organizational capacity.

The Occupy Twitter data suggest two things: that there is a core group of more active participants and that the contribution of this core is highlighted and amplified by the crowd. It is clear that some participants contributed more intensely than others to all parts of the organizational package. The roughly 27% of accounts in the #ows stream that were the most active dynamic integrators in linking to the most diverse types of resources across platforms (Figure 4) were also the most notable producers and curators: they posted and retweeted the most messages. Figure 7 displays the average Occupy-related tweet production from accounts contributing single vs. multiple links and shows that the more prolific the tweeting, the more diverse the production.[14] Figure 8 shows the curation rate by accounts producing single vs. multiple types of links measured as average retweet (the rate at which the account retweeted other messages). Both figures display a classic power law distribution headed by the producers of multiple-link types.

We have been able to determine that these core multi-resource contributors were overwhelmingly serious about and supportive of Occupy. This is not something that can be taken for granted – as noted earlier there was a fair amount of noise in the Occupy Twitter traffic. Nevertheless, as can be seen in Table 2, analysis of the top contributors in the #ows stream showed that spam accounts (malignant spam and product promotion) clustered amongst those tweeting only one or two link types.[15]Meanwhile, accounts that were explicitly negative toward Occupy were most visible around the middle linking mark.[16] Accounts tweeting supportive messages could be found throughout this sample, but dominated the category of most prolific producers in terms of tweets, retweets, and links type. It is noteworthy that while this top category included city camp accounts and an Occupy media platform account, it was dominated by unaffiliated (non-organization) pro-Occupy accounts.[17] Some of the unaffiliated accounts were used by activists (possibly several sharing an account) on the ground in demonstrations and city camps or who were highly engaged over time,[18]18. A couple of the top contributors already appear in the lists of top ten accounts in the study of Occupy-related Twitter in the summer of 2011 (Tremayne, 2013 but others had less clear off-Twitter credentials.

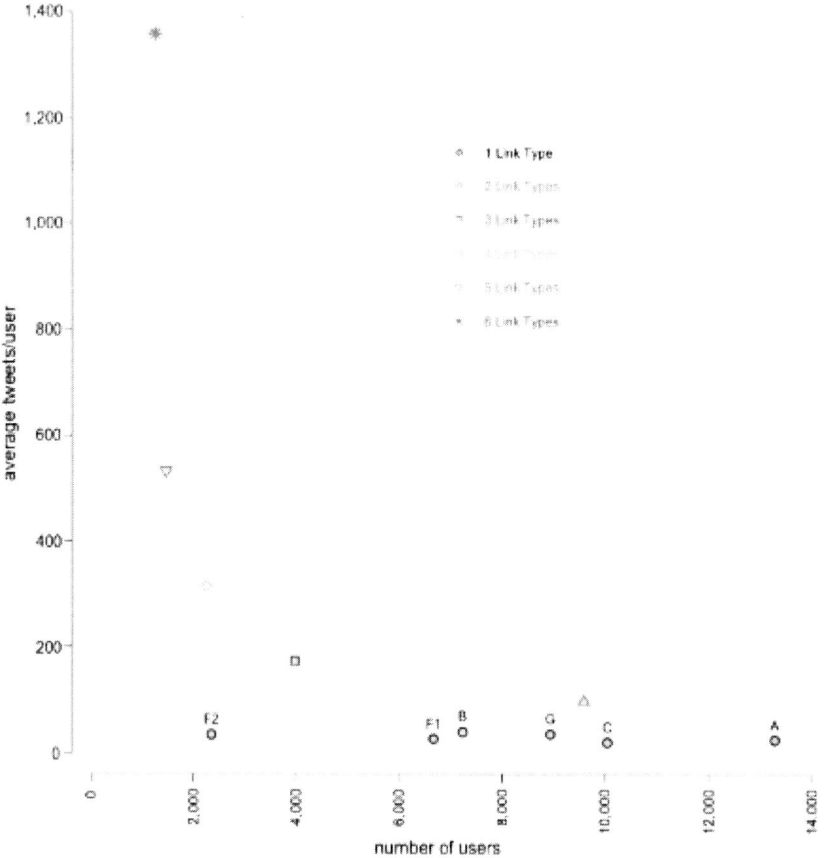

Figure 7. Average Occupy-related tweet production from accounts producing single vs. multiple types of links on #ows. Multiple linkers include combinations from two to six link types tweeted by the same users (e.g. 2 = two types of links only; 6 = links to all six resource types) during the period October–December 2011. *Y*-axis shows numbers of all tweets from those accounts in the whole Occupy dataset. The six key link types: A = news; B = commentary; C = specialty/hybrid sites; F1 = Occupy sites built on commercial platforms; F2 = custom Occupy sites; G = links to personal content.

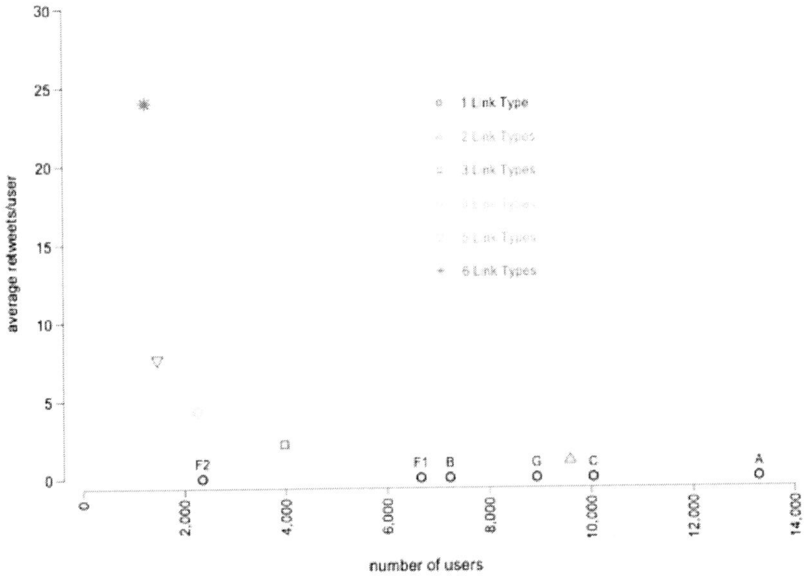

Figure 8. Average Occupy-related retweet production from accounts producing single vs. multiple types of links on #ows. Multiple link clusters refer to the combinations of different link types tweeted by the same users (e.g. 2 = two kinds of links only; 6 = links to all six) during the period October–December 2011. *Y*-axis shows average numbers of retweets from those users in the whole Occupy dataset. The six key link types: A = news; B = commentary; C = specialty/hybrid sites; F1 = Occupy sites built on commercial platforms; F2 = custom Occupy sites; G = links to personal content.

Table 2. Top contributors in #ows (October–December 2011) by numbers of links and type of account.

Account type	**1–2 links**	**3–4 links**	**5–6 links**
Occupy ORG/EQUIVALENT	2% (4)	4% (7)	8% (15)[a]
Pro-Occupy NON-ORG	23% (46)	63% (125)	89% (176)

Account type	1–2 links	3–4 links	5–6 links
Anti-Occupy NON-ORG	8% (16)	20% (40)	1% (2)
SPAM	56% (111)	4% (8)	0
No category	11% (21)	10% (18)	3% (5)

Note: Sample size in each link category = 99.

[a]This number should be approached with caution as it includes 8 bot accounts, which for the most part retweeted messages relating to a particular city camp. We cannot verify that they were city-related accounts. While their content suggests that they were straightforward bots, we have also observed a number of bot accounts in the data set mixing serious retweets with spam.

Accordingly, although Figures 7 and 8 display what one would expect to see in a distribution of contributions in online peer production (Wilkinson, 2008), the twist in this case is that the accounts with the largest following are in the tail of the curve while many of the accounts at the head have relatively small followings. In part, this can be explained by the fact that some of the producers of only one link type were established news sites or blogs with large followings. What is interesting is that even though the core contributors started out with less-explicit recognition, the crowd propelled them to greater visibility over time. As shown in Figure 9, the average increase of new followers was greatest for the accounts tweeting the most diverse array of links. This pattern continued as the protests tailed off: although the total number of active users declined, the numbers of followers for those tweeting all 6 core link types increased from an average of 1,908 per account in the Fall to over 3,000 from January to April 2012. In some cases the relative increase in followers is more extreme because the account was created for the protests and thus started with no followers at all. For example, the account producing six links using #ows that had the greatest follower change in the database was OccupyWallStNYC, which had a total of 100,346 followers by the end of December. However, as just noted, there were few recognizable especially created Occupy accounts amongst the top contributors, and the trend of increased followings indicated in Figure 9 holds also for the other top contributing accounts. Since 'following' an account is a form of curation that signals interest and enables stable resource flows from valued sources, this suggests that the crowd recognized the value of different producers' specific contributions.

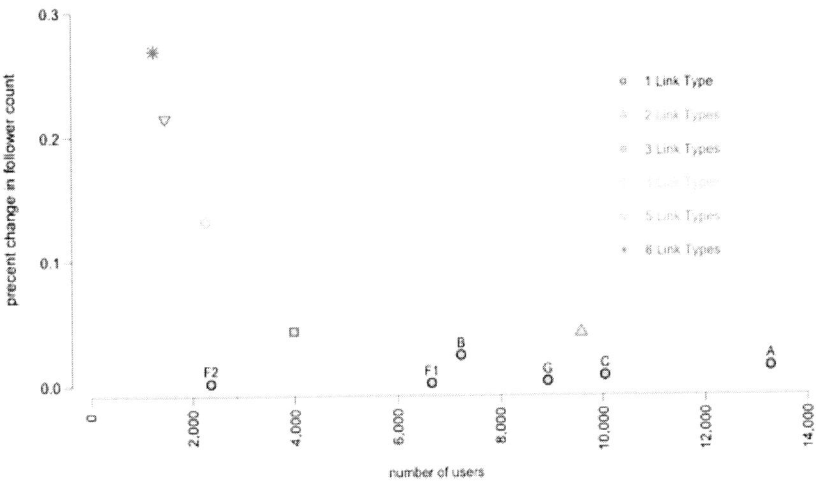

Figure 9. Percentage change in followers of accounts (in whole database) that linked to multiple link types in #ows sample October–December 2011.

Toward an analytical framework of crowd organization

It is important to understand how crowd-enabled networks of the kind that became prominent in recent protests work. One puzzle is that even when participants reject organizations and the bounded conditions conventionally associated with organizational capacity, some large-scale networks nonetheless seem to display coherent organization despite the ebb and flow of participation on personalized terms. We have introduced a preliminary framework for understanding peer-production organizing processes in crowd-enabled networks, and thereby the workings of communication as organization in this context. Further analysis is required to understand the potential and limits of crowd-enabled action in political contexts. This study is a first step.

Our analysis developed in two broad stages. First, we showed that technological stitching platforms play an important role in enabling large-scale crowds to achieve macro-level organization involving many different networks. Importantly, this point does not hang on a particular technology.

Twitter was critical for large-scale crowd-enabled organization in the Occupy movement, but other stitching technologies were also in play (e.g. SMS systems) and may be even more important in other cases (Livingston & Walter-Dropp, 2014). Moreover, many crowds are sustained by the interaction of virtual and physical participants who may move between different modes of participation. Second, we identified a basic package of peer-production processes that are integral for achieving the coherent organization that network stitching implies. All three categories – production, curation, and dynamic integration – are important. Deficits or poor implementation in any one undermines the capacity of the network to mobilize resources, respond to events, or adjust in periods of change. Without production, there is little substance in the network, and what there is will not be significantly co-produced among the crowd; without curation the network will be saturated with noise; without the micro-level connections of dynamic integration, groups and networks will scatter or become significantly dependent on key bridging actors. The indicators will differ across contexts, but the presence of the basic processes, along with routinized patterns of use, helps shape the degree to which large-scale crowd-enabled networks attain organization. The preliminary evidence suggests that engaging the full package is a complex achievement involving both core participants and the broader crowd.

At this point it is important to recall that this analysis isolates a slice of the crowd-enabled network that needs to be re-embedded in several respects in order to gain full traction on organization in the crowd. On the one hand, analysis of the stitching mechanism should be carefully re-integrated amongst the networks and parts it connects, with their various actor constellations and symbolic resources as well as conditioning principles, spatiality, and historical residues of practice. On the other hand, it is important to integrate and enrich this kind of analysis with insights drawn from other analytical and methodological approaches, such as those involved in social network analysis, ethnography, and platform studies. This will open deeper paths to analyzing what sustains and shapes organization in crowd-enabled networks, including how specific actor configurations facilitate framing processes across networks, how different kinds of resources circulate, and how stitching mechanisms change in different crowds and over time as new conditions emerge. Such development will help illuminate not only how crowd-enabled networks spread content and connections, but also how and when organizing mechanisms break down and fail. We also point to future studies of how particular stitching mechanisms (e.g. email, SMS, Twitter, Facebook) condition the character of the large-scale mediated crowd.

Finally, while our study focuses on mechanisms that build organizational capacity, we also note that many larger questions surround the crowd-enabled network as a political organizational form. Crowd networks may transition to other forms, but they do not necessarily represent a phase of emerging organization that must transform into more conventional forms or fail. The fundamental differences in organizational logics between collective and connective action as described by Bennett and Segerberg (2013) make holding crowds up to the standards of more conventional bureaucratic organization a misleading comparison. Since the conditions that produce crowd-enabled networks often do not support bureaucratic participation, we should be wary of using idealized organization or mobilization models as the only measure of political capacity or success. Moreover, although the study of political crowds has often been clouded by strongly diverging ideas about their nature and role, it cannot be assumed that their political character is predefined or that all crowds can be lumped together in terms of capacity, impact, or other normative generalizations (Borch, 2012; McClelland, 1989; McPhail, 1991). In this light, our work suggests the need for focused conceptual and empirical analyses of these organizational forms, along with the development of adequate standards for comparing them, and determining how they combine and clash with other forms of collective action. Such efforts will be important for understanding the workings of crowd-enabled networks across a variety of political contexts, and to learn how organized crowds fit into the democratic politics of claim-making, resistance, and dissent.

ACKNOWLEDGMENTS

This article benefited greatly from the comments received on an early version presented at the European Consortium for Political Research Joint Sessions in Mainz, March 2013. Our sincere thanks go to Elena Pavan, Aaron Shaw, Mayo Fuster Morell, as well as Thomas Poell, Benjamin Mako Hill, Bob Boynton, Michael Jensen, Henrik Bang, Steven Livingston, Helen Margetts, and the other participants. We also thank Sidney Tarrow and the two referees for their excellent comments.

FUNDING

Bennett and Segerberg gratefully acknowledge the support of Swedish Research Council grant [No. 421-2010-2303]. This project has also been supported by the US National Science Foundation INSPIRE grant [No. 1243170]: 'Tools, Models and Innovation Platforms for Research on Social Media'.

NOTES

1. This continuum of political network types (organizationally brokered, organizationally enabled, and crowd enabled) is from Bennett and Segerberg (2013).
2. As we argue elsewhere (Bennett & Segerberg, 2012, 2013), connective action hinges on intrinsically self-motivated – though not necessarily self-*centered* – action. Such personal offerings (messages, ideas, links to other sources, offers of help, stories, reflections, action etc.) may be shared through communication networks. Depending on how and whether those contributions are engaged by others, the result may be the production of valued outcomes for both the original contributors and others. This argument in part draws on peer production as conceived by Benkler (2006). We are however not suggesting that the broader crowd is consciously inspired by peer-production principles (although this has been shown to be the case for key participants in protests such as the 15-M, see Morell, 2012; Postill, 2013).
3. The history of Wikipedia according to Wikipedia contains a reference to its hybrid bureaucracy: http://en.wikipedia.org/wiki/Wikipedia. There is even a contested and dynamic discussion of what kind of bureaucrats Wikipedia has produced: http://en.wikipedia.org/wiki/Wikipedia:Bureaucrats.
4. This means that other nodes, whether individuals or other technology platforms, connected via different Twitter accounts and hashtags. Unlike campaign websites or Facebook pages, Twitter was not itself a singular, stable, or coherent node.
5. Determining the full population is problematic in large technology-enabled crowds. However, context-sensitive building of key search terms, and knowledge of different API streams helps secure reliable data, as does developing various measures of error (e.g., non-relevant vs. relevant tweets) and ways of addressing it. The dataset captured from October 2011 to August 2012 is remarkably complete and well matched to what is reported to be a nearly full population of Occupy tweets released through the commercially licensed GNIP PowerTrack (Driscoll & Walker, 2013).
6. The growth in number does not transparently imply Occupy movement activity (nor that all contributors supported Occupy). The continuing expansion of terms increases noise in a dynamic dataset. In order to check the relevance of tweets in the dataset, we coded a random sample of messages with and without Occupy-related hashtags (e.g., #ows). The coders achieved better than 90%

agreement on what distinguished a relevant Occupy tweet. Tweets with Occupy-related hashtags maintained a low rate of error (under 13%), while tweets without Occupy hashtags had a higher error rate, in part because they often pertained to issues other than Occupy. (For example, the search term 'strike' produced messages about strikes unrelated to Occupy.) Sampling on widely used Occupy hashtags enabled us to capture high levels of actual Occupy-related tweets (85–90%) compared to lower Occupy-relevancy rates for the general population of tweets containing more generic key terms (e.g. strike) and satirical references (e.g., Occupy Chipotle). While a percentage of the tweets came from spambots, it is not large or disproportionately distributed in any particular category of the coded data.

7. The added metadata included the expanded URL for any shortened URL; a list of hashtags in the tweet; a list of mentions in the tweet; the number of hashtags, URLs, and mentions in the tweet; a list of data collection keywords that matched within the tweet, and the location of the match (hashtag, @mention, text, or URL) in order to explain why each tweet made it into the archive.
8. Coding files were generated for each of the three hashtags by randomly sampling 225 non-retweeted tweets with URLs from every other day for the respective period studied for each hashtag. Samples were generated via the following process to ensure that each day sampled contained links that could subsequently be tracked through retweets and multiple appearances in other hashtags: 1) All non-retweeted tweets with at least one URL and the appropriate hashtag were extracted from the archive. 2) These tweets were randomly sorted. 3) They were sequentially examined to determine if the URL was part of a prior sample. If the URL was unique across all samples already generated, it was added to the coding sample and randomly assigned to one of three coders (who were not the authors). If the URL already appeared in the coding sample from a prior day, it was excluded. 4) This process was repeated until the coding sample for the day contained 225 URLs or the entire sample of tweets with the date and hashtag from the archive had been exhausted.
9. #ows tops the daily hashtag list over the six months that we checked activity levels (October 2011–March 2012), and surpassed 250,000 tweets per day during the height of police evictions of camps during mid-November. It appeared in over 3 million tweets in the time frame from which we drew our coding samples.
10. Coding reliability among 3 trained coders was .851 based on Krippendorf's Alpha and Fleiss Kappa, and .829 for pairwise

agreement using Cohen's Kappa. For details of the development of the coding categories and process we refer the reader to Agarwal et al. (forthcoming).

11. While some Twitter users retweet to show endorsement, some account profiles explicitly state that retweeting should not be regarded as positive endorsement.
12. Twitter states that it is 'best practice' to include a maximum of two hashtags in regular communication (http://www.twitter.com, 2013).
13. Here, as elsewhere in the study, it should be borne in mind that Twitter actors are in themselves a delimited part of the overall movement population.
14. The data in Figure 7 are from the peak activity period of October–December 2011. The pattern for January–April is similar, but the numbers of active accounts dwindles as the protests tailed off.
15. We selected the 99 most active accounts in the database archive in the 2011 time frame using #ows for each link diversity level (1–6) for the key codes A, B, C, F1, F2, and G, including RT and non RT links. The account type was categorized in terms of whether it represented an Occupy group or organization (e.g., city assembly, campaign, etc.) or was unaffiliated according to information in the account profile. Spam behavior and promotion, pro- or anti-Occupy stance was judged on message content. Journalists and news organizations were coded as 'no category', as were any cases in which it was not possible to establish category from the material available.
16. Accounts explicitly expressing complex support, e.g. for both Occupy and the Tea Party, clustered primarily in the upper middle category.
17. In terms of account profile description.
18. A couple of the top contributors already appear in the lists of top ten accounts in the study of Occupy-related Twitter in the summer of 2011 (Tremayne, 2013).

NOTES ON CONTRIBUTORS

W. Lance Bennett is Ruddick C. Lawrence Professor of Communication and Professor of Political Science at the University of Washington, where he is the founder and director of the Center for Communication and Civic Engagement. His books include *When the press fails: Political power and the news media from Iraq to Katrina* (Chicago, with Regina Lawrence and Steven Livingston) and *The logic of connective action: Digital media and the personalization of contentious politics* (Cambridge, with Alexandra

Segerberg). Current research areas include: press-government relations and the quality of public information, changing patterns of citizenship and civic engagement, and the uses of digital media in public life. [email: lbennett@u.washington.edu]

Alexandra Segerberg is a Research Fellow in the Department of Political Science at Stockholm University, and Associate Editor of ECPR Press, the publishing imprint of the European Consortium for Political Research. Her research interests revolve around political, philosophical and empirical theories of collective action. [email: alex.segerberg@statsvet.su.se]

Shawn Walker is a PhD Candidate and founding member of the Social Media (SoMe) Lab at the University of Washington Information School. His research focuses on methodologically sound techniques and research designs for the collection, curation, and analysis of social media data in social science research. This work takes place in the context of social movements utilizing social media. [email: stw3@uw.edu]

REFERENCES

1. Agarwal, S., Bennett, W. L., Johnson, C., & Walker, S. (forthcoming). A model of crowd-enabled organization: Theory and methods for understanding the role of Twitter in the Occupy protests.
2. Ancelovici, M. (2012). *Making sense of the Spanish Indignados.* Paper presented at the annual meeting of the Nineteenth Annual Conference of the Council for European Studies, Omni Parker House Hotel, Boston, MA, not available. August 2013. Retrieved from http://citation.allacademic.com/meta/p547131_index.html
3. Anduiza, E., Cristancho, C., & Sabucedo, J. (2013). Mobilization through online social networks: The political protest of the Indignados in Spain. *Information, Communication & Society.* doi:10.1080/1369118X.2013.808360
4. Barzilai-Nahon, K. (2008). Toward a theory of network gatekeeping: A framework for exploring information control. *Journal of the American Information Science and Technology*, 59, 1–20.
5. Barzilai-Nahon, K., Hemsley, J., Walker, S., & Hussain, M. (2011). Fifteen minutes of fame: The place of blogs in the life cycle of viral political information. *Policy & Internet*, 3(1), 1–28. Article 2.
6. Baym, G., & Shah, C. (2011). Circulating struggle: The on-line flow of environmental advocacy clips from the daily show and the colbert report. *Information Communication & Society*, 14(7), 1017–1038.

7. Benkler, Y. (2006). *The wealth of networks: How social production transforms markets and freedom*. New Haven: Yale University Press.
8. Benkler, Y., Roberts, H., Faris, R., Solow-Niederman, & Etling, B. (2013). Social mobilization and the networked public sphere: Mapping the SOPA-PIPA debate. The Berkman Center for Internet & Society Research Publication Series: Retrieved November 5, 2013, from http://cyber.law.harvard.edu/publications/2013/social_mobilization_and_the_networked_public_sphere
9. Bennett, W. L. (2003). Communicating global activism: Strengths and vulnerabilities of networked politics. *Information, Communication & Society*, 6(2), 143–168.
10. Bennett, W. L. (2012). The personalization of politics: Political identity, social media, and changing patterns of participation. *The Annals*, 644, 20–38.
11. Bennett, W. L., & Segerberg, A. (2012). The logic of connective action: Digital media and the personalization of contentious politics. *Information Communication & Society*, 15(5), 739–768.
12. Bennett, W. L., & Segerberg, A. (2013). *The logic of connective action: Digital media and the personalization of contentious politics*. New York: Cambridge University Press.
13. Bernstein, M. S., Monroy-Hernández, A., Harry, D., André, P., Panovich, K., & Vargas, G. (2011). 4chan and /b/: An analysis of anonymity and ephemerality in a large online community. In *Proceedings of AAAI Conference on Weblogs and Social Media (ICWSM)* (pp. 50–57). Menlo Park, CA: AAAI Press.
14. Bimber, B. (2003). *Information and American democracy: Technology in the evolution of political power*. New York, NY: Cambridge University Press.
15. Bimber, B., Flanagin, A., & Stohl, C. (2012). *Collective action in organizations: Interaction and engagement in an era of technological change*. New York: Cambridge University Press.
16. Borch, C. (2012). *The politics of crowds: An alternative history of sociology*. Cambridge: Cambridge University Press.
17. boyd, d., Golder, S., & Lotan, G. (2010). *Tweet, tweet, retweet: Conversational aspects of retweeting on Twitter*. Presented at HICSS-43, IEEE, Kauai, HI.
18. Bruns, A. (2008). *Blogs, wikipedia, second life, and beyond: From production to produsage*. New York: Peter Lang.
19. Bruns, A., & Burgess, J. (2011, August 25–27). *The use of Twitter hashtags in the formation of ad hoc publics*. Paper presented at the

European Consortium for Political Research General Conference, Reykjavik.

20. Butler, B., Joyce, E., & Pike, J. (2008). Don't look now, but we've created a bureaucracy: The nature and roles of policies and rules in Wikipedia. In *Proceedings of ACM Conference on Human Factors in Computing Systems (CHI)* (pp. 1101–1110). New York, NY: Association for Computing Machinery.
21. Calderaro, A. (2011, August 26). *New political struggles in the network society: The case of the free and open source (FOSS) Movement.* Paper presented at the European Consortium of Political Research (ECPR) General Conference, Reykjavik, Iceland.
22. Castells, M. (2009). *Communication power.* Oxford: Oxford University Press.
23. Castells, M. (2012). *Networks of outrage and hope: Social movements in the internet age.* Cambridge, UK: Polity.
24. Chadwick, A. (2013). *The hybrid media system: Politics and power.* Oxford: Oxford University Press.
25. Crowston, K., Wei, L., Howison, J., & Wiggins, A. (2012). Free/libre open-source software development: What we know and what we do not know. *ACM Computing Surveys*, 44(2), Article 7.
26. De Alfaro, L., Kulshreshtha, A., Pye, I., & Adler, B. T. (2011). Reputation systems for open collaboration. *Communications of the ACM*, 54(8), 81–87.
27. De Sanctis, G., & Monge, P. (1999). Introduction to the special issue: communication processes for virtual organizations. *Organization Science*, 10(6), 693–703.
28. Diani, M. (2011). Social movements and collective action. In J. Scott & P. Carrington (Eds.), *The sage handbook of social network analysis* (pp. 223–235). London: Sage Publications.
29. Donath, J., & boyd, d. (2004). Public displays of connection. *BT Technology Journal*, 22(4), 71–82.
30. Driscoll, K., & Walker, S. (2013). *Working within a black box: Transparency in the production of Big Twitter Data.* Unpublished manuscript.
31. Eckert, J., Walker, S., Hemsley, J., Nahone, K., & Mason, R. (2013). 'Opening the black box of social media research methods: SoMe Ways Forward'. Retrieved November 23, 2013, from http://somelab.net/wp-content/uploads/2013/01/HICSS_workshop_final29Dec.pdf
32. Fang, Y., & Neufeld, D. (2009). Understanding sustained participation in open source software projects. *Journal of Management Information Systems*, 25(4), 9–50.

33. Forte, A., & Lampe, C. (2013). Defining, understanding and supporting open collaboration: Lessons from the literature. *American Behavioral Scientist.*
34. Forte, A., Larco, V., & Bruckman, A. (2009). Decentralization in wikipedia governance. *Journal of Management Information Systems,* 26(1), 49–72.
35. Gaby, S., & Caren, N. (2012). Occupy online: How cute old men and Malcolm X recruited 400,000 US users to OWS on facebook. *Social Movement Studies,* 11(3–4), 367–374.
36. Gerbaudo, P. (2012). *Tweets and the streets: Social media and contemporary activism.* London: Pluto Press.
37. Gladwell, M. (2010, October 4). Small change: Why the revolution will not be tweeted, New Yorker. Retrieved August 19, 2013, from http://www.newyorker.com/reporting/2010/10/04/101004fa_fact_gladwell
38. González-Bailón, S. (2013). Online social networks and bottom-up politics, Rochester, NY: Social Science Research Network. SSRN Scholarly Paper. Retrieved from http://papers.ssrn.com/abstract=2246663
39. González-Bailón, S., Borge-Holthoefer, J., & Moreno, Y. (2013). Broadcasters and hidden influentials in online protest diffusion. *American Behavioral Scientist.*
40. Harkinson, J. (2012, April 13). How occupy co-opted moveon.org. Mother Jones. Retrieved from http://www.motherjones.com/politics/2012/04/99-spring-moveon-occupy-wall-street
41. Haythornthwaite, C. (2009, January 5–8). *Crowds and communities: Light and heavyweight models of peer production.* Proceedings of the Hawaii International Conference on System Sciences, Big Island, Hawaii.
42. Hemsley, J. (2011, December 9). Beginning to map social networks. Social Media Lab @ UW. Retrieved August 23, 2013, from http://somelab.net/2011/12/social-networks/
43. Hill, B. M., & Monroy-Hernández, A. (2013). *The cost of collaboration for code and art: Evidence from a remixing community.* Proceedings of the 2013 ACM Conference on Computer Supported Cooperative Work, San Antonio, TX, ACM 2013.
44. Hill, B. M., Shaw, A., & Benkler, Y. (n.d.). *Status, social signaling and collective action in a peer production community.* Unpublished manuscript.
45. Jenkins, H., Clinton, K., Purushotma, R., Robison, A., & Weigel, M. (2006). Confronting the challenges of participatory culture: Media education for the 21st century. The John D. and

Catherine T. MacArthur Foundation, Digital Media and Learning Initiative. Chicago, IL.

46. Jenkins, H., Ford, S., & Green, J. (2013). *Spreadable media: Creating value and meaning in a networked culture.* New York: New York University.
47. Jensen, M., & Bang, H. (2013). Occupy wall street: A new political form of movement and community? *Journal of Information Technology and Politics.*
48. Juris, J. (2005). The new digital media and activist networking within anti-corporate globalization movements. *The ANNALS of the American Academy of Political and Social Science,* 597, 189–208.
49. Kirkpatrick, D. D. (2011, February 9). Wired and shrewd, Young Egyptians Guide Revolt. *New York Times.* Retrieved August 14, 2013, from http://www.nytimes.com/2011/02/10/world/middleeast/10youth.html?pagewanted=all&_r=0
50. Kittur, A., & Kraut, R. (2008). Harnessing the wisdom of crowds in Wikipedia: Quality through coordination. Human-Computer Interaction Institute, Paper 99. Retrieved from http://repository.cmu.edu/hcii/99
51. Kittur, A., Pendleton, B., & Kraut, R. E. (2009). *Herding the cats: The influence of groups in coordinating peer production.* Proceedings of the 5th International Symposium on Wikis and Open Collaboration. ACM.
52. Kriplean, T., Beschastnikh, I., & McDonald, D. W. (2008). Articulations of wikiwork: Uncovering valued work in Wikipedia through barnstars. In *Proceedings of Computer-Supported Cooperative Work (CSCW)* (pp. 47–56). New York, NY: Association for Computing Machinery.
53. Kwak, H., Lee, C., Park, H., & Moon, S. (2010). What is Twitter, a social network or a news media? *Intl. WWW Conference,* Raleigh, NC, ACM, 591–600.
54. Lampe, C., Johnston, E., & Resnick, P. (2007). Follow the reader: Filtering comments on Slashdot. In *Proceedings of the ACM Conference on Human Factors in Computing Systems*(pp. 1253–1262). New York, NY: Association for Computing Machinery.
55. Lampe, C., & Resnick, P. (2004, April 24–29). *Slash(dot) and burn: Distributed moderation in a large online conversation space.* CHI 2004, Vienna, Austria.
56. Lievrouw, L. (2011). *Alternative and activist new media.* Cambridge: Polity.
57. Livingston, S., & Walter-Dropp, G. (Eds.). (2014). *Bits and atoms: Information and communication technology in areas of limited statehood.* Oxford: Oxford University Press.

58. Luther, K., Fiesler, C., & Bruckman, A. (2013). *Redistributing leadership in online creative collaboration*. Proceedings of the ACM Conference on Computer-Supported CooperativeWork (CSCW 2013), San Antonio, TX.
59. Margetts, H., John, P., Escher, T., & Reissfelder, S. (2012). Social information and political participation on the internet: An experiment. *European Political Science Review*, 3(3), 321–344.
60. McClelland, J. S. (1989). *The crowd and the mob: From Plato to Canetti*. London: Unwin Hyman.
61. McPhail, C. (1991). *The myth of the madding crowd*. New York: A. de Gruyter.
62. Meraz, S., & Papacharissi, Z. (2013). Networked gatekeeping and networked framing on #egypt. *The International Journal of Press/Politics*, 18(2), 138–166.
63. Morell, M. F. (2012). The free culture and 15M movements in Spain: Composition, social networks and synergies. *Social Movement Studies*, 11(3), 386–392.
64. Mueller, M. (2010). *Networks and states: The global politics of Internet governance*. Cambridge, MA: MIT Press.
65. Murthy, D. (2013). *Twitter: Social communication in the Twitter age*. Cambridge: Polity.
66. Olson, M. (1965). *The logic of collective action: Public goods and the theory of groups*. Cambridge, MA: Harvard University Press.
67. Ostrom, E. (1990). *Governing the commons: The evolution of institutions for collective action*. Cambridge: Cambridge University Press.
68. Pavan, E. (2012). *Frames and connections in the governance of global communications: A network study of the internet governance forum*. Plymouth, UK: Lexington Books.
69. Penney, J., & Dadas, C. (2013). (Re)tweeting in the service of protest. Digital composition and circulation in the Occupy Wall Street movement. *New Media & Society*.
70. della Porta, D. (2005). Multiple belongings, flexible identities, and the construction of 'another politics': Between the European Social Forum and local social fora. In D. della Porta & S. Tarrow (Eds.), *Transnational protest and global activism* (pp. 175–202). Lanham, MD: Rowman and Littlefield.
71. Postill, J. (2013). The uneven convergence of digital freedom activism and popular protest: A global theory of the new protest movements. Melbourne: RMIT University. Retrieved October 7, 2013, from: http://rmit.academia.edu/JohnPostill

72. Powell, W. (1990). Neither market nor hierarchy: Network forms of organization. In B. Staw (Ed.), *Research in organizational behavior* (pp. 295–336). Greenwich, CT: JAI Press.
73. Reagle, J. (2007). Do as I do: Authorial leadership in Wikipedia. In *Proceedings of the 2007 International Symposium on Wikis* (pp. 143–156). New York, NY: Association for Computing Machinery.
74. Schweik, C., & English, R. (2007). Tragedy of the FOSS commons? Investigating the institutional designs of free/libre and open source software projects, *First Monday*, [S.l.], feb. 2007. ISSN 13960466. Retrieved August 23, 2013, from http://firstmonday.org/ojs/index.php/fm/article/view/1619/1534.
75. Shaw, A. (2012). Centralized and decentralized gatekeeping in an open online collective. *Politics & Society*, 40(3), 349–388.
76. Starbird, K., & Palen, L. (2012). *(How) will the revolution be retweeted? Information diffusion and the 2011 Egyptian Uprising.* CSCW 2012. ACM.
77. Suh, B., Hong, L., Pirolli, P., & Chi, E. H. (2010). 'Want to be retweeted? Large scale analytics on factors impacting retweet in Twitter networks'. In *IEEE Intl Conference on Social Computing* (pp. 177–184), IEEE.
78. Thorson, K., Driscoll, K., Ekdale, B., Edgerly, S., Gamber Thompson, L., Schrock, A., ... Wells, C. (2013). YouTube, Twitter and the Occupy movement: Connecting content and circulation practices. *Information, Communication & Society*,
79. Tremayne, M. (2013). Anatomy of protest in the digital era: A network analysis of Twitter and Occupy Wall Street. *Social Movement Studies: Journal of Social, Cultural and Political Protest.*
80. Turner, F. (2009). Burning Man at Google: A cultural infrastructure for new media production. *New Media & Society*, 11(1&2), 73–94.
81. Vasi, I. B., & Suh, C. S. (2013). *Protest in the internet age: Public attention, social media, and the spread of 'Occupy' protests in the United States.* Unpublished manuscript. Retrieved November 24, 2013, from http://politicsandprotest.ws.gc. cuny.edu/files/2012/07/PPW-2-Vasi.pdf
82. Weber, S. (2004). *The success of open source.* Cambridge, MA: Harvard University Press.
83. Welser, H. T., Cosley, D., Kossinets, G., Lin, A., Dokshin, F., Gay, G., & Smith, M. (2011). Finding social roles in wikipedia. In *Proceedings of the 2011 iConference* (pp. 122–129). New York, NY: Association for Computing Machinery.
84. Wilkinson, D. (2008). Strong regularities in online peer production. In *Proceedings of the 9th ACM Conference on Electronic*

Commerce (pp. 302–309). New York, NY: Association for Computing Machinery.

85. Zhu, H., Kraut, R., & Kittur, A. (2012). Effectiveness of shared leadership in online communities. In *Proceedings of the ACM Conference on Computer Supported Cooperative Work* (pp. 407–416). New York, NY: Association for Computing Machinery.

Index